Power Electronics-Enabled Autonomous Power Systems

Power Electronics-Enabled Autonomous Power Systems

Next Generation Smart Grids

Qing-Chang Zhong
Illinois Institute of Technology & Syndem LLC
Chicago, USA

Registered Offices
John Wiley & Sons, Inc., 111 River Street, Hoboken, NJ 07030, USA
John Wiley & Sons Ltd, The Atrium, Southern Gate, Chichester, West Sussex, PO19 8SQ, UK

Editorial Office
The Atrium, Southern Gate, Chichester, West Sussex, PO19 8SQ, UK

For details of our global editorial offices, customer services, and more information about Wiley products visit us at www.wiley.com.

Wiley also publishes its books in a variety of electronic formats and by print-on-demand. Some content that appears in standard print versions of this book may not be available in other formats.

Library of Congress Cataloging-in-Publication data applied for

HB ISBN: 9781118803523

Cover Design: Wiley
Cover Images: Electric Car © Nerthuz/Getty Images, Computer Technology © RoyFWylam/Getty Images, Led light bulb © ppart/Shutterstock, Industrial electric motor © scanrail/Getty Images, Hydro dam water © ChrisGorgio/Getty Images, Solar Panel © filo/Getty Images, Energy © Voyagerix/Shutterstock, Nuclear power plant © TTstudio/Shutterstock, A field of wind turbines © linearcurves/Getty Images

Set in 9.5/12.5pt STIXTwoText by SPi Global, Chennai, India

Printed and bound by CPI Group (UK) Ltd, Croydon, CR0 4YY

10 9 8 7 6 5 4 3 2 1

To

Ms. Lihua Luo, my first-grade teacher, who told me:

"you have nothing but potential. Keep moving forward. Never stop."
and

Ms. Xiufen Lin, my third-grade teacher, who told me:

"you have nothing but do not envy others."

Contents

List of Figures

List of Tables

Foreword

When I first met and interviewed Qing-Chang Zhong at the Illinois Institute of Technology in Chicago, I already knew a lot about the innovative synchronous converter technologies he'd invented to harmonize future power grids. Qing-Chang's invention is a game changer for the grid. It is the sort of breakthrough – like the touch screen in smart phones – that helps to push an industry from one era to the next.

What I learned about that day, and on ensuing visits between our families, was Qing-Chang's towering ambition, keen intelligence, and vision for humanity. He has intimate knowledge of the complexity of generating consistent and affordable power to the masses, and the necessity to get this invention into everyday use in time to save the planet.

This book is a comprehensive guide to Qing-Chang's path-breaking thinking and work over the last two decades for next-generation smart grids. It explores the theoretical foundations of his skilled engineering and offers case studies of their application. The book offers a technical solution to implement the lateral power envisioned by futurist Jeremy Rifkin that underpins the Third Industrial Revolution.

Qing-Chang's goal is to help develop an electrical supply and distribution network that gives more access to various supplies and loads with more influence in its management and operation and advance global energy freedom for billions of people with access to low-cost clean energy.

Qing-Chang envisions a near future of complete, 100% generation from distributed energy sources to achieve what he calls "synchronized and democratized (SYNDEM) smart grid." This unifies and harmonizes distributed energy resources and loads with a common law of synchronization. As a result, his work is designed to promote local resiliency, avoid wide-area blackouts, and secure the grid from terrorist attacks.

Qing-Chang's SYNDEM grid architecture, in effect, is an essential tool for achieving one of the most technically challenging industrial transformations in history. The architecture eases stress on contemporary electric transmission grids as coal, nuclear, and gas-fired centralized generating plants give way to thousands of decentralized and distributed renewable energy stations being installed offshore, in fields, and on ridges and rooftops all over the world. It does so by automatically sensing and instantly reacting to fluctuations in electricity production, demand, and voltage across the grid. That's a real trick for an electrical supply industry diversifying its generating sources so rapidly.

A warm, earnest, accomplished engineer who's studied and taught at prestigious universities on three continents, Qing-Chang is sincere about his goal. You'll see that in this book's straightforward prose. The development of his autonomous synchronized smart grid speeds our evolution to cleaner energy sources. It is as essential to human progress as the often-punishing switch from sailing schooners and wood, to steamships and coal. What Qing-Chang describes is a pathway and state-of-the-art tools that will make our transition to low-carbon power much more reliable and cleaner, and far less disruptive and expensive.

I've spent four decades reporting on access and demand for energy for the New York Times, the Los Angeles Times, National Geographic, ProPublica, Energy News Network, and other prominent news organizations. My work on six continents is defined by the ability to discover vital trends in energy, science, the environment, and economics, elevate them to national and global attention, and prompt action.

I reported on the Three Mile Island nuclear accident in 1979. A decade later I led a team of New York Times reporters that uncovered the decrepit and dangerous conditions inside America's nuclear weapons manufacturing plants spread across 12 states.

I was among the first journalists to reveal how water scarcity in China's Yellow River Basin was harming the country's energy sector, affecting harvests, and weakening the Chinese economy. That reporting compelled the Chinese Central Government to warn about the impending barrier to development in its 2012 five-year plan, and to propose solutions. In 2014 the US and China agreed to limit their carbon emissions. The US–China climate agreement includes six provisions. Two of them focused on the confrontation between rising energy demand and declining freshwater reserves. The agreement also was the diplomatic breakthrough that led to the Paris Climate Accord in December 2015.

In other words, I recognize how urgency inspires innovation. Researchers, engineers, professors, students, policy makers, regulatory and legislative organizations will benefit from this book, which thoroughly explores Qing-Chang's theoretical framework and technologies to make it much easier to develop decentralized and clean sources of electrical energy. As he says, "in this way, we can work together to save the planet before it is too late."

Keith Schneider
New York Times correspondent since 1982

Preface

Electrification is the greatest engineering achievement of the 20th century and power systems are arguably the most important infrastructure that underpins social life and economic growth. The generation of electricity has been dominated by centralized large power plants, but the landscape of power systems is fast changing. Large centralized power plants are being replaced with incompatible, non-synchronous, and relatively small distributed energy resources (DERs), electric vehicles (EVs), and storage systems, which is often referred to as the democratization of power systems. This is comparable to the great historical event of personal computers replacing mainframes in the technology domain or republics replacing monarchies in the political domain. The number of active players, including DER units and storage units etc., is rapidly growing and could easily reach millions. This brings unprecedented challenges to grid stability, reliability, security, and resiliency. The fundamental challenge is that future power systems will be power electronics based, instead of electrical machines based, with millions of non-synchronous heterogeneous players. This is less of a *power* problem but more of a *systems* problem – power systems are going through a paradigm shift.

Adding an ICT (information and communications technology) system into power systems, hence the birth of smart grids, has emerged as a potential solution. However, this could lead to serious reliability concerns. On 1 April 2019, a computer outage of Aerodata affected all major US airlines, causing 3000+ flight cancellations or delays. This sends a clear warning to the power industry: ICT systems may be a single point of failure and similar system-wide outages could happen to power systems. Moreover, when the number of active players reaches a certain level, the management of ICT systems is itself a challenge. What is even worse is that adding ICT systems to power systems *opens the door for potential cyber-attacks by anybody, at any time, from anywhere.* There are numerous reports about this and, on 23 December 2015, hackers compromised ICT systems of three distribution companies in Ukraine and disrupted the electricity supply to 230 000 customers. The reliance of power systems operation on ICT systems has become *a fundamental systemic flaw.*

Another fundamental systemic flaw of current power systems is that a local fault can lead to cascading failures. On 16 June 2019, a blackout originating in Argentina's northeast struck all of Argentina and Uruguay, and parts of Brazil, Chile and Paraguay. There is a pressing demand to prevent local faults from cascading into wide-area blackouts.

This book summarizes the author's profound thinking into these problems over the last 18 years and presents a theoretical framework, together with its underpinning technologies and case studies, for future power systems with up to 100% penetration of DERs to achieve harmonious interaction, to prevent local faults from cascading into wide-area blackouts, and to operate autonomously without relying on ICT systems. Thus, it is possible to completely avoid cyber-attacks. The theoretical framework, referred to as SYNDEM (meaning synchronized and democratized), adopts the synchronization mechanism of synchronous machines, which has underpinned power systems operation and growth for over 100 years, as its rule of law. Hence, the SYNDEM framework brings backward compatibility to current power systems. The underpinning technologies will turn power electronic converters into virtual synchronous machines (VSMs) and achieve legal equality among millions of non-synchronous heterogeneous players, including flexible loads that have power electronic converters at the front end. This unifies and harmonizes their interface and integration with the grid, making it possible to achieve autonomous operation for power systems without relying on ICT systems. Hence, the SYNDEM theoretical framework and its underpinning technologies will significantly smooth the paradigm shift of power systems from the *centralized control* of a small number of large facilities to *democratized interaction* of a large number of relatively small generators and flexible loads, accelerating the large-scale adoption of renewables while enhancing grid stability, reliability, security, and resiliency.

This book consists of one introductory chapter (Chapter 1) and five parts: Theoretical Framework (Chapters 2–3), First-Generation VSMs (Chapters 4–14), Second-Generation VSMs (Chapters 15–20), Third-Generation VSMs (Chapter 21), and Case Studies (Chapters 22–25). In the introductory chapter, the outline of the book and the evolution of power systems are presented. In Part I, the SYNDEM theoretical framework and the ghost power theory are presented. In Part II, the first-generation VSMs (synchronverters) are presented, together with their application in wind power, solar power, flexible loads, STATCOM, and motor drives, the removal of PLLs, the improvement to bound frequency and voltage, and the reconfiguration of virtual inertia, virtual damping, and fault ride-through. In Part III, the equivalence between droop control and the synchronization mechanism is established at first. Then, the second-generation VSMs (robust droop controllers) are presented, followed by universal droop control, the removal of PLLs, droop-controlled flexible loads to provide continuous demand response, and current limiting converters to prevent cascading failure. In Part IV, the third-generation VSM, which achieves passivity of itself and the whole system if it is open-loop passive, is presented. In Part V, four case studies are presented, including a single-node system based on a reconfigurable SYNDEM smart grid research and educational kit, a 100% power electronics based SYNDEM smart grid testbed, a home grid, and the Panhandle wind power system.

This book is suitable for graduate students, researchers, and engineers in control engineering, power electronics, and power systems. The SYNDEM theoretical framework chapter is also suitable for policy makers, legislators, entrepreneurs, commissioners of utility commissions, utility personnel, investors, consultants, and attorneys. The book covers various challenging problems in the control of power electronic converters, distributed generation, wind power integration, solar power integration, microgrids, smart grids, flexible AC transmission systems, etc. It can be also adopted as a textbook for relevant graduate programs. Actually, it has been adopted as the textbook for ECE 537–Next Generation Smart Grids at Illinois Institute of Technology since 2018.

Acknowledgments

It is Father's Day as I am writing this. I cannot help but miss my parents, who live 12,000 km away from me. Needless to say, it is they who have planted the seeds for everything in this book. They nurtured me to keep being upright and taught me to serve the society wherever I am. They are like a mountain in my home town, constantly providing me with energy, courage, and calmness. There are many words I can say to thank them but, whatever I say, it is never enough to express my gratitude for what they have given me. There are many things I can do to thank them but, whatever I do, it is never enough to show my appreciation for them. The only thing that makes me feel slightly better is to call them once a week while I keep moving forward every day – I have been doing this for nearly 20 years. Many teachers and professors have taught me with their life-long experiences, valuable wisdom, and profound knowledge, which have directly contributed to the development of the technologies described in this book. "Teacher for one day, father forever," as a Chinese proverb says. I will always remember them. This book is dedicated to two of them, my first-grade teacher Ms. Luo and my third-grade teacher Ms. Lin for their inspirational words. This book holistically summarizes the intensive and extensive thinking I have had about future power systems since 2001. Over the last 18 years, I have had the great opportunity to advance this line of research with collaborators, postdoctoral researchers, and PhD students, including Mohammad Amin, Frede Blaabjerg, Dushan Boroyevich, Jianyun Chai, Chengxiu Chen, Joseph M. Guerrero, Tomas Hornik, George Konstantopoulos, Miroslav Krstic, Fred Lee, Hong Li, Zijun Lyu, Zhenyu Ma, Wen-Long Ming, Long Nguyen, Beibei Ren, Tiancong Shao, Wanxing Sheng, Márcio Stefanello, Yeqin Wang, George Weiss, Yu Zeng, Xiaochao Zhang, Yangyang Zhao, Qionglin Zheng, and Yuanfeng Zhou, just to name a few. Their contributions to the work included in this book are greatly appreciated. I would like to thank Royal Academy of Engineering, U.K. and the Leverhulme Trust for the award of a Senior Research Fellowship during 2009–2010. Their visionary decision has made a significant impact on my research in this area. I would also like to thank the Engineering and Physical Sciences Research Council, UK for their support (under Grant No. EP/J01558X/1 and EP/J001333/2), which has facilitated me to make some major breakthroughs in this area. It is a great honor for me to be invested as the Max McGraw Endowed Chair Professor of Energy and Power Engineering and Management at Illinois Institute of Technology. Max McGraw (1883–1964) founded McGraw Electric Company in 1900 when he was 17 years old and acquired Thomas A. Edison, Inc. in 1957 to form the McGraw–Edison Company, which employed 21,000 people in 1985 when acquired by Cooper Industries. He never retired. He was so visionary and established McGraw Foundation in 1948 with the mission to provide financial assistance for educational and charitable purposes in furtherance of the public good and promoting the well-being of all humanity. He will be remembered forever. The support of our industrial partners and advisors has always been instrumental for our research. I am particularly grateful to Phillip Cartwright and Kevin Daffey (Rolls-Royce), James Carlson (Carlson Wireless Technologies), Gene Frantz (former Principal Fellow of Texas Instruments), Mark Harral (GroupNire), Aris Karcanias (FTI Consulting), Tony Lakin (Turbo Power Systems), Jim MacInnes (Crystal Mountain), Brian MacCleery (National Instruments), Robert Owen (Texas Instruments), and Zhenyu Yu (Texas Instruments). It has been a great pleasure to work with the colleagues of John Wiley & Sons, Ltd and IEEE Press. The support and help from Steven Fassioms (the Project Editor), Michelle Dunckley (the Editorial Assistant), Karthika Sridharan (the Production Editor), and Sandra Grayson (the Commissioning Editor) are greatly appreciated. The visionary decision of Peter Mitchell in signing the contract for publishing this book should be acknowledged as well. I would also like to thank the readers of this book in advance. I believe the SYNDEM grid architecture together with the underpinning technologies will play a key role in making power systems worldwide stable, reliable, sustainable, secure, and resilient. We need your help, support, and dedication. Spread the word and save the planet! I would like to thank my wife Shuhong Yu for her support, patience, love, and sacrifice. She has taken over all the responsibilities in taking care of the family, which has allowed me to focus on thinking and research. I would also like to thank my wonderful daughters Rui and Lisa for being supportive, considerate, active, and forward-looking. Last but not least, I would also like to take this opportunity to thank those I have not mentioned above but have had a direct or indirect impact on me or my research, good or bad. You have made me better and stronger.

Thank you all!

Qing-Chang Zhong
zhongqc@ieee.org
IEEE Fellow, IET Fellow

Founder & CEO, Syndem LLC
Chicago, USA
http://www.syndem.com

Max McGraw Endowed Chair Professor
　in Energy and Power Engineering
Department of Electrical and Computer
　Engineering
Illinois Institute of Technology
10 W 35th Street
Chicago, USA
http://peac.ece.iit.edu/

16 June 2019

About the Author

Qing-Chang Zhong, an IEEE Fellow and an IET Fellow, holds the Max McGraw Endowed Chair Professor in Energy and Power Engineering and Management at Illinois Institute of Technology, Chicago, USA and is the Founder and CEO of Syndem LLC, Chicago, USA. He was educated at Imperial College London (PhD, 2004), Shanghai Jiao Tong University (PhD, 2000), Hunan University (MSc, 1997), and Hunan Institute of Engineering (Diploma, 1990).

He served as a Distinguished Lecturer for IEEE Power Electronics Society, IEEE Control Systems Society, and IEEE Power and Energy Society. He has (co-)authored four research monographs: *Control of Power Inverters in Renewable Energy and Smart Grid Integration* (Wiley-IEEE Press, 2013), *Control of Integral Processes with Dead Time* (Springer-Verlag, 2010), *Robust Control of Time-Delay Systems* (Springer-Verlag, 2006), and this book, *Power Electronics-Enabled Autonomous Power Systems: Next Generation Smart Grids* (Wiley-IEEE Press, 2020). He proposed the SYNDEM grid architecture for the next-generation smart grids based on the synchronization mechanism of synchronous machines, which unifies and harmonizes the interface and interaction of power system players with the grid to achieve autonomous operation without relying on communication networks. This line of research has been featured by *IEEE Power Electronics Magazine* as a cover story, by Energy News Network as Game Changer for Grid, by *IEEE Spectrum* as a vision for a harmonious grid, and by IEEE PES Task Force on Primary Frequency Control as the Path to the Future.

He is truly globalized. Before moving to Chicago, he spent about 14 years in the UK, as Research Associate at Imperial College London, Senior Lecturer at University of Glamorgan and University of Liverpool, Chair Professor in Control Engineering at Loughborough University, and Chair Professor in Control and Systems Engineering at The University of Sheffield, and one year in Israel at Technion – Israel Institute of Technology as a postdoctoral researcher. He has delivered 200+ invited talks in 20+ countries, including delivering a semiplenary talk on SYNDEM smart grids at the 20th IFAC World Congress, which is the world's largest conference in control and systems engineering.

He served on the Steering Committee of IEEE Smart Grid and is a Vice-Chair of IFAC TC on Power and Energy Systems. He was a Senior Research Fellow of the Royal Academy of Engineering/Leverhulme Trust, UK and the UK Representative to the European Control Association. He served as an Associate Editor for *IEEE Transactions on Automatic Control*, *IEEE Transactions on Power Electronics*, *IEEE Transactions on Industrial Electronics*, *IEEE Transactions on Control Systems Technology*, *IEEE Journal of Emerging and Selected Topics in Power Electronics*, *IEEE Access*, and *European Journal of Control*.

Here is a version written by Keith Schneider, a *New York Times* correspondent since 1982.

Zhong did not advance to the uppermost echelons of global energy innovation by privilege of birth or happy accident. Raised in a rural area of Sichuan, China, his family was poor. His prospects for attending a university were slim. For this reason, his family sent him to a technical secondary school in Xiangtan instead of high school. Having his trajectory pointed in that direction for him, his chances of getting a college education were practically zero.

During the technical school, though, Zhong excelled and was promoted to attend college level courses. After graduation, he began his career as a technician. He soon started his first entrepreneurial endeavor by filing for a patent in 1991. Zhong successfully commercialized it – manufactured and sold 120 units quickly. The invention stirred enough market interest that the radio station of Xiangtan City promoted the product for one month for free.

Zhong realized that he could achieve even more than he thought and developed these six words as his mantra. *Control yourself. Challenge yourself. Excel yourself.*

In 1994, Zhong returned to university to study for a master's degree. One of the entrance requirements was passing a national exam that included knowledge of English. His first foreign language was Japanese, so learning English was nearly Mission Impossible. He tried unsuccessfully to find an English tutor, but in the end he taught himself. He mastered the language portion and was admitted to Hunan University, the best university in the province, with the highest total score among all applicants to the university that year.

In 1997, Zhong temporarily left his wife, Sue, and his one-year-old daughter in Hunan and went to Shanghai to earn a PhD degree. After receiving his doctorate in 2000, Dr. Zhong went to Israel for a postdoctoral position. This time, he was accompanied by his family.

After completing the contract in Israel, the Zhongs moved to England for a second post-doctoral position at Imperial College London. Dr. Zhong envisioned becoming a full professor in 10 years. In order to accomplish that goal he studied at night for a second PhD degree while working as a postdoctoral researcher by day. Nine years later, Dr. Zhong was appointed to Chair Professor in Control Engineering at Loughborough University. Just as important is what occurred two years ago. He made a breakthrough in his research and invented the synchronverter, the first-generation virtual synchronous machine. He knew this was revolutionary. While advancing the research, including building a $5 million lab at The University of Sheffield, he started exploring various ways for commercialization.

Dr. Zhong eventually realized that the best way to achieve his vision is to operate at a global stage. After nearly 14 years of work in the UK, he decided to move one more time, this time to the US. He was recruited to Chicago, a global hub, where he assumed Max McGraw Endowed Chair Professor at Illinois Institute of Technology.

After settling down, Dr. Zhong founded Syndem LLC in 2017. The company focuses on developing virtual synchronous machines to accelerate large-scale adoption of distributed energy resources and to make power systems autonomous without relying on communication networks. Within one year, Dr. Zhong accomplished the rare – almost impossible – task of leading his team to nearly break even in their first year while launching two products and securing contracted revenue for another year of operation.

Dr. Zhong lives in a Chicago suburb with Sue, herself an engineer, and their two daughters. The older one, Rui, works for a top investment bank in the green technology sector and is an accomplished visual artist. The younger, Lisa, is an energetic reader, dancer, and gymnast, eagerly awaiting attending high school. She read over 800 books in 2018.

List of Abbreviations

1G	First-generation
2G	Second-generation
3G	Third-generation
AC	Alternating current
AF	Active filtering
APFM	Amplitude phase frequency model
APM	Amplitude phase model
BA	Balancing authority
CA	Control area
C-inverter	Inverter with dominantly capacitive impedance
CM	Common mode
CSI	Current source inverters
CSM	Cyber synchronous machines
CVCF	Constant voltage constant frequency
DC	Direct current
DERs	Distributed energy resources
DFIG	Doubly fed induction generator
DPC	Direct power control
DQ, dq	Direct quadrant
DSM	Demand-side management
DSP	Digital signal processing
ERCOT	Electric Reliability Council of Texas
EPLL	Enhanced PLL
EV	Electric vehicle
FACTS	Flexible AC transmission systems
FERC	Federal Energy Regulatory Commission
GPS	Global positioning system
GS	Grid-support mode
GSC	Grid-side converter
HVAC	Heating, ventilation, and air conditioning
HVDC	High-voltage DC
ICT	Information and communications technology
IGBT	Insulated gate bipolar transistor

IM	Induction machine
ISS	Input-to-state stable
LED	Light-emitting diodes
L-inverter	Inverter with dominantly inductive impedance
LPF	Low-pass filter
MODEM	Modulation and demodulation
MPPT	Maximum power point tracking
NEC	National Electrical Code
NS	No-support mode
PCC	Point of common coupling
PFR	Primary frequency response
PH	Port Hamiltonian
PI	Proportional integral
PLL	Phase-locked loop
PMSG	Permanent magnet synchronous generator
POI	Point of interconnection
PWM	Pulse width modulation
PV	Photovoltaic
QoS	Quality of service
R-inverter	Inverter with dominantly resistive impedance
RMS	Root mean square
RoCoF	Rate of change of frequency
RSC	Rotor-side converter
SC	Synchronous condensers
SCADA	Supervisory control and data acquisition
SCIM	Squirrel-cage induction machine
SG	Synchronous generator
SLL	Sinusoid-locked loop
SM	Synchronous machine
SSPS	Solid state power substations
STA	Sinusoid tracking algorithm
STATCOM	Static synchronous compensator
SVC	Static var compensators
SYNDEM	Synchronized and democratized, or generally synchroniz- and democratiz-
THD	Total harmonic distortion
USB	Universal serial bus
VCO	Voltage-controlled oscillator
VOC	Voltage oriented control
VSD	Variable speed drives
VSG	Virtual synchronous generators
VSI	Voltage-source inverter
VSM	Virtual Synchronous Machines
WLDS	Ward Leonard drive systems
WPGS	Wind power generation systems

1

Introduction

1.1 Motivation and Purpose

Electrification is the greatest engineering achievement of the 20th century (NAE 2000) and the power system is regarded as the largest and most complex machine engineered by humankind (Kundur 1994). Power systems are arguably the most important infrastructure that underpins social life and economic growth because "energy regimes shape the nature of civilizations –how they are organized, how the fruits of commerce and trade are distributed, how political power is exercised, and how social relations are conducted" (Rifkin 2011).

The generation of electricity has been dominated by large centralized facilities that burn fossil fuels, such as coal, oil, and gas. While fossil fuels have greatly contributed to the world's civilization, two challenging consequences are emerging: one is that fossil fuels are not sustainable and the other is that the combustion of fossil fuels emits greenhouse gases, which is a major cause for climate change – *one of today's most pressing global challenges* (Hutt 2016). The large-scale adoption of renewable distributed energy resources (DERs) has been widely accepted as a promising means to tackle these two challenges. The United Nations have put the Paris Agreement into force to reduce greenhouse gas emissions (UN 2018) and many countries have set strategic plans to utilize renewable energy and make a transition to a low-carbon economy. For example, France and the UK plan to close all coal plants by 2023 (England 2016) and 2025 (Vaughan 2018), respectively. Many states in the US, including Hawaii, California, and New York, have decided to generate 100% carbon-free electricity by around 2050. As a result, the number of DER units is rapidly growing and could easily reach millions, even hundreds of millions, in a power system. This is often referred to as the democratization of power systems (Farrell 2011). Integrating a small number of DERs into the grid is not a problem, but integrating millions of DERs into the grid brings unprecedented challenges to grid stability, reliability, security, and resiliency (Zhong and Hornik 2013). For example, it has been reported that Hawaii's solar push strains the grid (Fairley 2015) and renewable energy could leave you mired in blackouts (Brewer 2014). Fundamentally speaking, this is less of a *power* problem but more of a *systems* problem.

Adding an ICT system into the power system, hence the birth of the smart grid, has emerged as a potential solution to make power systems more efficient, more resilient to threats, and friendlier to the environment (Amin 2008; Amin and Wollenberg 2005).

Power Electronics-Enabled Autonomous Power Systems: Next Generation Smart Grids,
First Edition. Qing-Chang Zhong.

However, this could also lead to serious reliability concerns (Eder-Neuhauser et al. 2016; Overman et al. 2011). If the ICT system breaks down then the whole power system could crash. On 1 April 2019, a computer outage of Aerodata caused 3000+ flight cancellations or delays, affecting all major US airlines including Southwest, American, Delta, United, Alaska, and JetBlue (Gatlan 2019). This sends a clear warning signal to the power industry: ICT systems may become a single point of failure for power systems and similar system-wide outages could happen to power systems if their operation relies on ICT infrastructure. Moreover, when the number of players reaches a certain level, how to manage the ICT system is itself a challenge. What is even worse is that adding ICT systems to power systems opens the door for cyber-attacks by anybody, at any time, from anywhere. On 23 December 2015, hackers compromised the ICT systems of three energy distribution companies in Ukraine and temporarily disrupted the electricity supply to approximately 225,000 customers (Lee et al. 2016). There are numerous reports about cyber-attacks to power systems (Carroll 2019) and, on 15 June 2019, *The New York Times* reported that *US Escalates Online Attacks on Russia's Power Grid* (Sanger and Perlroth 2019). There are certainly issues beyond engineering and technologies but the reliance of power system operation on ICT systems has become a fundamental systemic flaw. It is the responsibility of engineers to correct this systemic flaw and make power systems secure and reliable. Actually, there is no other choice left. On 27 June 2019, the US Senate passed a bipartisan bill, the *Securing Energy Infrastructure Act*, with the aim of removing vulnerabilities that could allow hackers to access the energy grid. There is a pressing demand to correct this systemic flaw and avoid potential cyber-attacks.

Another fundamental systemic flaw of current power systems is that a local fault can lead to cascading failures (Schäfer et al. 2018; Yang et al. 2017). On 28 September 2016, tornadoes with high-speed winds damaged and tripped two 275 kV transmission lines in South Australia (SA), causing six voltage dips in 2 min. Eight wind farms exceeded the preset number of voltage dips and tripped, losing 456 MW wind generation in 7 s. The imported power through the Victoria–SA Heywood Interconnector (510 MW) significantly increased and forced it to trip within 0.7 s, islanding SA from the National Electricity Market and leaving an imbalance of 1 GW. Subsequently, all gas generators tripped and all supply to SA was lost, leading to a state-wide blackout. About 850,000 customers lost electricity (AEMO 2017). On 16 June 2019, all of Argentina and Uruguay, and parts of Brazil, Chile, and Paraguay in South America were hit by a massive blackout, affecting approximately 48 million people (Regan and McLaughlin 2019). The blackout originated at an electricity transmission point between the power stations at Argentina's Yacyreta Dam and Salto Grande in the country's northeast. On 9 August 2019, a major blackout struck England and Wales, affecting almost a million homes and forcing trains to a standstill around the UK (Ambrose 2019). Again, it was caused by something not uncommon, the loss of a gas-fired power plant and an offshore wind farm. There is a pressing demand to correct this systemic flaw and prevent local faults from cascading into wide-area blackouts.

The purpose of this book is to summarize the author's profound thinking over the last 18 years on these problems, which he anticipated in 2001, and present a theoretical framework, together with its underpinning technologies and case studies, for future power systems with up to 100% penetration of distributed energy resources to achieve harmonious interaction, to prevent local faults from cascading into wide-area blackouts, and to operate autonomously without relying on ICT systems and completely avoid cyber-attacks.

It is purely a coincidence that *The New York Times* report on cyber-attacks on Russia's power grid, the South America blackout, the US legislation on preventing cyber-attacks to the grid, and the UK blackout all happened within the last two months when this book was being finalized, signaling the right time to complete this book.

1.2 Outline of the Book

As shown in Figure 1.1, this book contains this introductory chapter (Chapter 1) and five parts: Part I: Theoretical Framework (Chapters 2 and 3), Part II: First-Generation VSMs (Chapters 4–14), Part III: Second-Generation VSMs (Chapters 15–20), Part IV: Third-Generation VSMs (Chapter 21), and Part V: Case Studies (Chapters 22–25). Most of the chapters include experimental results or real-time simulation results, as indicated with a large or small triangle tag at the bottom-right corner of the corresponding chapter box in Figure 1.1, and, hence, the technologies can be applied in practice with minimum effort.

In this introductory chapter, in addition to the outline of the book, the evolution of power systems is briefly presented to set the stage for the following five parts.

Part I: Theoretical Framework contains two chapters. Chapter 2 presents the SYNDEM theoretical framework for next-generation smart grids – power electronics-enabled autonomous power systems, covering the concept of SYNDEM smart grids, the rule of law that governs SYNDEM smart grids, the legal equality for all SYNDEM active players to equally take part in grid regulation, the architecture of SYNDEM smart grids, a brief description of potential technical routes, and the roots of the SYNDEM concept. Chapter 3 introduces a new operator, called *the ghost operator g*, to physically construct the ghost of a (sinusoidal) signal and, further, the ghost of a system with sinusoidal inputs. Moreover, the reactive power of an electrical system is shown to be the real power of the ghost system with its input being the ghost of the input to the original system. This is then applied to define the reactive power for mechanical systems, completing the electrical-mechanical analogy, and, furthermore, generalizing to any dynamic system that can be described by a port-Hamiltonian (PH) system model, establishing a significantly simplified instantaneous power theory, referred to as the ghost power theory. This can be applied to any dynamic system, single phase or poly-phase, with or without harmonics.

Part II: First-Generation VSMs contains 11 chapters related to the first-generation VSMs (synchronverters). Chapter 4 presents the synchronverter (1G VSM) technology to operate an inverter to mimic a synchronous generator (SG) after directly embedding the mathematical model of synchronous generators into the controller. The real and reactive power delivered by synchronverters connected in parallel can be automatically shared with the well-known frequency and voltage droop mechanism. Synchronverters can also be easily operated in the standalone mode. Chapter 5 describes a control strategy to operate a PWM rectifier to mimic a synchronous motor. Two controllers, one to directly control the power exchanged with the grid and the other to control the DC-bus voltage, are discussed. At the same time, the reactive power can be controlled as well. Chapter 6 applies the synchronverter technology to control the back-to-back PWM converters of

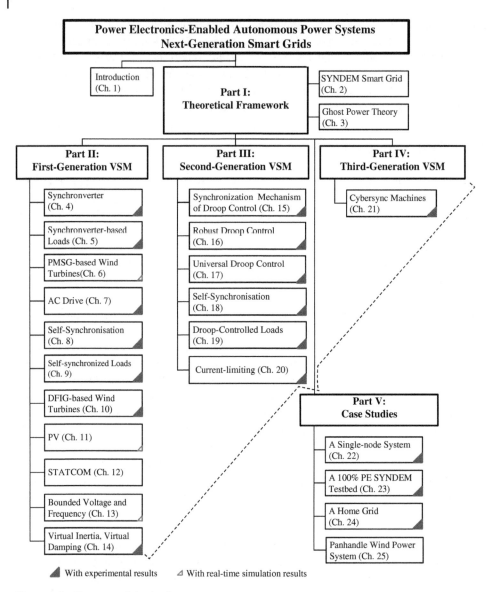

Figure 1.1 Structure of the book.

PMSG based wind turbines. Both converters are operated as synchronverters. Different from the common practice, the rotor-side converter (RSC) is controlled to maintain the DC-link voltage and the grid-side converter (GSC) is controlled to inject the maximum power available to the grid. Since both converters are operated with 1G VSM technology, the whole wind power system is able to support the grid frequency and voltage when there is a grid fault, making it friendly to the grid. Chapter 7 applies the 1G VSM to control the speed of an AC machine in four quadrants, via powering the AC machine with a VSM that generates a variable-voltage-variable-frequency supply. This is a natural and mathematical, rather than physical, extension of the conventional Ward Leonard drive systems for DC machines to AC machines. If the rectifier providing the DC bus for the AC drive is controlled as a virtual synchronous motor according to Chapter 5, then an AC drive is equivalent to a motor–generator–motor system. This facilitates the analysis of AC drives and the introduction of some special functions. Chapter 8 takes a radical step to improve the synchronverter as a self-synchronized synchronverter by removing the dedicated synchronization unit, which is often a phase-locked loop (PLL). It can automatically synchronize itself with the grid before connection and maintain synchronization with the grid after connection without using a PLL. Experimental results show that this improves the performance of frequency tracking by more than 65%, the performance of real power control by 83% and the performance of reactive power control by about 70%. Chapter 9 discusses the removal of the dedicated synchronization unit from synchronverter based loads (rectifiers). Two controllers are presented: one is to directly control the real power exchanged with the grid and the other is to control the DC-bus voltage. At the same time, the reactive power can be controlled as well. Chapter 10 reveals the analogy between differential gears and DFIG and then operates a DFIG based wind turbine as a VSM. An electromechanical model is presented to represent a DFIG as a differential gear that links a rotor shaft driven by a prime mover (the wind turbine), a virtual stator shaft coupled with a virtual synchronous generator and a virtual slip shaft coupled with a virtual synchronous motor. Moreover, an AC drive, which consists of an RSC and a GSC, is adopted to regulate the speed of the slip virtual synchronous motor. Then, the whole DFIG–converter system is operated as one VSG with the stator shaft synchronously rotating at the grid frequency, even when the rotor shaft speed changes, through controlling the virtual slip shaft to synchronously rotate at the slip frequency. Following the concept of the AC Ward Leonard drive systems discussed in Chapter 7, the RSC is controlled as a virtual synchronous generator to generate an appropriate slip voltage, via regulating the real power and reactive power sent to the grid. In order to facilitate this, the GSC is controlled as a virtual synchronous motor to maintain a stable DC-bus voltage without drawing reactive power in the steady state. A prominent feature is that there is no need to adopt a PLL – either on the grid side or on the rotor side. The system is not only able to provide the kinetic energy stored in the turbine/rotor shaft as inertia to support the grid frequency, but it is also able to provide reactive power to support the grid voltage. Chapter 11 applies the 1G VSM to a transformerless PV system, which consists of an independently controlled neutral leg and an inversion leg. The presence of the neutral leg enables the direct connection between the ground of PV panels and the neutral line of the grid, removing the need to have an isolation transformer. This significantly reduces leakage currents because the stray capacitance between the PV panels and the grid neutral line (ground) is bypassed. Another benefit is that the voltage of the PV is only required to be higher than the peak value of the grid voltage, which is the same as that required by conventional full bridge inverters. The resulting PV inverter is also grid-friendly. Chapter 12 applies the 1G VSM to operate a STATCOM as a synchronous generator in the condenser mode, without using a dedicated synchronization unit, such as a PLL. In addition to the conventional voltage regulation mode (or the V-mode in short) and the direct Q control mode (or the Q-mode in short), a third operation mode, i.e. the voltage droop control mode (or the V_D-mode in short), is introduced to the operation of the STATCOM. This allows parallel-operated STATCOMs to share reactive power properly and provides more control flexibility. Chapter 13 presents an improved 1G VSM to make sure that its frequency and voltage always stay within given ranges. Furthermore, its stability region is analytically characterized according to system parameters so that the improved synchronverter can be always stable and converges to a unique equilibrium as long as the power exchanged at the terminal is kept within this region. Chapter 14 explains that the concept of inertia has two aspects of meaning: the inertia time constant that characterizes the speed of the frequency response and the inertia constant that characterizes the amount of energy stored. It is then shown that, while the energy storage aspect of the virtual inertia of a VSM can be met by storage units, the inertia time constant that can be provided by a VSM may be limited because a large inertia time constant may lead to oscillatory frequency responses. A VSM with reconfigurable inertia time constant is then introduced by adding a low-pass filter to the real power channel. Moreover, a virtual damper is introduced to provide the desired damping ratio, e.g., 0.707, together with the desired inertia time constant. Two approaches are presented to implement the virtual damper: one through impedance scaling with a voltage feedback controller and the other through impedance insertion with a current feedback controller. A by-product from this is that the fault ride-through capability of the VSM can be designed as well.

Part III: Second-Generation VSMs contains six chapters related to the 2G VSM. First, Chapter 15 reveals that the widely adopted droop control mechanism structurally resembles a PLL or the synchronization mechanism of synchronous machines, making droop control a potential technical route to implement SYNDEM smart grids. A conventional droop controller for inverters with inductive impedance is then operated to behave as a PLL, without a dedicated synchronization unit. Chapter 16 reveals the fundamental limitations of the conventional droop control scheme at first and then presents a robust droop controller to achieve accurate proportional sharing without these limitations for R-, L- and C-inverters. The load voltage can be maintained within the desired range around the rated value. The strategy is robust against numerical errors, disturbances, noises, feeder impedance, parameter drifts, component mismatches, etc. The only error comes from measurement, which can be controlled by using sensors with required tolerance. Chapter 17 shows that there exists a universal droop control principle for impedance having a phase angle between $-\frac{\pi}{2}$ rad and $\frac{\pi}{2}$ rad and it takes the form of the droop control for R-inverters. In other words, the robust droop control for R-inverters is universal and can be applied to inverters with different types of impedance having a phase angle from $-\frac{\pi}{2}$ rad to $\frac{\pi}{2}$ rad. Chapter 18 removes the PLL from the universal droop controller to achieve self-synchronization without a PLL. Chapter 19 presents a general framework based on the universal droop control for a rectifier-fed load to continuously take part in the regulation of grid voltage and frequency without affecting the operation of the DC load. As a result, such a load can provide a primary frequency response, excelling the FERC requirement on newly integrated generators to provide primary frequency response. It can automatically change the power consumed to support the grid, without affecting the normal operation of the load. This is a critical technology that prevents local faults from cascading into wide-area blackouts via releasing the full potential of loads to regulate system frequency and voltage. Chapter 20 presents a current-limiting universal droop controller to operate a grid-connected inverters under both normal and faulty grid conditions without damage by adopting an advanced nonlinear control strategy. This is another critical technology that help prevent local faults from cascading into wide-area blackouts, via maintaining connection without trip-off when there is a fault unless itself is faulty.

Part IV: Third-Generation VSMs contains Chapter 21, which briefly touches upon the third generation VSMs that are expected to be able to guarantee the stability of a power system with multiple power electronic converters. A generic control framework is presented to render the controller of a power electronic converter passive by using the PH systems theory and the ghost operator. The controller consists of two symmetric control loops and an engendering block. With the critical concepts of the *ghost signal* and the *ghost system* introduced in Chapter 3, the engendering block is augmented as a lossless interconnection between the control block and the plant pair that consists of the original plant and its ghost plant. The whole system is then passive if the plant pair is passive. Moreover, some practical issues, such as controller implementation, power regulation and self-synchronization without a dedicated synchronization unit, are also discussed.

Part V: Case Studies contains four chapters. Chapter 22 describes a single-node system implemented with a SYNDEM Smart Grid Research and Educational Kit, which is reconfigurable to obtain over 10 different topologies, covering DC/DC conversion and single-phase/three-phase DC/AC, AC/DC, and AC/DC/AC conversion. Hence, it is ideal for carrying out research, development, and education of SYNDEM smart grids. It adopts the widely used Texas Instrument (TI) C2000 ControlCARD and is equipped with the automatic code generation tools of MATLAB®, Simulink®, and TI Code Composer Studio™ (CCS), making it possible to quickly turn computational simulations into physical experiments without writing any code. The single-node system is equipped with 2G VSM technology and additional functions so that it can autonomously blackstart, regulate voltage and frequency, detect the presence of the public grid, self-synchronize with the grid, connect to the grid, detect the loss of the grid, and island it from the grid. Chapter 23 presents a 100% power electronics based SYNDEM smart grid testbed with eight nodes of VSMs connected to the same AC bus to demonstrate the operation of a SYNDEM smart grid. Experimental results are presented to show that the SYNDEM smart grid framework is very effective and all the VSM nodes, including wind power, solar power, DC loads, AC loads, and an energy bridge, can work together to collectively regulate the SYNDEM grid frequency and voltage, without relying on ICT systems for control. Chapter 24 presents a practical home grid based on the SYNDEM framework. It consists of four 3 kW solar inverters, one 3 kW wind inverter, and one 3 kW energy bridge for interconnection with the public grid. The home grid can be operated in the islanded mode or the grid-tied mode if needed. Chapter 25 discusses the Texas Panhandle wind power system, which suffers from the severe problem of exporting the wind power generated to load centers far away. It is shown that the SYNDEM smart grid architecture and its underpinning technologies could remove the export limit imposed on the wind farms in Panhandle so that they can export the wind power generated at full capacity without causing problems to the grid.

1.3 Evolution of Power Systems

Electricity is the workhorse of the modern world and has been in existence for over 100 years.

1.3.1 Today's Grids

A power system today typically consists of facilities to generate, transmit, distribute, and utilize electrical power (Karady and Holbert 2004). Power plants generate electricity from energy sources, such as fossil fuels, hydro, nuclear, etc., and are often far from load centers because of associated pollution and risks or geographical limitations. In order to reduce losses and investment costs, the generation of electricity is currently dominated by centralized large power plants and the electricity generated is transformed into high or ultra-high voltage for transmission and then transformed to low voltage for utilization at load centers. At the transmission level, there are often interconnections in order to form a strong grid, to which massive power plants are connected. Distribution networks are often radial. Generation facilities and loads are generally separated by transmission and distribution networks and electricity normally flows unidirectionally from generation to loads. In particular, the electricity flow in distribution networks is unidirectional.

There is a need to maintain balance between generation and load demand in power systems. Otherwise, the frequency and/or voltage may vary in a wide range, which may cause damage. In current power systems, the system stability is maintained by regulating a small number of large generators to meet the balance between generation and demand. Most loads in the system do not actively take part in system regulation.

1.3.2 Smart Grids

The large-scale utilization of DERs, including renewables, EVs and energy storage systems, brings unprecedented challenges to grid stability, reliability, security, and resiliency (Zhong and Hornik 2013). The conventional centralized control paradigm is no longer feasible for power systems with millions of relatively small and distributed generators.

Adding an ICT system into power systems, hence the birth of smart grids, has emerged as a potential solution to make power systems friendlier to the environment as well as more efficient (Amin 2008; Amin and Wollenberg 2005). The main characteristics of smart grids in comparison to today's grids are shown in the second and the third columns of Table 1.1, with a prominent intention to address all problems via adding an ICT system. More details about smart grids can be found from the literature, e.g. (Amin and Wollenberg 2005; DOE 2009; Ekanayake et al. 2012; Fang et al. 2012; Farhangi 2010; Momoh 2012).

1.3.3 Next-Generation Smart Grids

As mentioned before, adding an ICT system into power systems does not solve the problem of how active players, such as DER units and flexible loads, interact with the grid at the physical level. This could also lead to serious concerns about reliability if their operation has to rely on the ICT infrastructure (Eder-Neuhauser et al. 2016; Overman et al. 2011). If the ICT system breaks down then the whole power system could crash. Moreover, when the number of active players reaches a certain level, how to manage the ICT system is itself a challenge. What is even worse is that adding ICT systems to power systems opens the door for cyber-attacks by anybody, at any time, from anywhere, making

Table 1.1 Comparison of today's grids, smart grids, and next-generation smart grids.

Characteristic	Today's grid	Smart grid	Next-generation smart grid
Role of ICT	Constantly growing	Tendency to introduce bidirectional ICT into every part and corner of power systems	Unidirectional ICT for monitoring and management but not for control to prevent cyber-attacks and single point of failures
Load participation in system regulation	Limited, non-participative (passive)	Binary (ON/OFF) demand responses enabled by ICT	Autonomous, continuous demand responses for fully active regulation of frequency and voltage, without the need of ICT
Generation mix and DER integration	Dominated by central generation with limited but growing current-controlled DER units	Diversifying, with mostly current-controlled DERs, strong tendency of relying on ICT	DER dominated or even 100% renewable with grid-friendly, compatible DERs, autonomous system regulation without the need of ICT
Cascading failures and blackouts	Inherent, systemic flaw	Significantly reduced number of incidents enabled by ICT but catastrophic when it happens	Built-in capability of preventing local faults from cascading into wide-area blackouts
Resiliency against faults and natural disaster	Vulnerable to faults, natural disasters, and the resulting cascading failures	Significant improvement enabled by ICT	Fully autonomous, mutually supported, fast recovery, reduced burden on utilities, without the need of ICT
Cyber-security	Vulnerable to malicious cyber-attacks	Significantly improved but still a systemic flaw	No access to the system through ICT; no more cyber-attacks to power systems
Power quality	Focus on outages, sometimes low power quality	A priority with a variety of quality/price options	Built-in control flexibility and functions to fundamentally reduce faults and outages and improve voltage quality
System-wide efficiency	Limited means for improvement	Significant improvement enabled by ICT but causing data explosion and ever-increasing complexity	Organic and harmonious operation underpinned by optimally designed components and system architecture

it a systemic flaw. While ICT systems could certainly bring many benefits to the operation and management of power systems, there is a need to confine the role of ICT.

It is envisioned that the next-generation smart grids will be power-electronics-enabled autonomous power systems without relying on ICT systems, underpinned by the synchronization mechanism of synchronous machines that brings backward compatibility to current power systems (Zhong 2013b, 2016b, 2017e,f).

Table 1.1 outlines the main characteristics of the next-generation smart grids with comparison to today's grids and smart grids. The prominent features of the next-generation smart grids include:

- That the role of ICT is defined to be unidirectional for monitoring and management only, excluding control, to prevent cyber-attacks and single point of failures.
- That all active players are unified with the same rule of law, which is backward compatible with today's grid, to achieve autonomous operation without relying on ICT and prevent wide-area blackouts.
- That it is governed by a harmonious system architecture that allows the grid to expand as well as to decompose when needed, e.g., in the case of faults to prevent local faults from cascading into wide-area blackouts.

1.4 Summary

In this chapter, the motivation and purpose of this book are briefly presented before describing the structure and contents of the book. Then, the evolution of power systems is outlined, highlighting the prominent features of next-generation smart grids with comparison to today's grids and the current-generation smart grids. The next-generation smart grid has three prominent features: (1) the role of ICT is defined to be unidirectional for monitoring and management only, excluding control, in order to prevent cyber-attacks and single point of failures; (2) all active players are unified with the same rule of law, which is backward compatible with today's grid, in order to achieve autonomous operation without relying on ICT and to prevent wide-area blackouts; and (3) it is governed by a harmonious system architecture that allows the grid to expand as well as to decompose when needed, e.g., in the case of faults to prevent local faults from cascading into wide-area blackouts.

Parts I–V will present the theoretical framework for next-generation smart grids, its underpinning technical routes, and case studies in detail.

Part I

Theoretical Framework

2

Synchronized and Democratized (SYNDEM) Smart Grid

In this chapter, a theoretical framework referred to as the SYNDEM smart grid is presented for the next-generation smart grids – power electronics-enabled autonomous power systems. This covers the concept of SYNDEM smart grids, the rule of law that governs SYNDEM smart grids, the legal equality for all SYNDEM active players to take part in grid regulation, the SYNDEM grid architecture and its potential benefits, a brief description of technical routes, its primary frequency response, and the roots of the SYNDEM concept.

2.1 The SYNDEM Concept

The paradigm shift of power systems has been recognized by many visionary thinkers. President Pöttering of the European Parliament says that "this is no Utopia, no futuristic vision: in twenty-five years' time, we will be able to construct each building as its own 'mini power station producing clean and renewable energy for its own needs, with the surplus being made available for other purposes." Jeremy Rifkin calls it the transition from hierarchical to lateral power in his book *The Third Industrial Revolution* (Rifkin 2011) and stresses that the lateral power will have the same kind of transformative effect on society as steam power and the printing press first had, followed by electric power and television. John Farrell calls it the democratization of the electric system (Farrell 2011).

Democracy is a well known political concept that empowers all eligible individuals to play an equal role in decision making. In recent years, this has been widely applied to other areas. For example, Thomas Friedman argues in (Friedman 2005) that the era of globalization has been characterized by the democratization of technology, the democratization of finance, and the democratization of information. These ideas have already started impacting the way commerce is conducted, societies are governed, children are educated, etc.

The most fundamental features of a democratized society include the rule of law and legal equality. The rule of law implies that every individual is subject to the law and legal equality implies that all individuals are equal and should be treated equally. In order to realize the democratization of power systems, there is a need to realize these two features while taking into account the fundamentals of power systems and democratized societies.

For power systems, one important fact is that it is not feasible to rebuild existing power systems from scratch because huge investment has been made into power systems over the last 100+ years. The ultimate solution should be to make millions of newly added

Power Electronics-Enabled Autonomous Power Systems: Next Generation Smart Grids,
First Edition. Qing-Chang Zhong.
© 2020 John Wiley & Sons Ltd. Published 2020 by John Wiley & Sons Ltd.

heterogeneous players *compatible* with the grid and follow the fundamental principles of the current power systems. Hence, this is a compatibility problem, which is not much talked about in power systems but is common sense in information technology. For example, the invention of the MODEM solved the compatibility problem of computers with telephone systems and revolutionized the access of computers to telephone systems. The widespread usage of the USB interface made many devices, such as cameras, printers and phones, compatible with computers. There is a need to solve the compatibility problem for power systems, in order to accelerate the paradigm shift of power systems from centralized control of a small number of large facilities to democratized interaction of a large number of relatively small generators and flexible loads (Zhong 2017f). In other words, the rule of law for future power systems should follow the fundamental principle that has been established in current power systems.

For democratized societies, one fundamental fact is that individuals can have different or even divisive opinions (Brennan 2016; Grayling 2017). Figure 2.1 illustrates two such examples: the 2016 UK Brexit Referendum and the 2016 US presidential election. There was no consensus in each case. If the democratization of power systems is implemented based on the current principles of democratized societies, it would sow the seeds of a systemic flaw for future power systems that may lead to system-wide consequences. Arrow's Impossibility Theorem (Arrow 1951, 2012), named after the economist and Nobel laureate Kenneth Arrow, states that, when there are more than two distinct candidates, there does not exist a voting (democratic) system that can convert individual preferences about the candidates into a transitive, i.e. strictly ranked, order of community preferences under the following four fairness criteria:

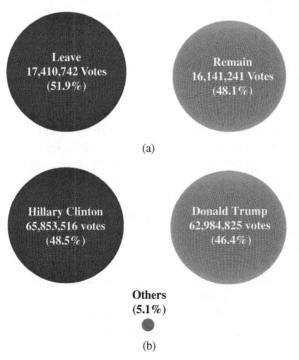

Figure 2.1 Examples of divisive opinions in a democratic society. (a) 2016 UK Brexit Referendum. (b) 2016 US presidential election.

Leave
17,410,742 Votes
(51.9%)

Remain
16,141,241 Votes
(48.1%)

(a)

Hillary Clinton
65,853,516 votes
(48.5%)

Donald Trump
62,984,825 votes
(46.4%)

Others
(5.1%)

(b)

(1) Universal admissibility: each individual can have any set of rational preferences.

(2) Unanimity: if every individual prefers candidate A to candidate B, then the community prefers candidate A to candidate B.

(3) Freedom from irrelevant alternatives: if every individual prefers candidate A to candidate B, then any change that does not affect this relationship must not affect the community preference for candidate A over candidate B.

(4) Non-dictatorship: no particular individual can dictate the community preference with his own preference independent of others.

When each criterion is considered separately, the four criteria seem to be perfectly reasonable but, when all the four conditions are required at the same time, no voting systems that guarantee a rational consequence exist and nonsensical or clearly undemocratic consequences may appear. More dramatically, the first three conditions imply that there will be a dictatorship if a transitive order of group preferences is demanded. For this reason, the theorem is also known as the *dictator theorem*.

This offers a high-level theoretical explanation why current power systems "dictated" by large power plants are stable. This also indicates that, after large power plants disappear, a democratized power system *cannot* guarantee its stability without introducing an additional mechanism for this. It is vital for all individuals to synchronize with each other to reach a consensus, i.e. to maintain frequency and voltage stability. Hence, future power systems will not only be democratized but also should be synchronized. This leads to the concept of synchronized and democratized (in short, SYNDEM) smart grids (Zhong 2017f). For such a SYNDEM grid, all active players, large or small, conventional or renewable, supplying or consuming, would follow the same rule of law and play the same equal role to achieve the same common goal, that is to maintain grid stability.

It is shown in (Zhong 2013b) that the synchronization mechanism of synchronous machines, which has underpinned the organic growth and stable operation of power systems for over 100 years, can continue serving as the rule of law for SYNDEM smart grids. Moreover, the power electronic converters that are adopted to integrate different players can be equipped with the synchronization mechanism through control to achieve legal equality.

2.2 SYNDEM Rule of Law – Synchronization Mechanism of Synchronous Machines

There are different power plants, such as coal-fired, nuclear, and hydro power plants, in existing power systems. However, electricity generation is dominated by only one type of electrical machine – synchronous machines. The reason why the industry has decided to adopt synchronous machines while there are different types of electric machines is because synchronous machines have an inherent synchronization capability. This can be understood by looking at the mathematical model of synchronous machines.

A synchronous generator, which is a synchronous machine operated as a generator, is governed by the swing equation

$$\ddot{\theta} = \frac{1}{J}(T_{\mathrm{m}} - T_{\mathrm{e}} - D_{\mathrm{p}}\dot{\theta}), \tag{2.1}$$

where θ is the rotor angle; $\dot{\theta}$ is the angular speed of the machine; T_m is the mechanical torque applied to the rotor; J is the moment of inertia of all the parts rotating with the rotor; D_p is the friction coefficient; and T_e is the electromagnetic torque

$$T_e = pM_f i_f \langle i, \ \widetilde{\sin} \ \theta \rangle. \tag{2.2}$$

Here, p is the number of pole pairs of the magnetic field and can be assumed to be 1 without loss of generality; i is the stator current; i_f is the field excitation current; M_f is the maximum mutual inductance between the stator windings and the field winding; and $\langle \cdot, \ \cdot \rangle$ denotes the conventional inner product. The vectors $\widetilde{\sin} \ \theta$ and $\widetilde{\cos} \ \theta$ are defined, respectively, as

$$\widetilde{\cos} \ \theta = \begin{bmatrix} \cos \theta \\ \cos(\theta - \frac{2\pi}{3}) \\ \cos(\theta - \frac{4\pi}{3}) \end{bmatrix}, \quad \widetilde{\sin} \ \theta = \begin{bmatrix} \sin \theta \\ \sin(\theta - \frac{2\pi}{3}) \\ \sin(\theta - \frac{4\pi}{3}) \end{bmatrix}.$$

The three-phase generated voltage e and the reactive power Q are, respectively,

$$e = E\widetilde{\sin} \ \theta, \tag{2.3}$$

$$Q = -E \langle i, \ \widetilde{\cos} \ \theta \rangle, \tag{2.4}$$

with $E = \dot{\theta} M_f i_f$ being the amplitude of the voltage. Assume that the terminal voltage is v. Then the stator current is

$$i = \frac{e - v}{sL + R}, \tag{2.5}$$

where $sL + R$ is the impedance of the stator windings. Note that i, e and v in (2.5) are the Laplace transforms of the corresponding signals. It should be clear whether a signal is in the time domain or in the frequency domain from the context.

The mathematical model of a synchronous machine described in (2.1)–(2.5) for the single-phase case is shown in Figure 2.2, after adding one integrator to zero the output of the D_p block and two low-pass filters $H(s)$ to remove the ripples in the torque and the reactive power. This is actually an enhanced phase-locked loop called the sinusoid-locked loop (Zhong and Hornik 2013; Zhong and Nguyen 2012). The core of the upper part of Figure 2.2 represents the swing equation (2.1) and the torque (2.2), which is a conventional phase-locked loop that can synchronize the frequency and the phase with those of the terminal voltage. The lower part is an amplitude channel to synchronize the amplitude of e with the terminal voltage. In the steady state, when $T_m = 0$ and the reference for Q is 0, there are $T_e = 0$ and $Q = 0$, which means $i = 0$ and $e = v$, achieving frequency, phase and amplitude synchronization. In other words, synchronous machines have the inherent mechanism of synchronization, which allows them to synchronize with each other or the grid autonomously.

The synchronization mechanism of synchronous machines is the mechanism that has underpinned and facilitated the organic growth and stable operation of power systems for over 100 years. In order to guarantee the compatibility of millions of heterogeneous players with the grid, this mechanism should be followed and adopted as the rule of law for SYNDEM smart grids. In this way, the synchronization mechanism also guarantees that all individuals could synchronize with each other to reach a consensus, i.e. for the voltage

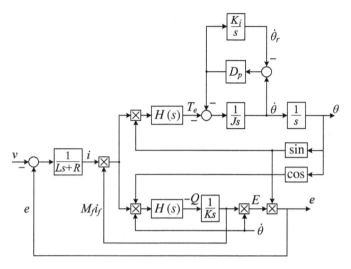

Figure 2.2 The sinusoid-locked loop (SLL) that explains the inherent synchronization mechanism of a synchronous machine.

and the frequency to stay around the rated values, e.g. 230 V voltage and 50 Hz frequency in Europe and 120V voltage and 60 Hz frequency in the US, so that the system stability is maintained. Moreover, this can be achieved without relying on a dedicated communication network. The function of communication is achieved based on the inherent synchronization mechanism of synchronous machines through the electrical system. As a result, the communication system in a SYNDEM smart grid can be released from low-level controls and adopted to focus on high-level functions, e.g. information monitoring, management, electricity market, etc.

As a matter of fact, the tendency to synchronize, or to act simultaneously, is probably the most mysterious and pervasive phenomenon in nature, from orchestras to GPS, from pacemakers to superconductors, from biological systems to communication networks (Strogatz, 2004). The observations that organisms adapt their physiology and behavior to the time of the day in a circadian fashion have been documented for a long time. For example, Chuang Tzu, who was an influential Chinese philosopher, a follower and developer of Taoism in the 4th century BC, wrote in his book *Chuang Tzu* (Chuang Tzu 2016) "to go to work at sunrise and go to rest at sunset," pointing out the importance of synchronizing human activities with the sun. The synchronization phenomenon has intrigued some of the most brilliant minds of the 20th century, including Albert Einstein, Richard Feynman, and Norbert Wiener. In 2017, the Nobel Prize in Physiology or Medicine was awarded to Jeffrey C. Hall, Michael Rosbash and Michael W. Young for their discoveries of molecular mechanisms that control circadian rhythms (Nobelprize.org 2017). They uncovered the internal clocks that synchronize cellular metabolism and organismal behavior to the light/dark cycle to generate biological rhythms with 24 h periodicity.

Hence, adopting the synchronization mechanism of synchronous machines as the rule of law to govern SYNDEM smart grids is probably also the most natural option.

2.3 SYNDEM Legal Equality – Homogenizing Heterogeneous Players as Virtual Synchronous Machines (VSM)

Figure 2.3 illustrates the approximate electricity consumption in the US, according to the US Electric Power Research Institute. Although there are many different loads, there are four main load types: motors that consume over 50% of electricity, internet devices that consume over 10% of electricity, lighting devices that consume about 20%, and other loads that consume the remaining 20% of electricity. It is well known that the adoption of variable-speed motor drives, which are equipped with power electronic rectifiers to convert AC electricity into DC electricity at the front-end, is able to significantly improve the efficiency of motor applications (Bose 2009). Hence, the 50% of electricity consumed by motors could actually be consumed by power electronic rectifiers. Internet devices consume DC electricity so the 10% of electricity consumed by internet devices is consumed by power electronic rectifiers as well. As to lighting devices, there is a clear trend in the lighting market to adopt LED lights, which also include power electronic rectifiers at the front end. Hence, in the future, the majority of electricity will be consumed by rectifiers, whatever the end function is.

On the supply side, most DERs are connected to the grid through power electronic inverters. For example, wind turbines generate more electricity at variable speeds, which means the electricity generated is not compatible with the grid and power electronic converters are needed to control the generation and interaction with the grid. Solar panels generate DC electricity, which needs to be converted into AC electricity to make it compatible with the grid as well. Similarly, electric vehicles and energy storage systems also require power electronic converters to interact with the grid.

In transmission and distribution networks, more and more power electronic converters, such as HVDC (high-voltage DC) links (Arrillaga 2008) and FACTS (flexible AC transmission systems) devices (Hingorani and Gyugyi 1999), are being added to electronically, rather than mechanically, control future power systems (Hingorani 1988) in order to reduce power losses and improve controllability. The US Department of Energy is developing a roadmap to strategically adopt solid state power substations (SSPS) to provide enhanced capabilities and support the evolution of the grid (Taylor et al. 2017).

Putting all the above together, future power systems will be power electronics based, instead of electric machines based, with a huge number of relatively small and non-synchronous players at the supply side, inside the network, and at the demand side. Although these players are heterogeneous, they are all integrated with the transmission and distribution network through power electronic converters that convert electricity

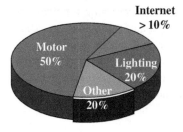

Figure 2.3 Approximate electricity consumption in the US.

Figure 2.4 A two-port virtual synchronous machine (VSM).

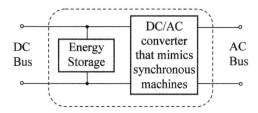

between AC and DC. If all these power electronic converters could be controlled to behave in the same way, then millions of heterogeneous players could be homogenized and equalized (in the sense of per unit, i.e. in proportion to the capacity), achieving legal equality. Even better, if these power electronic converters could be controlled to behave like synchronous machines, then they would possess the inherent synchronization mechanism of synchronous machines as well. Such converters are called virtual synchronous machines (VSMs) (Zhong 2016b) or cyber synchronous machines (CSMs) as coined in (Zhong 2017e).

A VSM is a DC/AC converter that mimics synchronous machines with a built-in energy storage unit connected on the DC bus, as illustrated in Figure 2.4 or on an additional port, as illustrated in Figure 19.1. The capacity of the energy storage unit can be large or small, depending on the magnitude and length of the power imbalance between the DC bus and the AC bus to be handled. For some applications, it is enough just to use the capacitors on the DC bus without adding an extra energy storage unit. The electricity can flow from the AC side to the DC side as a rectifier or, vice versa, as an inverter.

It is worth highlighting that operating power electronic converters as VSMs should not stop at the stage of simply mimicking conventional SMs but should advance to transcend conventional SMs. Power electronic converters can respond much faster than conventional SMs and possess much better controllability, which makes it possible for a VSM to transcend conventional SMs. This will be described in detail in later chapters.

2.4 SYNDEM Grid Architecture

2.4.1 Architecture of Electrical Systems

After homogenizing all heterogeneous players to achieve legal equality and equipping them with the synchronization mechanism of synchronous machines (SM) as a rule of law, the SYNDEM grid architecture is obtained and shown in Figure 2.5. All conventional power plants, including coal-fired, hydro and nuclear power plants, are integrated to the transmission and distribution network through SM as normally done without any major changes. All DERs that need power electronic inverters to interface with the grid are controlled to behave as VSMs, more specifically, as virtual synchronous generators, to interact with the grid and all loads that have rectifiers at the front end are controlled to behave as VSMs, more specifically, as virtual synchronous motors. For HVDC links, the power electronic converters at both ends are controlled as VSMs, one as a virtual synchronous generator and the other as a virtual synchronous motor. This presents a unified, harmonized, and scalable architecture for future power systems. It is applicable to a system with one generation node and one load node or a system with millions of nodes.

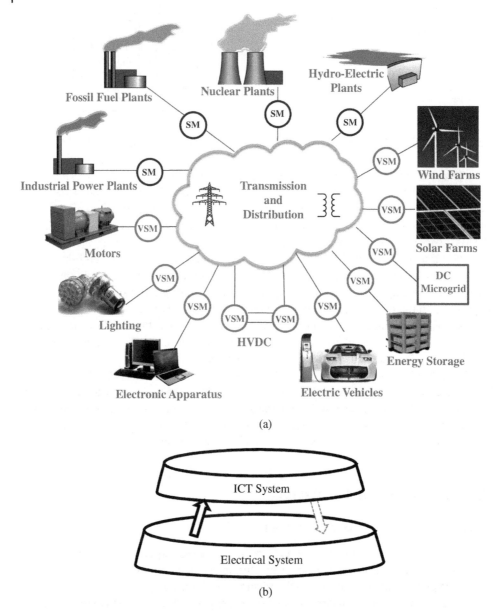

Figure 2.5 SYNDEM grid architecture based on the synchronization mechanism of synchronous machines (Zhong 2016b, 2017e). (a) Electrical system. (b) Overall system architecture highlighting the relationship between the electrical system and the ICT system, where the dashed arrow indicates that it may not exist.

In such a power system, all power electronics based players, at the supply side, inside the network, and at the demand side, are empowered to actively regulate system stability in the same way as conventional power plants do, unlike in current power systems where only a small number of large generators regulate the system stability. Hence, this architecture is able to achieve the paradigm shift of power systems from *centralized control* to *democratized interaction.* Because the synchronization mechanism of SMs are inherently embedded inside all active players, they autonomously interact with each other via exchanging power through the electrical system. This paves the way for the autonomous operation of power systems, which means minimal human intervention is needed to maintain system operation within the designed frequency and voltage boundaries. For example, when a coal-fired power plant is tripped off, the system frequency drops. All SMs/VSMs that take part in the autonomous regulation of system stability on the supply side would quickly and autonomously respond to the frequency drop and increase their power output in order to balance the power shortage. At the same time, all VSMs that take part in the autonomous regulation of system stability on the demand side would autonomously decrease power consumption to balance the power shortage. As a result, the frequency drop is reduced, which helps reduce the number of loads to be tripped off. If a new power balance cannot be reached after all generators reach the maximum capacity then some VSMs that serve non-critical loads would further reduce the power intake until a new power balance is reached. Similarly, if a heavy load is turned off, all SMs/VSMs that take part in the autonomous regulation of system stability on the supply side would quickly and autonomously reduce the power output and all VSMs that can take part in the autonomous regulation of system stability on the demand side would autonomously increase power consumption to help reach a new power balance. The change of load power can be of short term or long term, depending on the types and functions of loads. Similarly, the variability of DERs can be taken care of by the players as well. As a result, a SYNDEM smart grid can prevent local faults from cascading into wide-area blackouts, correcting the systemic flaw of power systems about local faults cascading into wide-area blackouts.

The deployment of the SYNDEM grid architecture depends on the flexibility of power systems in generation and consumption. This is not a problem. Indeed, power systems worldwide are designed to be very flexible. For example, the UK Grid Code (National Grid, 2016) dictates that the system frequency shall be controlled within 49.5–50.5 Hz, i.e. $\pm 1\%$ around the nominal frequency, and that generators and apparatus should be capable of operating continuously when the system frequency is within 49.0–51.0 Hz. For a 2% frequency droop slope, a 0.5 Hz change of frequency is equivalent to having additional reserve at the level of 50% of the system capacity. Moreover, the SYNDEM grid architecture is able to release the inertia in wind turbines and large motors etc., which further increases the system inertia. If the reserve/inertia is still not enough, storage systems can be added. Note that the fast reaction of power electronic converters could also reduce the required level of inertia. Hence, it is envisioned that the flexibility of a SYNDEM smart grid is not a problem at all. Similarly, the normal operating range for voltage is $\pm 5\%$ for 400 kV and $\pm 10\%$ for 275 kV and 132 kV in the UK. There is plenty of flexibility in reactive power and voltage. Thus, the SYNDEM grid architecture offers a means to fully release the potential of the flexibility already in power systems, improving system stability and reliability.

The architecture shown in Figure 2.5(a) empowers all players to directly take part in the regulation of system stability, which enhances system autonomy. This is consistent with the worldwide trend of increasing autonomy and declining hierarchy in different areas (Anderson and Brown 2010; Friedman 2005; Moore 2011).

The architecture shown in Figure 2.5(a) is lateral rather than hierarchical, although it allows the addition of hierarchical management layers on top of it if needed. Hence, it offers

Table 2.1 Machines that power the industrial revolutions.

Industrial revolution	Machines that power the revolution
The first	Steam engines (mechanical)
The second	Electric machines (electro-mechanical)
The third	Virtual synchronous machines (electrical)?

a technical solution to realize the lateral power envisioned in (Rifkin 2011). It has been widely recognized that steam engines are the machines that powered the first industrial revolution and electric machines are the machines that powered the second industrial revolution. It is likely that virtual synchronous machines are the machines that will power the third industrial revolution because it is fair to say that steam engines are mechanical, electric machines are half-electrical and half-mechanical, while virtual synchronous machines are fully electrical. This is summarized as shown in Table 2.1.

2.4.2 Overall Architecture

As is well known, what makes a grid smart is the introduction of an ICT system to the electrical system of a power system. However, any unnecessary or improper use of ICT systems may lead to serious concerns on, e.g. cyber security and reliability, because cyber-attacks could be launched by anybody, at any time, from anywhere. This is a systemic flaw of the current-generation smart grid. Needless to say, the electrical system is primary and the ICT system is secondary. The electrical system is the foundation for providing electrical services while the role of the ICT system is to offer added value without causing additional problems. The electrical system should be able to work without the ICT system. Otherwise, the interdependence of electrical and ICT systems may open the door to unknown problems. In a SYNDEM grid, all SMs and VSMs have the same synchronization mechanism so there is no need to rely on an additional communication network for low-level control and system regulation. Figure 2.5(b) illustrates the overall SYNDEM grid architecture, with the electrical system at the low level and the ICT system at the high level. It has clarified the relationship of the ICT system and the electrical system: various data from the electrical system can be passed on to the ICT system as indicated by the solid arrow but the instructions from the ICT system may or may not be passed to the electrical system as indicated by the dashed arrow, depending on the required level of security. The ICT system can be released from low-level control to focus on high-level functions, such as SCADA and market operations.

It is pertinent to clarify the unidirectional role of the ICT system in a SYNDEM smart grid and decouple it from the low-level or local control of devices and equipment because the SYNDEM smart grid architecture inherently empowers all players to actively take part in the regulation of the grid. Each device or equipment is able to operate autonomously without relying on external communication, while having the capability to receive reference set-points and instructions if needed. This significantly reduces the need for a high-speed communication infrastructure and mitigates the concern of cyber security.

The ICT system is no longer essential for the basic operation of the electrical system. Even if the ICT system at the high level is not working, the electrical system at the low level is able to operate and provide basic services autonomously. This provides the foundation to achieve cyberattack-free, autonomous, renewable electric (CARE) power systems and completely solves the systemic flaw of power systems with regarding to cyber security.

The focus of this book is on the SYNDEM electrical system. For ICT systems, see e.g., (Ye et al. 2017).

2.4.3 Typical Scenarios

The SYNDEM grid architecture is scalable and can be applied to power systems at different scales, from single-node systems to million-node systems.

When there is a need, small-scale systems can be connected together, through a device called an *energy bridge*. Depending on the need, an energy bridge can be simply a circuit breaker, a transformer, a back-to-back power electronic converter, or their combination. Its main function is to connect and isolate two circuits. Its secondary functions may include protection, voltage conversion, and/or frequency conversion. If it is a back-to-back power electronic converter, then both sides can be operated as a VSM. If a part of the system is faulty, then it can be disconnected to isolate the faulty part of the system; after the fault is cleared, it can be re-connected. Hence, while the SYNDEM grid architecture allows small-scale grids to merge and form large-scale grids, it also naturally allows large grids to break into small ones, making future power systems self-organizable and autonomous.

Some typical scenarios of SYNDEM smart grids are outlined below.

2.4.3.1 Home Grid

Figure 2.6 illustrates a SYNDEM home grid, which consists of a wind turbine VSM, a solar VSM, and some load VSMs. It can be operated with or without a utility grid, making it an ideal solution to advance energy freedom. It is possible to power homes with several solar panels or small wind turbines. This is particularly useful for disaster relief and remote areas without public utilities. As a result, the SYNDEM grid architecture could eventually bring energy freedom to individuals, providing the highest resiliency to each home.

2.4.3.2 Neighborhood Grid

The home grids in a neighborhood can be connected together to form a neighborhood grid through energy bridges, as illustrated in Figure 2.7. This will allow neighbors to support each other when there is a fault on the public utility grid, providing the highest resiliency to neighborhoods.

2.4.3.3 Community Grid

The neighborhood grids can be connected together to form a community grid through energy bridges, as illustrated in Figure 2.8. These energy bridges often have higher current ratings than the ones used to connect/isolate individual homes. Since an important function of the energy bridge is to isolate a part of the grid when there is a fault without affecting the normal part, this provides the highest resiliency to the community.

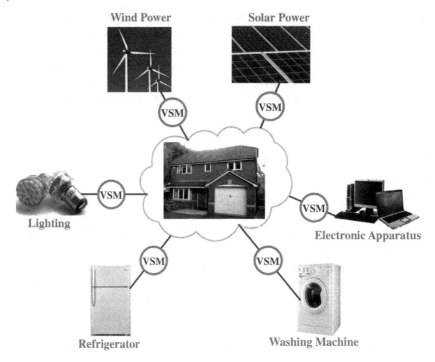

Figure 2.6 A SYNDEM home grid.

2.4.3.4 District Grid
The community grids in a district can be connected together to form a district grid through energy bridges, as illustrated in Figure 2.9. These energy bridges often step up the voltage to reduce losses. Again, any part of the grid that is faulty can be isolated and reconnected after removing the fault, providing the highest resiliency.

2.4.3.5 Regional Grid
The district grids in a region can be connected together to form a regional grid through energy bridges, as illustrated in Figure 2.10. These energy bridges often step up the voltage further. Again, any part of the grid that is faulty can be isolated and reconnected after clearing the fault, providing the highest resiliency.

2.5 Potential Benefits

All the SMs and VSMs have the same intrinsic mechanism of synchronization so there is no need to rely on ICT systems to achieve low-level control. In other words, ICT systems can be released from low-level control to focus on high-level functions of power systems, e.g. SCADA and market operations (Wu et al. 2005), significantly reducing the investment in the ICT infrastructure. This also helps enhance the cyber security of the system because reduced or even no access to low-level controllers is provided to malicious attackers,

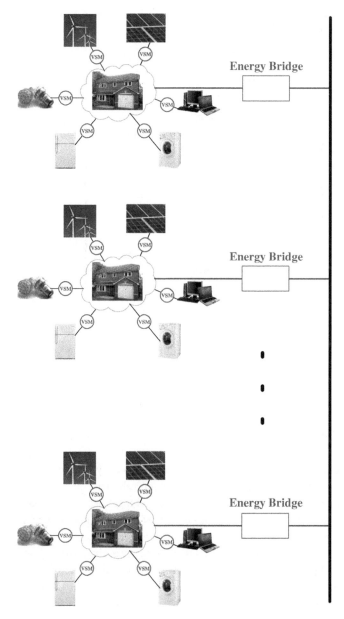

Figure 2.7 A SYNDEM neighbourhood grid.

significantly reducing the investment in cyber security and eliminates a systemic flaw for power systems with regarding to cyber security.

The SYNDEM architecture turns all loads with a power electronic converter at the front end into active and responsible players to maintain system stability and achieves a continuous demand response. This prevents critical customers from suffering complete loss of electricity. Instead, all such loads make a small, often negligible, amount

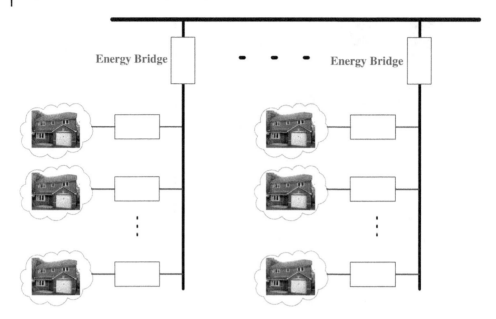

Figure 2.8 A SYNDEM community grid.

of contributions. This improves quality of service (QoS) and prevents local faults from cascading into wide-area blackouts, correcting another systemic flaw of power systems. Because the synchronization mechanism of synchronous machines is the key principle that has underpinned the growth and operation of power systems for over 100 years, the transition from today's grid into tomorrow's SYNDEM grid is evolutionary rather than revolutionary.

The SYNDEM grid architecture is scalable and can be applied to power systems at different scales, from single-node systems to million-node systems, from vehicles and aircraft to public grids. When there is a need, small systems can be connected together. When a part of the grid is faulty, it can be disconnected; after the fault is cleared, it can be re-connected. If HVDC links are used to link AC systems at different frequencies, then the AC systems can be operated together or, if needed, independently. Hence, while the architecture allows small grids to merge and form large-scale power grids, it also naturally allows large grids to break into small ones. Hence, this may offer the technical foundation to turn a move in China that broke up the Chinese Southern AC grid (Fairley 2016) into a natural trend worldwide.

The deployment of SYNDEM grids could considerably reduce infrastructure investment in generation, transmission and distribution networks. Take the UK power grid that has the capacity of about 70 GW as an example. If all loads contribute 2% in a contingency, which is within the tolerance for most individual loads, then the total contribution from all loads is about 1.4 GW. This is higher than the capacity of any current UK nuclear power plants, which means the UK grid is able to cope with the trip-off of any nuclear power plant if all loads consumes 2% less electricity. The architecture also allows all inertia that already exists in the system, e.g., those in wind turbines, large industrial motors and electric vehicles, to be

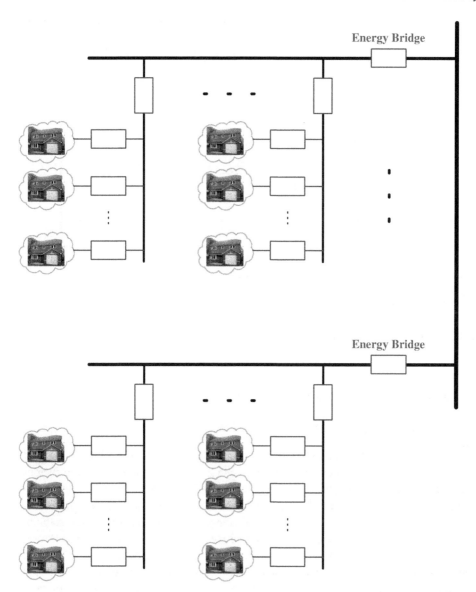

Figure 2.9 A SYNDEM district grid.

released, which helps considerably reduce the operational cost because of the fast reaction of power electronic converters. For example, if there are 10 million laptops plugged in a grid and each contributes 10 W when needed, then the equivalent reserve amounts to 100 MW, which is at the level of power generated by 50 wind turbines rated at 2 MW.

Because of the intrinsic synchronization mechanism embedded into each VSM, it is less likely for a VSM to disconnect from the grid under grid variations. This will increase the uptime of renewable generators and hence the yield, bringing more revenue to the owner.

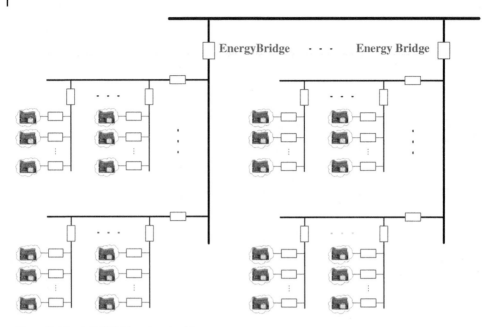

Figure 2.10   A SYNDEM regional grid.

2.6   Brief Description of Technical Routes

2.6.1   The First-Generation (1G) VSM

Different options to implement VSMs are available in the literature. The VISMA approach (Beck and Hesse 2007; Chen et al. 2011) controls the inverter current to follow the current reference generated according to the mathematical model of synchronous machines, which makes inverters behave like controlled current sources. Since power systems are dominated by voltage sources, this may bring detrimental impact, in particular on system stability (Dong et al. 2013; Sun 2011; Wen et al. 2015). The approach proposed in (Gao and Iravani 2008) follows the mathematical model of SMs but it requires the measurement of the grid frequency, which is often problematic in practice (Dong et al. 2015). The approach proposed in (Karimi-Ghartemani 2015) controls the voltage but it also requires the measurement of the grid frequency for the real power frequency droop control. The synchronverter approach (Zhong and Weiss 2009, 2011; Zhong et al. 2014) directly embeds the mathematical model of synchronous machines into the controller to control the voltage generated, even without the need for measuring the grid frequency or a phase-locked loop (Zhong et al. 2014). The synchronverter has been further developed for microgrids (Ashabani and Mohamed 2012), HVDC applications (Aouini et al. 2016; Dong et al. 2016), STATCOM (Nguyen et al. 2012), PV inverters (Ming and Zhong 2014), wind power (Zhong et al. 2015), motor drives (Zhong 2013a), and rectifiers (Ma et al. 2012; Zhong et al. 2012b). The synchronverter technology offers a promising technical route to implement SYNDEM smart grids and is described in detail in Part II. Because it offers a basic and conceptual implementation of VSMs, it is classified as the first-generation (1G) VSM.

2.6.2 The Second-Generation (2G) VSM

It is well known that synchronous machines have inductive output impedances because of the stator windings. However, the output impedance of power electronic converters changes with the hardware design and the controller and could be inductive (denoted L-converters), resistive (denoted R-converters) (Guerrero et al. 2005; Zhong 2013c), capacitive (denoted C-converters) (Zhong and Zeng 2011, 2014), resistive-inductive (denoted R_L-converters), resistive-capacitive (denoted R_C-converters) or complex around the fundamental frequency. The impedance of converters plays an important role in system stability (Sun 2011; Vesti et al. 2013; Wang et al. 2015b; Zhong and Zhang 2019). For inverters with different types of output impedance, the widely adopted droop control appears in different forms (Zhong and Hornik 2013), which means converters with different types of impedance connected together could lead to instability. The droop controller adopted in synchronverters and conventional power systems implicitly assumes that the impedance is dominantly inductive. There is a need to develop a SYNDEM technical route that is applicable to converters with different types of impedance, while possessing the synchronization mechanism of synchronous machines.

Since a droop controller structurally resembles an enhanced PLL (Zhong and Boroyevich 2013, 2016), it also has the intrinsic synchronization mechanism of synchronous machines and can provide a potential technical route to implement VSMs. The robust droop controller (Zhong 2013c), initially proposed for R-inverters to achieve accurate power sharing and tight voltage regulation, has been proven to be universal and applicable to inverters with output impedance having an impedance angle between $-\frac{\pi}{2}$ rad and $\frac{\pi}{2}$ rad (Zhong and Zeng 2016). Moreover, it can be equipped with a self-synchronization mechanism without a PLL (Zhong et al. 2016). Hence, the robust droop controller offers another, actually better, technical route to implement SYNDEM smart grids. A VSM based on the robust droop controller is classified as a second-generation (2G) VSM.

2.6.3 The Third-Generation (3G) VSM

There are many challenges during this paradigm shift of power systems. The iceberg of power system challenges and solutions is illustrated in Figure 2.11. On the surface, the challenges are seen as the change from centralized generation to distributed generation or democratization with many heterogeneous players, which are homogenized after operating power electronic converters as VSMs. However, the system stability is still a question because an interconnected system may become unstable even for individually stable systems. This relies on the synchronization mechanism (somewhat hidden). Both the first and the second generations of VSMs are equipped with the inherent synchronization mechanism, but it is not straightforward to rigorously prove the system stability. The third-generation (3G) of VSMs are expected to guarantee the stability of each individual VSM and also the stability of a system with multiple VSMs interconnected together.

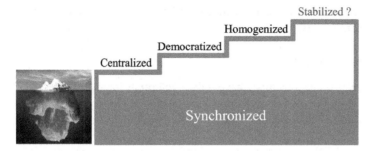

Figure 2.11 The iceberg of power system challenges and solutions.

2.7 Primary Frequency Response (PFR) in a SYNDEM Smart Grid

Maintaining the stability of frequency is a top priority in power systems operation. There are three types of overlapping frequency control (Illian H. et al. 2017):

(1) Primary frequency control, which is any action provided on an interconnection to stabilize frequency in response to a change frequency. Primary frequency control comes from an automatic generator governor response (also known as speed regulation) and a load response (load damping), typically from motors and other devices that provide an immediate response based on local (device-level frequency responsive) control systems or device characteristics.

(2) Secondary frequency control, which is any action provided by an individual control area (CA), balancing authority (BA), or its reserve sharing group to correct the resource–load imbalance that creates the original frequency deviation, and restores both scheduled frequency and primary frequency responsive reserves. Secondary control comes from either manual or automated dispatch from a centralized control system to correct frequency error.

(3) Tertiary frequency control, which is any action provided by control areas on a balanced basis that is coordinated so there is a net zero effect on the area control error (ACE). Examples of tertiary control include the dispatching of generation to serve native load, economic dispatch to affect interchange, and the re-dispatching of generation. Tertiary control actions are intended to restore secondary control reserves by reconfiguring reserves.

The US Federal Energy Regulatory Commission (FERC) requires newly interconnecting large and small generating facilities, both synchronous and non-synchronous, to install, maintain, and operate equipment capable of providing primary frequency response as a condition of interconnection (FERC 2018). Since all SMs and VSMs can provide primary frequency control or PFR against frequency excursions, a SYNDEM smart grid naturally meets this FERC requirement. Moreover, the loads interfaced with power electronic converters are able to provide PFR as well. It is expected that this will be required by the regulatory commissions in the US and other countries in the near future.

It is envisioned that, eventually, no secondary or tertiary frequency control is needed for a SYNDEM grid because all players can autonomously take part in system regulation.

2.7.1 PFR from both Generators and Loads

In a SYNDEM smart grid, all non-synchronous active participants, both generators and loads, are equipped with the intrinsic synchronization mechanism of synchronous machines. The generators are all turned into frequency-responsive synchronous participants and can provide (virtual) kinetic balancing inertia to improve frequency stability, in the same way as conventional synchronous machines. This stops the trend of decreasing inertia due to the penetration of DERs. The loads interfaced through power electronic converters also become frequency-responsive and can automatically respond to the change of frequency, quickly regulating power consumption without impacting user experience. This frequency-responsive characteristic makes these customer-side loads play a similar role to PFR on the generator side. It stops the trend of decreasing load damping because of the increasing adoption of loads controlled independently of frequency. See Chapters 5 and 19 for more detail about PFR provided by loads.

2.7.2 Droop

Droop control plays an important role in PFR. The droop settings of individual participants specify the slope and the amount of the PFR. In a SYNDEM smart grid, VSMs associated with different types of suppliers and loads, according to their nature, can be configured with different droop coefficients addressing critical levels, economic benefits, frequency conditions, and other factors. For example, a wind generator may not be able to provide enough PFR for a low frequency condition but can easily provide PFR for a high frequency condition. It is normally not a problem to completely shut down heating, ventilation, and air conditioning (HVAC) systems for several minutes or to shift washing machine and dishwasher use by a couple of hours, even up to 24 h. For economic reasons, it is also possible to set a large droop, e.g. 10% for small frequency conditions and a small droop, e.g. 3%, for large frequency conditions. In this way, achieving the maximum PFR without significantly affecting the quality of service is possible in many cases. Moreover, shifting the peak load reduces the peak/normal load ratio and the PFR needed as well.

2.7.3 Fast Action Without Delay

VSMs are inherently power electronic converters and act upon frequency changes without delay. Since any delayed response increases the maximum frequency change in the event of disturbances, the fast action of VSMs reduces the amount of balancing inertia required before the frequency change is arrested in the event of disturbances. For short frequency spikes, the impact on the system is small because of the relatively large system inertia. A VSM acts upon frequency spikes quickly but also returns to normal quickly after the spikes.

2.7.4 Reconfigurable Virtual Inertia

The kinetic balancing inertia of a conventional synchronous machine does not vary. However, the virtual inertia of a VSM is reconfigurable. This provides more flexibility for the PFR. Moreover, the virtual inertia of a VSM does not involve estimation of system frequency or the rate of change of frequency (RoCoF). This is very different from the synthetic inertia

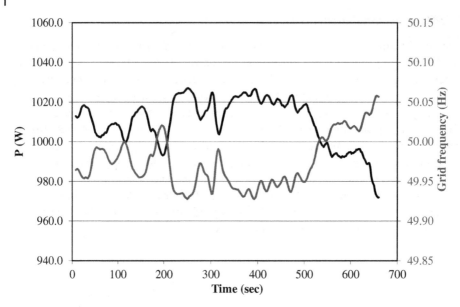

Figure 2.12   The frequency regulation capability of a VSM connected the UK public grid.

that changes the power according to the measured frequency and its rate of change, which causes noise amplification due to the measurement of df/dt (Duckwitz and Fischer 2017; Zhu et al. 2013). Another important factor related to system instability is damping. The damping of a VSM can be reconfigured to improve stability as well. See Chapter 14 for details about reconfiguration of virtual inertia and virtual damping.

2.7.5   Continuous PFR

A prominent feature of the SYNDEM grid architecture is that some active participants, both supply and load, can provide PFR. They can continuously adjust the output or the intake according to the system frequency in an autonomous manner. Figure 2.12 illustrates this capability of a VSM connected to the UK grid. The real power output changed autonomously according to the changing frequency, demonstrating excellent PFR. Some non-essential loads, e.g., HVAC systems and pumping systems, can also provide this continuous PFR at a very high level. In other words, these loads can provide continuous demand response, autonomously, instead of the conventional on–off demand response. See Chapters 5 and 19 for more details about continuous PFR provided by loads.

2.8   SYNDEM Roots

2.8.1   SYNDEM and Taoism

The principle of synchronization and democratization is inspired by, and has deep roots in, the Chinese classic philosophical text *Tao Te Ching*; see, e.g., (Lao Tzu 2016a,b). Written by Lao Tzu some 2500 years ago, it is believed to be the second most translated work in the world after the *Bible*. It is a fundamental text for both philosophical and religious Taoism, with strong influences on other schools, such as Legalism, Confucianism, and Chinese Buddhism. It describes the *Tao* (principles) of nature and *Te* (virtue) for the human race, with emphases on

(1) The harmony between nature and the human race. Nature is vast and there are many people. However, nature and the human race can and should live in harmony. Excessive human activities may damage nature, which can in turn affect or even destroy the human race.

(2) The criticality of following natural principles. An artificial solution may work for a short period but in the long run it could be catastrophic if it does not follow natural principles. It is critical to *let things happen* rather than *make things happen*.

(3) Everything is nothing and nothing is everything. If natural principles are not followed, whatever developed by the human race may get destroyed. "When natural principles are followed, even if there is nothing, one would appear. Then, one generates two; two generates three; and three generates many."

(4) Simplicity and minimum action. "Managing a country is like cooking a small dish." Not much needs to be done. Complex solutions have no practical benefit. A solution for a challenging problem has to be simple and non-intrusive.

The deep roots of SYNDEM smart grids in *Tao Te Ching* is evidenced by the following[1]:

(1) The newly added renewable energy sources and flexible loads are harmonized to behave in the same way as conventional power plants.

(2) The most natural principle of synchronization is followed by all players in SYNDEM smart grids. They actively interact with the grid to maintain system stability. They are not forced or led through commands given over communication networks.

(3) A SYNDEM smart grid can work with one, two, three or many active players, and when needed a large power system can decompose into small regional grids or even microgrids in residential homes. It is extremely scalable and the architecture is very simple.

(4) Managing a SYNDEM smart grid becomes very easy because all suppliers and loads in a grid can autonomously take part in the regulation of the grid. In other words, a SYNDEM smart grid manages itself with minimum human intervention.

2.8.2 SYNDEM and Chinese History

The concept of SYNDEM smart grids also has deep roots in Chinese history[2]. During the Spring–Autumn period (approximately 771 to 476 BCE) and the Warring States period (approximately 475 to 221 BCE) of ancient China, China was in the form of multiple states. It was an era of great cultural and intellectual expansion in China, when hundreds of schools of thought and philosophies flourished. Among them, the most influential thoughts included:

- Taoism – established by Lao Tze (571–471 BCE) with the core principle of living in harmony with nature.
- Confucianism – established by Confucius (551–479 BCE) with the core principle of being inclusive and kind.

---

1 www.syndem.com
2 https://www.youtube.com/watch?v=nk6F2jQJdDg

- Legalism – established by Li Kui (455–395 BCE) with the core principles of governing with laws.
- Mohism – established by Mo Tze (470–c. 391 BCE) with the core principle of thinking with logic.
- School of Yin-Yang – established by Zou Yan (305–240 BCE) with the attempt to explain the universe in terms of basic forces of Yin-Yang and the Five Elements of metal, wood, water, fire, and earth. Later, this was adopted by Taoism, Confucianism, and the Chinese medical framework.

It is fair to say that China was highly democratized during the Spring–Autumn and Warring States period, for over 500 years. People could move freely from one state to another and enjoyed freedom of speech. The most powerful states often held regular conferences to decide important matters. The thoughts and ideas refined during this era then became the cultural foundation of China and other East Asian countries.

However, it was also an era filled with chaos and bloody battles. During the Spring–Autumn period, there were nearly 500 battles, i.e. two battles a year on average. *The Art of War*, an ancient Chinese military treatise of strategy and tactics, was written by Sun Tze around 515–512 BCE, during the Spring–Autumn period. During the Warring States period, as the name of the period indicates, there were also many major battles among the states.

In 221 BCE, Emperor Qin (Chin) united China and harmonized Chinese characters, currency, trade (weights and measures), and communication (transportation). This laid a structurally stable foundation for China for many years to come, demonstrating the power of harmonization and unification. Harmony has then become the most important principle of Chinese culture, cultivating peace, stability, prosperity, and sustainability.

What future power systems need is a structurally stable foundation to enable natural growth and operation. This is exactly what the SYNDEM theoretical framework offers.

2.9 Summary

Based on (Zhong 2013b, 2017e,f), this chapter presents the SYNDEM theoretical framework for the next-generation smart grids – power electronics-enabled autonomous power systems, including its concepts, rules of law, legal equality, architecture, benefits, technical routes, and roots. It is highlighted that the synchronization mechanism of synchronous machines can continue underpinning the next-generation smart grids and that power electronic converters can be equipped with this mechanism. As a result, all players can actively take part in the grid regulation in a synchronized and democratized manner, achieving Cyberattack-free, Autonomous, Renewable Electric (CARE) power systems. The SYNDEM grid framework is able to correct the systemic flaws of cyber security and cascading failures in the current power systems.

3

Ghost Power Theory

The concepts of real power and reactive power play a vital role in power systems. The physical meaning of real power is very clear: it represents the power (energy) that does the real work, either generated or consumed. However, the physical meaning of reactive power is not clear. It is an imaginary mathematical concept. In this chapter, a new operator, called *the ghost operator g*, is introduced to physically construct the ghost of a (sinusoidal) signal and further the ghost of a system with sinusoidal inputs. The ghost operator satisfies $g^2 = -1$ but it is different from the imaginary operator. With the concept of the ghost system, the reactive power of an electrical system is the real power of the ghost system with its input being the ghost of the input to the original system. This is then extended to define the reactive power for mechanical systems, which completes the electrical-mechanical analogy. Furthermore, this is generalized to any dynamic system that can be described by a port-Hamiltonian (PH) system model. This establishes a significantly simplified instantaneous power theory, referred to as the ghost power theory. It is applicable to any dynamic system, single phase or poly-phase, with or without harmonics.

3.1 Introduction

In electrical engineering, real power and reactive power are two well known concepts. The physical meaning of the real power is very clear: it represents the real power (energy) that is consumed. However, the physical meaning of the reactive power is still not clear. It is a mathematical concept and has different interpretations; see e.g. (Akagi et al. 2007; de Leon et al. 2012; Montanari and Gole 2017; Peng and Lai 1996) and the references therein. It reflects the quantity of power due to the phase mismatch between the current and the voltage caused by reactive elements like capacitors and inductors. The reactive power is conventionally defined as a component of the instantaneous (real) power. It is sometimes regarded as the power oscillating in the system. However, for balanced three-phase systems there is no oscillating power; for single-phase resistive systems there is no reactive power although the power pulsates. Many attempts have been made to reveal the physical meaning of reactive power, e.g., by using the vector product (Peng and Lai 1996), the Clarke transformation (Akagi et al. 2007), the Poynting theorem (de Leon et al. 2012), and the *mno*

Power Electronics-Enabled Autonomous Power Systems: Next Generation Smart Grids,
First Edition. Qing-Chang Zhong.
© 2020 John Wiley & Sons Ltd. Published 2020 by John Wiley & Sons Ltd.

transformation (Montanari and Gole 2017). However, these only reveal some aspects of the reactive power. The physical meaning of reactive power is not fully established. It is still imaginary.

Actually, this phenomenon is quite common in science and engineering: many imaginary concepts are introduced without a physical meaning. Reactive power is just one of them. Another famous example is the imaginary number, as the name itself indicates. An imaginary number is a complex number that can be written as a real number multiplied by the imaginary unit j, which is defined by $j^2 = -1$. Imaginary numbers do not exist physically but have been playing a fundamental role in science and engineering. The searching for the physical meaning of these imaginary concepts has been ongoing for years.

The mathematical operator called *the ghost operator* coined in (Zhong, 2017b), in short the g-operator, shifts the phase of a sine or cosine function by 90° leading. It is similar to the imaginary operator because it satisfies $g^2 = -1$ but it does not return imaginary numbers. Here, by $g^2 = -1$, it means applying the ghost operator g twice to a sine or cosine function returns the opposite of the sine or cosine function. The ghost operator can be applied to *physically construct* the ghost of a signal and furthermore the ghost of a system. Interestingly and surprisingly, by adopting the PH system theory (van der Schaft and Jeltsema 2014), it is proven that the ghost of a system behaves exactly in the opposite way to the original system if the input to the ghost is the ghost of the input to the original system. An important property of the g-operator is that the operator $\begin{bmatrix} 0 & -g \\ g & 0 \end{bmatrix}$ transforms the signal and its ghost into themselves. With this, it is revealed that the instantaneous reactive power of a system is the instantaneous real power of its ghost system with its input being the ghost of the input to the original system.

It is well known that electrical systems and mechanical systems are dual to each other (Evangelou et al. 2006; Jiang and Smith 2011). While the role and importance of reactive power is well recognized for electrical systems, it is rarely mentioned for mechanical systems. There are only very limited attempts in the literature, e.g. (Mizoguchi et al. 2013; Rengifo et al. 2012), about the role of reactive power and power factor in mechanical systems. Following the understanding of the physical meaning of reactive power for electrical systems, the electrical-mechanical analogy is reviewed and a missing term, the reactive power, is identified. The ghost operator is then applied to define the reactive power for mechanical systems, completing the electrical-mechanical analogy. As a matter of fact, this can be generalized to other energy systems. This leads to an instantaneous power theory, referred to as the ghost power theory, for the sake of ease of reference.

3.2 Ghost Operator, Ghost Signal, and Ghost System

3.2.1 The Ghost Operator

Definition 1. The *ghost operator*, in short the g-operator, is the operator that shifts the phase of a sine or cosine function by $\frac{\pi}{2}$ rad leading.

Lemma 2. The ghost operator satisfies $g \sin \theta = \cos \theta$, $g \cos \theta = -\sin \theta$ and $g^2 = -1$.

Proof: It is straightforward to see that

$$g \sin \theta = \sin(\theta + \frac{\pi}{2}) = \cos \theta,$$

$$g \cos \theta = \cos(\theta + \frac{\pi}{2}) = -\sin \theta.$$

Applying the g-operator once more, then

$$g^2 \sin \theta = g \cos \theta = \cos(\theta + \frac{\pi}{2}) = -\sin \theta,$$

$$g^2 \cos \theta = -g \sin \theta = -\sin(\theta + \frac{\pi}{2}) = -\cos \theta.$$

Hence, $g^2 = -1$. This completes the proof. □

Note that $g \sin \theta$ means "applying" the ghost operator g to the function $\sin \theta$, instead of "multiplying" it with $\sin \theta$, and $g^2 = -1$ means applying the ghost operator g to a sine or cosine function twice returns the opposite of the sine or cosine function. Other notation involving the ghost operator g is similar. Apparently, $g^3 = -g$ and $g^4 = 1$. It seems that the ghost operator is very similar to the commonly used imaginary operator j, which satisfies $j^2 = -1$, but they are actually very different. The ghost operator is applicable to a sine or cosine function but the imaginary operator is applicable to any (complex) number. Moreover, applying the g-operator to a sine or cosine function always returns a real function (value) but applying the imaginary operator to a real number returns an imaginary number. Figure 3.1 illustrates the operation of the two operators. When both operators are applied to $\cos \theta$, respectively, the imaginary operator returns $j \cos \theta$ but the ghost operator returns $g \cos \theta = -\sin \theta$. When both are applied to $\sin \theta$, respectively, the imaginary operator returns $j \sin \theta$ but the ghost operator returns $g \sin \theta = \cos \theta$.

The ghost operator is also different from the Hilbert transform (Bracewell 1999; Weisstein 1999), which is defined for a real function $f(t)$ in L^2 as

$$\hat{f}(t) = \frac{1}{\pi} PV \int_{-\infty}^{\infty} \frac{f(\tau)}{\tau - t} d\tau \tag{3.1}$$

when the integral exists, where PV denotes the Cauchy principal value, i.e. $\lim_{\delta \to 0} \int_{-\infty}^{t-\delta} + \int_{t+\delta}^{\infty}$. This definition follows the widely used mathematical symbolic computation program Wolfram Mathematica and Mathworld (Weisstein 1999) and also (Bracewell 1999), using $\tau - t$ in the denominator instead of $t - \tau$ as in much other literature, e.g. (Johansson 1999; King 2009), with the difference being a negative sign. The real functions $f(t)$ and $\hat{f}(t)$ are orthogonal and, moreover, form a unique analytic function $f(t) - j\hat{f}(t)$, where j is the imaginary operator. When the Hilbert transform is applied twice, it transforms $f(t)$ into $-f(t)$, which is the same as the ghost operator. However, it transforms $\cos \omega t$ into $-\sin \omega t$ and $\sin \omega t$ into $\cos \omega t$ *only* when the frequency ω is positive. The ghost operator always transforms $\cos \omega t$ into $-\sin \omega t$ and $\sin \omega t$ into $\cos \omega t$ for *any* frequency ω.

3.2.2 The Ghost Signal

Without loss of generality, for a sinusoidal signal

$$e = E \sin(\omega t + \phi),$$

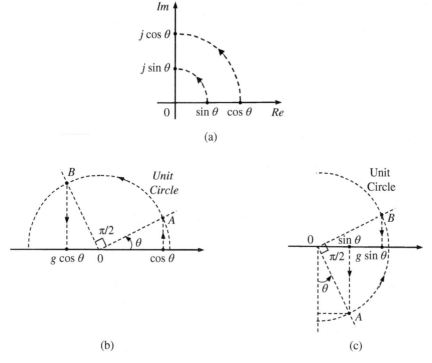

Figure 3.1 Illustrations of the imaginary operator and the ghost operator. (a) The imaginary operator applied to cos θ and sin θ. (b) The g-operator applied to cos θ. (c) The g-operator applied to sin θ.

its ghost signal is defined as

$$e_g = ge = Eg\sin(\omega t + \phi) = E\cos(\omega t + \phi).$$

It leads the signal e by 90°.

Lemma 3. The signal e and its ghost e_g satisfy

$$\begin{bmatrix} e_g \\ e \end{bmatrix} = \begin{bmatrix} 0 & g \\ -g & 0 \end{bmatrix}\begin{bmatrix} e_g \\ e \end{bmatrix} \quad \text{or} \quad \begin{bmatrix} e \\ e_g \end{bmatrix} = \begin{bmatrix} 0 & -g \\ g & 0 \end{bmatrix}\begin{bmatrix} e \\ e_g \end{bmatrix}.$$

Proof: It is straightforward to show these. □

In other words, $\begin{bmatrix} 0 & g \\ -g & 0 \end{bmatrix}$ and $\begin{bmatrix} 0 & -g \\ g & 0 \end{bmatrix}$ are identity operators, which transform the signal e and its ghost e_g into themselves without any change. This is a significant property because the eigenvalues of a skew-symmetric matrix always appear in pairs as $\pm\lambda$ (plus an unpaired 0 eigenvalue in the odd-dimensional case) and the non-zero eigenvalues of a real skew-symmetric matrix are all purely imaginary. The fact that the eigenvalues of the skew-symmetric operators $\begin{bmatrix} 0 & g \\ -g & 0 \end{bmatrix}$ and $\begin{bmatrix} 0 & -g \\ g & 0 \end{bmatrix}$ are all equal to 1 may open a door for many new applications.

3.2.3 The Ghost System

Systems take signals in the form of inputs and generate signals in the form of outputs. There are many ways to describe a system. In this chapter, the PH framework is adopted because of its direct link with the concept of power.

The PH system theory (van der Schaft and Jeltsema 2014) offers a systematic mathematical framework for structural modeling, analysis, and control of complex networked multi-physics systems with lumped and/or distributed parameters. It combines the historical Hamiltonian modeling approach in geometric mechanics (Holm, 2011a,b; Holm et al. 2009) and the port-based network modeling approach in electrical engineering (Carlin 1967, 1969; Tellegen 1952), via geometrically associating the interconnected network with a Dirac structure (van der Schaft and Jeltsema 2014), which is power-conserving. The Hamiltonian dynamics is defined with respect to the Dirac structure and the Hamiltonian representing the total stored energy. PH systems are open dynamical systems and interact/interconnect with their environment through ports.

For a dynamical system Z, if (i) there are no algebraic constraints between the state variables, (ii) the interconnection port power variables can be split into input and output variables, and (iii) the resistive structure is linear and of the input–output form, then the system Z can be described in the usual input u-state x-output y format (Jeltsema and Doria-Cerezo 2012; Maschke et al. 2000; van der Schaft and Jeltsema 2014) as

$$\dot{x} = [J(x) - R(x)]\frac{\partial H}{\partial x}(x) + G(x)u, \tag{3.2}$$

$$y = G^T(x)\frac{\partial H}{\partial x}(x), \tag{3.3}$$

where $x \in R^{n\times1}$ is the state vector, and u and $y \in R^{m\times1}$ are the input and the output, $H(x)$ is the Hamiltonian representing the total energy of the system and $\frac{\partial H(x)}{\partial x} \in R^{n\times1}$ is its gradient, $J(x) = -J^T(x)$ is a skew-symmetric matrix representing the network structure, $R(x)$ is a positive semi-definite symmetric matrix representing the resistive elements of the system, and $G(x)$ is the input matrix. All these matrices depend smoothly on the state x. Note that the Hamiltonian $H(x)$ is not necessarily non-negative nor bounded from below. A more general case is that there is a feed-through term $D(x, u)$ on the right-hand side of the output equation.

In general, the input to a system can be arbitrary but in this chapter it is assumed that the input u is periodic and hence can be described by the sum of a series of sinusoidal signals. This covers a wide range of engineering systems.

The time-derivative of the Hamiltonian $H(x)$ is

$$\frac{dH}{dt}(x(t)) = -\frac{\partial^T H}{\partial x}(x)R(x)\frac{\partial H}{\partial x}(x) + y^T u, \tag{3.4}$$

which characterizes the power conservation/balance property of PH systems. The product $y^T u$ is called the supply rate and has the unit of power. As a result, the Hamiltonian always satisfies

$$H(x(t)) \leq H(x(0)) + \int_0^t y^T u\, dt \tag{3.5}$$

because of the dissipated energy associated with $R(x)$. Moreover, if the Hamiltonian $H(x)$ is bounded from below by $C > -\infty$, then the system is passive (Byrnes et al. 1991)

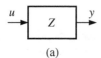

(a)

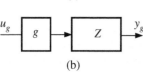

(b)

Figure 3.2 The system pair that consists of the original system and its ghost. (a) The original system Z. (b) The ghost system Z_g.

with $H_s(x) = H(x) - C$ being a non-negative storage function; the system is lossless if $\frac{dH_s}{dt}(x(t)) = y^T u$ (van der Schaft and Jeltsema 2014).

Definition 4. The ghost of a system is a mathematical object that contains an exact model of the system, including the same initial conditions, cascaded with the ghost operator at its input.

The system Z described in (3.2) and (3.3), and its corresponding ghost Z_g, are illustrated in Figure 3.2; both together form a system pair (Z, Z_g). Note that the line on the left side of the g-block does not have an arrow. This is to indicate that the application of the g-operator to the signal u_g instead of multiplying with u_g. Denote the state of the ghost system Z_g as x_g and the output as y_g. Note that u_g, x_g, and y_g are the input, the states, and the output of the ghost system Z_g, which are not necessarily the ghost signals of the input, the states, and the output of the system Z.

Lemma 5. If the input to the ghost system Z_g is the ghost u_g of the input u to the original system Z, i.e. $u_g = gu$, then the ghost state x_g is the opposite of the system state x, i.e. $x_g = -x$, and the ghost output is the opposite of the system output, i.e. $y_g = -y$.

Proof: Since the ghost system contains a model of the system Z, for the input $u_g = gu$, its state equation satisfies

$$\dot{x}_g = (J - R)\frac{\partial H}{\partial x_g} + Ggu_g$$

$$= (J - R)\frac{\partial H}{\partial x_g} - Gu, \tag{3.6}$$

where the property $g^2 = -1$ from Lemma 2 is invoked. Multiplying both sides by -1, then

$$-\dot{x}_g = (J - R)\frac{\partial H}{\partial(-x_g)} + Gu.$$

Substituting $x_g = -x$ recovers the state equation of the system given in (3.2). Note that the Hamiltonian H satisfies $H(x_g) = H(-x) = H(x)$ so there is no need to differentiate it for the system and its ghost.

Substituting $x_g = -x$ into the output equation of the ghost

$$y_g = G^T \frac{\partial H}{\partial x_g},$$

then

$$y_g = -G^T \frac{\partial H}{\partial(-x_g)} = -y. \tag{3.7}$$

This completes the proof. □

Remark 6. The ghost system behaves symmetrically (or oppositely) as the original system with respect to the origin but does not exist in reality and, hence, the name.

3.3 Physical Meaning of Reactive Power in Electrical Systems

Assume that the system Z described in (3.2) and (3.3) is an AC electrical system with the input voltage u and the output current i, i.e. $y = i$. Then the supply rate to the system Z is the instantaneous power

$$p = y^T u = i^T u. \tag{3.8}$$

For a single-phase system with

$$u = \sqrt{2}U \sin(\omega t), \qquad i = \sqrt{2}I \sin(\omega t - \phi),$$

where U and I represent the root-mean-square (rms) values of the voltage and current, respectively, there is

$$p = UI \cos \phi - UI \cos(2\omega t - \phi), \tag{3.9}$$

which can be rewritten as

$$p = UI \cos \phi(1 - \cos(2\omega t)) - UI \sin \phi \sin(2\omega t).$$

The first term $UI \cos \phi(1 - \cos(2\omega t))$ is non-negative and the second term $UI \sin \phi \sin(2\omega t)$ has a zero mean over the period $\tau = \frac{2\pi}{\omega}$, both oscillating at twice the frequency. The average value of p over one period τ is the same as the average value of the first term over one period, which is known as the real power P, i.e.

$$P = \frac{1}{\tau} \int_0^\tau p \, dt = \frac{1}{\tau} \int_0^\tau i^T u \, dt = UI \cos \phi.$$

It is also often called the active power or the average power. The maximum of the second term $UI \sin \phi \sin(2\omega t)$ is defined as the reactive power Q, i.e.

$$Q = UI \sin \phi.$$

The physical meaning of the real power has been very clear: it is the power that is actually consumed or does the real work. However, the physical meaning of the reactive power is not clear and it is just a mathematical formulation. It is known as an unwanted but unavoidable part of AC electric circuits, which is often regarded as the oscillating power in the system. However, this is not generally true (Akagi et al. 2007). For example, for balanced three-phase systems, the power p is not oscillating.

Now, the puzzle is solved. The reactive power is the power of the ghost system Z_g!

Indeed, the instantaneous power of the ghost system is the product of its voltage and current given by

$$q = u_g^T y_g = u_g^T i_g = i_g^T u_g = -i^T u_g, \tag{3.10}$$

which is

$$q = -i^T u_g$$
$$= -\sqrt{2}I \sin(\omega t - \phi) \times \sqrt{2}U \cos(\omega t)$$
$$= UI \sin \phi - UI \sin(2\omega t - \phi). \tag{3.11}$$

Its average over one period τ is the average reactive power

$$Q = \frac{1}{\tau} \int_0^\tau q \, dt = -\frac{1}{\tau} \int_0^\tau i^T u_g \, dt = UI \sin \phi.$$

For balanced three-phase systems, the oscillating terms in (3.9) and (3.11) all disappear, and the instantaneous real and reactive power are all equal to the average real and reactive power, respectively.

Comparing (3.9) and (3.11), it is easy to see that

$$\begin{bmatrix} p \\ q \end{bmatrix} = \begin{bmatrix} 0 & g \\ -g & 0 \end{bmatrix} \begin{bmatrix} p \\ q \end{bmatrix}.$$

Moreover, there is

$$\begin{bmatrix} P \\ Q \end{bmatrix} = \begin{bmatrix} 0 & g \\ -g & 0 \end{bmatrix} \begin{bmatrix} P \\ Q \end{bmatrix}.$$

This is a very interesting result: $P = gQ$ and $Q = -gP$. Once again, it is shown that $\begin{bmatrix} 0 & g \\ -g & 0 \end{bmatrix}$ is an identity operator.

Actually, the reactive power is widely known as the imaginary power. The ghost system does not exist in reality so its power is of course imaginary. Since the definition of reactive power is now dual to the real power, it is applicable to systems under different scenarios like the real power: with any number of phases and any number of harmonics, and without any intermediate transformation.

Assume that, for a general system Z, the voltage and current are, respectively,

$$u = \Sigma_n \sqrt{2}U_n \sin(n\omega t - \alpha_n),$$
$$i = \Sigma_m \sqrt{2}I_m \sin(m\omega t - \beta_m).$$

The corresponding ghost voltage is

$$u_g = gu = \Sigma_n \sqrt{2}U_n \cos(n\omega t - \alpha_n).$$

Then, the instantaneous real power and reactive power are, respectively,

$$p = i^T u$$
$$= \Sigma_m \sqrt{2}I_m \sin(m\omega t - \beta_m) \cdot \Sigma_n \sqrt{2}U_n \sin(n\omega t - \alpha_n),$$
$$q = -i^T u_g$$
$$= -\Sigma_m \sqrt{2}I_m \sin(m\omega t - \beta_m) \cdot \Sigma_n \sqrt{2}U_n \cos(n\omega t - \alpha_n).$$

Integrating both over one period, then the average real power and reactive power can be obtained as

$$P = \Sigma_n U_n I_n \cos \phi_n,$$
$$Q = \Sigma_n U_n I_n \sin \phi_n,$$

where $\phi_n = \beta_n - \alpha_n$, because the integration of the cross-frequency terms with $m \neq n$ over one period is 0. This is consistent with the conventional definition for reactive power (and real power) (Erlicki and Emanuel-Eigeles 1968). The apparent power S is defined to reflect the capacity of the system as

$$S = UI,$$

where $U = \sqrt{\Sigma_n U_n^2}$ and $I = \sqrt{\Sigma_m I_m^2}$ are the rms values of the voltage and the current, respectively. The difference between S^2 and $P^2 + Q^2$ is then characterized by the distortive power D, also called the harmonic power, via

$$S^2 = P^2 + Q^2 + D^2.$$

Finally, the power factor of the system is defined as $\frac{P}{S}$ to reflect the utilization of the capacity.

What is described above is actually an instantaneous power theory for AC electrical systems that is significantly simpler than existing power theories, e.g. the commonly used instantaneous p-q power theory (Akagi et al. 2007), which involves the transformation of the voltages and currents into the $\alpha\beta 0$ coordinates through the Clarke transformation. Simplicity is beauty.

3.4 Extension to Complete the Electrical-Mechanical Analogy

Voltage, current, flux linkage and charge are fundamental concepts in electrical systems. The voltage is defined as the change in the flux linkage and the current is defined as the change in the charge. The other relationships among these four concepts are then characterized by the four basic electric elements: resistor (voltage ~ current), inductor (flux linkage ~ current), capacitor (charge ~ voltage) and memristor (flux linkage ~ charge) (Chua 1971). The term impedance is defined in the frequency domain as the ratio between the voltage and the current, with its inverse called admittance. The term power is defined as the product of the voltage and the current in the time domain, with its average called the real power. For AC electrical systems, there is also the term reactive power as discussed in the previous section.

It is well known that electrical systems and mechanical systems are analogous (Firestone 1933; Jiang and Smith 2011). The fundamental concepts in both fields are summarized in Table 3.1 to demonstrate this duality. Here, the force–current analogy instead of the commonly used force–voltage analogy is adopted, which has led to the discovery of the two-terminal mechanical device (called the inerter) corresponding to the capacitor (Smith, 2002). Most of the dual concepts in Table 3.1 are well known, except the charge–momentum and flux linkage–displacement pairs. According to Newton's second law, the rate of change of the momentum of a particle is proportional to the force acting on it, which makes the momentum dual to the charge in electric systems. For rotational

Table 3.1 The electrical-mechanical analogy based on the force–current analogy.

Electrical systems	Translational mechanical systems	Rotational mechanical systems
Resistor	Damper	Damper
Inductor	Spring	Spring
Capacitor	Inerter (mass)	Moment of inertia
Flux linkage	Displacement	Angle
Voltage	Velocity	Angular velocity
Current	Force	Torque
Charge	Momentum	Angular momentum
Impedance	Impedance	Impedance
Admittance	Admittance	Admittance
Real power	(Real) power	(Real) power
Reactive power	? *(Reactive power)*	? *(Reactive power)*

mechanical systems, the corresponding term is the angular momentum. As to the flux linkage–displacement analogy, it is clear from the fact that the derivative of the flux linkage is the voltage and the derivative of the displacement is the velocity. For rotational systems, the corresponding term is the angle. Note that the fourth electric element, the memristor that describes the relationship between the charge and the flux linkage, is not included in the table. As a side note, for the sake of completeness of the electrical-mechanical analogy, there should exist a mechanical element that describes the relationship between the momentum and the displacement, corresponding to the memristor. However, no known mechanical device exists and efforts should be made to identify this device. Anyway, this is not the focus here so it is not discussed further. The term impedance is defined as the ratio of velocity to force for translational systems and as the ratio of angular velocity to torque for rotational systems. Note that this is consistent with (Firestone 1933; Smith 2002) but it is not with those adopting the force–voltage analogy. The inverse of impedance is called admittance. As a result, a mechanical system can be analyzed or synthesized in the same way as its dual electrical system, if one more term – the reactive power – is properly defined.

For mechanical systems, the term work, which is the integral of power, is used more often than the term power. For translational systems, the work done is the product of the force and the displacement; for rotational systems, the work done is the product of the torque and the angle. This leads to the power defined as the product of the force and velocity for translational systems and as the product of the torque and the angular velocity for rotational systems. Apparently, the power defined in this way represents the real power.

For AC electrical systems, there are real power and reactive power as described in the previous section. However, for mechanical systems, there does not exist a well-accepted way of defining the reactive power. The understanding about the role of reactive power and power factor in mechanical systems is very limited (Mizoguchi et al. 2013; Rengifo et al. 2012). Actually, the reactive power in mechanical systems can be defined dually as the real power, following the electrical-mechanical analogy and the newly introduced ghost operator.

One barrier for this might be due to the fact that mechanical motions are caused by forces and torques, which are dual to current sources, but electrical motions are often caused by voltage sources. In order to better understand reactive power in mechanical systems, consider the case of an inductor L driven by a current source $i = -\sqrt{2}I \sin(\omega t - \frac{\pi}{2}) = \sqrt{2}I \cos(\omega t)$ at first. The voltage induced is

$$u = L\frac{di}{dt} = -\sqrt{2}I\omega L \sin(\omega t) = -\sqrt{2}\omega L I \sin(\omega t),$$

and its ghost is

$$u_g = gu = -\sqrt{2}\omega L I \cos(\omega t).$$

According to (3.10), the (instantaneous) reactive power of the system is the power of its ghost given by

$$
\begin{aligned}
q = i_g^T u_g &= -i^T u_g \\
&= \sqrt{2}I \cos(\omega t) \times \sqrt{2}\omega L I \cos(\omega t) \\
&= \omega L I^2 - \omega L I^2 \sin(2\omega t),
\end{aligned}
\tag{3.12}
$$

with the average reactive power being

$$Q = \omega L I^2.$$

This is consistent with the value obtained from the conventional way.

Now, consider a rigid body Z with the moment of inertia J_m driven by the torque $y = \sqrt{2}T \cos(\omega t)$, as described in Figure 3.2(a). The resulting angular velocity is $u = \sqrt{2}\frac{T}{\omega J_m} \sin(\omega t)$, with its ghost angular velocity being $u_g = gu = \sqrt{2}\frac{T}{\omega J_m} \cos(\omega t)$. As a result, dual to (3.10), the instantaneous reactive power is the power of the ghost illustrated in Figure 3.2(b), given by the product of $y_g = -y$ and the ghost angular velocity u_g as

$$
\begin{aligned}
q = y_g^T u_g &= -y^T u_g \\
&= -\sqrt{2}T \cos(\omega t) \times \sqrt{2}\frac{T}{\omega J_m} \cos(\omega t) \\
&= -\frac{T^2}{\omega J_m} + \frac{T^2}{\omega J_m} \sin(2\omega t),
\end{aligned}
$$

with its average reactive power being

$$Q = -\frac{T^2}{\omega J_m}.$$

The negative sign indicates that the rigid body "generates" reactive power, similar to the case of a capacitor. The second example to be considered is a spring with stiffness K subject to the force $y = \sqrt{2}F \cos(\omega t)$. The resulting velocity is $u = -\sqrt{2}\frac{\omega F}{K} \sin(\omega t)$, with its ghost velocity being $u_g = gu = -\sqrt{2}\frac{\omega F}{K} \cos(\omega t)$. As a result, dual to (3.10), the instantaneous reactive power is the power of the ghost given by the product of $y_g = -y$ and the ghost velocity u_g as

$$q = y_g^T u_g = -y^T u_g$$
$$= -\sqrt{2}F\cos(\omega t) \times (-\sqrt{2}\frac{\omega F}{K}\cos(\omega t))$$
$$= \frac{\omega F^2}{K} - \frac{\omega F^2}{K}\sin(2\omega t),$$

with its average reactive power being

$$Q = \frac{\omega F^2}{K}.$$

The positive sign indicates that the spring "consumes" reactive power, similar to the case of an inductor.

Because of the electrical-mechanical analogy, the two examples illustrated above can be generalized to any mechanical system. That is, the reactive power of a mechanical system is the power of its ghost system, i.e.

$$q = y_g^T u_g = -y^T u_g,$$

where y_g and u_g are, respectively, the output and input of the ghost system. For translational systems, they are related to the force and velocity of the system; for rotational systems, they are related to the torque and angular velocity of the system.

Note that for the real power defined in the normal way as

$$p = u^T y = y^T u$$

there are

$$q = -gp \qquad \text{and} \qquad p = gq.$$

Furthermore, the average reactive power and the power factor can also be defined for mechanical systems in a similar way as for electrical systems. This completes the electrical-mechanical analogy, as shown in Table 3.1.

It is well known that reactive power plays a critical role in electrical systems. However, the role of reactive power has not been well recognized for mechanical systems. The definition introduced here is expected to help understand the role of reactive power in mechanical systems and opens a new door for the analysis and synthesis of mechanical systems, which may in turn advance the understanding of electrical systems further.

3.5 Generalization to Other Energy Systems

As a matter of fact, the instantaneous power theory described above can be generalized to any dynamical system Z that is described by the PH model (3.2)–(3.3) with a periodic input u. Its instantaneous real power and reactive power are

$$p = u^T y = y^T u,$$

and

$$q = y_g^T u_g = -y^T u_g,$$

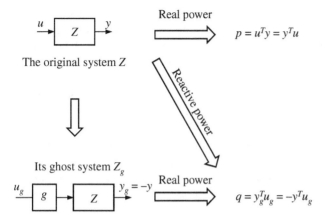

Figure 3.3 Illustration of the ghost power theory.

where $u_g = gu$ is the ghost of the input u to the original system Z and $y_g = -y$ is the output of the ghost. The average real power and reactive power can be easily obtained by taking the averages of p and q over one period. Since the PH system (3.2)–(3.3) can be applied to model complex networked multi-physics systems with lumped and/or distributed parameters (van der Schaft and Jeltsema 2014), the above interpretation/definition for instantaneous real power and reactive power is very generic and can be applied to other systems involving energy or power conversion as well, including fluid, thermal, magnetic, and chemical systems, in addition to the electrical and mechanical systems illustrated above. For the convenience of future references, this power theory is called the ghost power theory and is summarized as illustrated in Figure 3.3.

3.6 Summary and Discussions

Based on (Zhong 2017b), a new mathematical operator called the ghost operator g has been introduced for sine or cosine functions, followed by the physical construction of the ghost for a signal and the ghost for a system. The operator satisfies $g^2 = -1$ but it is different from the well known imaginary operator j. The g-operator does not return an imaginary number but the imaginary operator j does when both are applied to $\cos\theta$ or $\sin\theta$ with θ being a real number. The ghost of a system behaves exactly in the opposite way as the original system when its input is the ghost of the input to the original system, showing perfect symmetry with respect to the origin. With this, the physical meaning of reactive power has become very clear: it is the real power of the ghost system with its input being the ghost of the input to the original system. Moreover, it has been shown that this can be applied to introduce reactive power for mechanical systems, providing a missing concept in the electrical-mechanical analogy. Furthermore, this has been generalized to any dynamic system that can be modeled as a PH system, establishing the ghost power theory.

Note that the ghost operator is different from the Hilbert transform (Bracewell 1999; Weisstein 1999), although the application of both the ghost operator and the Hilbert transform leads to the opposite of the original signal.

The Chinese Yin–Yang philosophy describes how seemingly opposite or contrary objects may actually be complementary, interconnected, and interdependent, and how they may give rise to each other as they interrelate to one another. The ghost of a system and the original system actually form a perfect pair of Yin–Yang. Some interesting results may emerge with further research on this.

It is also conjectured that there exists a mechanical device dual to the memristor, which describes the relationship between the momentum and the displacement.

Since the control of power electronic converters heavily relies on the accurate calculation of real and reactive power (Montanari and Gole 2017; Rezaei et al. 2016; Zhong and Hornik 2013), the ghost power theory is expected to play an important role in SYNDEM smart grids.

One thing that has not been proved at the time when this book goes for printing is whether the ghost of a passive system is passive. There seems to be a gap in the proof in (Zhong 2017b).

Part II

1G VSM: Synchronverters

4

Synchronverter Based Generation

In this chapter, the 1G VSM, i.e. the synchronverter (Zhong and Weiss 2011, 2009), is presented to operate an inverter to mimic a synchronous generator (SG), after directly embedding the mathematical model of SGs into the controller. The implementation and operation of a synchronverter are described in detail. The real and reactive power delivered by synchronverters connected in parallel can be automatically shared with the well known frequency and voltage droop mechanism. Synchronverters can also be easily operated in the standalone mode and hence they provide an ideal solution for microgrids and smart grids. Extensive simulation and experimental results are presented.

4.1 Mathematical Model of Synchronous Generatorss

The model of an SG can be found in many sources, e.g., (Fitzgerald et al. 2003; Grainger and Stevenson 1994; Kundur 1994; Walker 1981). Most of the references make various assumptions, e.g. steady state and/or balanced sinusoidal voltages/currents, to simplify the analysis. Here, a model that is a passive dynamic system without any assumptions on the signals is established, from the perspective of system analysis and controller design, for a round rotor machine (without damper windings), with p pairs of poles per phase (and p pairs of poles on the rotor) and with no saturation effects in the iron core.

Since an SG is an electrical machine, it has an electrical part and a mechanical part.

4.1.1 The Electrical Part

The details on the geometry of the windings can be found in (Fitzgerald et al. 2003; Walker 1981). The field and the three identical stator windings are distributed in slots around the periphery of the uniform air gap. The stator windings can be regarded as concentrated coils having self-inductance L and mutual inductance $-M$ ($M > 0$ with a typical value $\frac{1}{2}L$, the negative sign is due to the $\frac{2\pi}{3}$ phase angle), as shown in Figure 4.1. The field (or rotor) winding can be regarded as a concentrated coil having self-inductance L_f. The mutual inductance

Power Electronics-Enabled Autonomous Power Systems: Next Generation Smart Grids,
First Edition. Qing-Chang Zhong.
© 2020 John Wiley & Sons Ltd. Published 2020 by John Wiley & Sons Ltd.

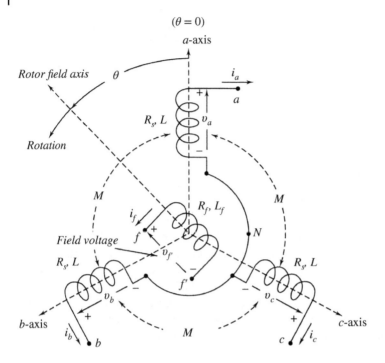

Figure 4.1 Structure of an idealized three-phase round-rotor synchronous generator with $p = 1$, modified from (Grainger and Stevenson 1994, figure 3.4).

between the field coil and each of the three stator coils varies with the (electrical) rotor angle θ as follows:

$$M_{af} = M_f \cos(\theta),$$
$$M_{bf} = M_f \cos(\theta - \frac{2\pi}{3}),$$
$$M_{cf} = M_f \cos(\theta - \frac{4\pi}{3}).$$

The flux linkages of the windings are

$$\Phi_a = Li_a - Mi_b - Mi_c + M_{af}i_f,$$
$$\Phi_b = -Mi_a + Li_b - Mi_c + M_{bf}i_f,$$
$$\Phi_c = -Mi_a - Mi_b + Li_c + M_{cf}i_f,$$
$$\Phi_f = M_{af}i_a + M_{bf}i_b + M_{cf}i_c + L_f i_f,$$

where i_a, i_b and i_c are the stator phase currents and i_f is the rotor excitation current. Denote

$$\Phi = \begin{bmatrix} \Phi_a \\ \Phi_b \\ \Phi_c \end{bmatrix}, \quad i = \begin{bmatrix} i_a \\ i_b \\ i_c \end{bmatrix}$$

and

$$\widetilde{\cos}\,\theta = \begin{bmatrix} \cos\theta \\ \cos(\theta - \frac{2\pi}{3}) \\ \cos(\theta - \frac{4\pi}{3}) \end{bmatrix}, \quad \widetilde{\sin}\,\theta = \begin{bmatrix} \sin\theta \\ \sin(\theta - \frac{2\pi}{3}) \\ \sin(\theta - \frac{4\pi}{3}) \end{bmatrix}.$$

Assume for the moment that the neutral line is not connected, then

$$i_a + i_b + i_c = 0.$$

It follows that the stator flux linkages can be rewritten as

$$\Phi = L_s i + M_f i_f \widetilde{\cos}\,\theta, \tag{4.1}$$

where $L_s = L + M$, and the field flux linkage can be rewritten as

$$\Phi_f = L_f i_f + M_f \langle i,\ \widetilde{\cos}\,\theta \rangle, \tag{4.2}$$

where $\langle \cdot,\ \cdot \rangle$ denotes the conventional inner product. The second term $M_f \langle i,\ \widetilde{\cos}\theta \rangle$ is constant if the three phase currents are sinusoidal (as functions of θ) and balanced. Assume that the resistance of the stator windings is R_s, then the phase terminal voltages $v = \begin{bmatrix} v_a & v_b & v_c \end{bmatrix}^T$ can be obtained from (4.1) as

$$v = -R_s i - \frac{d\Phi}{dt} = -R_s i - L_s \frac{di}{dt} + e, \tag{4.3}$$

where $e = \begin{bmatrix} e_a & e_b & e_c \end{bmatrix}^T$ is the back emf due to the rotor movement given by

$$e = M_f i_f \dot\theta \widetilde{\sin}\,\theta - M_f \frac{di_f}{dt} \widetilde{\cos}\,\theta. \tag{4.4}$$

Similarly, according to (4.2), the field terminal voltage is

$$v_f = -R_f i_f - \frac{d\Phi_f}{dt}, \tag{4.5}$$

where R_f is the resistance of the rotor winding. However, this is not used here because the field current i_f, instead of v_f, is used as an adjustable constant input. This completes modeling the electrical part of the machine.

4.1.2 The Mechanical Part

The mechanical part of the machine is governed by

$$J\ddot\theta = T_m - T_e - D_p\dot\theta, \tag{4.6}$$

where J is the moment of inertia of all parts rotating with the rotor, T_m is the mechanical torque, T_e is the electromagnetic toque and D_p is a damping factor. T_e can be found from the energy E stored in the machine magnetic field, i.e.

$$\begin{aligned} E &= \frac{1}{2}\langle i,\ \Phi \rangle + \frac{1}{2} i_f \Phi_f \\ &= \frac{1}{2}\langle i,\ L_s i + M_f i_f \widetilde{\cos}\,\theta \rangle + \frac{1}{2} i_f (L_f i_f + M_f \langle i,\ \widetilde{\cos}\,\theta \rangle) \\ &= \frac{1}{2}\langle i,\ L_s i \rangle + M_f i_f \langle i,\ \widetilde{\cos}\,\theta \rangle + \frac{1}{2} L_f i_f^2. \end{aligned}$$

From simple energy considerations (see, e.g., (Ellison 1965; Fitzgerald et al. 2003)), there is

$$T_e = \frac{\partial E}{\partial \theta_m}\bigg|_{\Phi, \ \Phi_f \text{constant.}}$$

It is not difficult to verify (using the formula for the derivative of the inverse of a matrix function) that this is equivalent to

$$T_e = -\frac{\partial E}{\partial \theta_m}\bigg|_{i, \ i_f \text{constant.}}$$

Since the mechanical rotor angle θ_m satisfies $\theta = p\theta_m$,

$$T_e = -p\frac{\partial E}{\partial \theta}\bigg|_{i, \ i_f \text{constant}}$$

$$= -pM_f i_f \left\langle i, \ \frac{\partial}{\partial \theta}\widetilde{\cos}\ \theta \right\rangle$$

$$= pM_f i_f \langle i, \ \widetilde{\sin}\ \theta \rangle. \tag{4.7}$$

Note that if $i = i_0\widetilde{\sin}\ \varphi$ (as would be the case in the sinusoidal steady state), then

$$T_e = pM_f i_f i_0 \langle \widetilde{\sin}\ \varphi, \ \widetilde{\sin}\ \theta \rangle = \frac{3}{2}pM_f i_f i_0\ \cos(\theta - \varphi).$$

Note also that if i_f is constant (as is usually the case), then (4.7) with (4.4) yields

$$T_e\dot\theta_m = \langle i, \ e \rangle.$$

4.1.3 Presence of a Neutral Line

The above analysis is based on the assumption that the neutral line is not connected. If the neutral line is connected, then

$$i_a + i_b + i_c = i_N,$$

where i_N is the current flowing through the neutral line. Then, the formula for the stator flux linkages (4.1) becomes

$$\Phi = L_s i + M_f i_f \widetilde{\cos}\ \theta - \begin{bmatrix} 1 \\ 1 \\ 1 \end{bmatrix} M i_N$$

and the phase terminal voltage (4.3) becomes

$$v = -R_s i - L_s\frac{di}{dt} + \begin{bmatrix} 1 \\ 1 \\ 1 \end{bmatrix} M\frac{di_N}{dt} + e,$$

where e is given by (4.4). Other formulae are not affected.

The presence of a neutral line makes the system model somewhat more complicated. However, in a synchronverter to be designed in the next section, M is a design parameter that can be chosen to be 0. The physical meaning of this is that there is no magnetic coupling between the stator windings, which does not happen in a physical synchronous generator but can be easily implemented in a synchronverter. When a neutral line is needed, it is advantageous to make $M = 0$ and then to provide an independently controlled neutral line, e.g. with the strategies discussed in (Zhong and Hornik 2013). The choice of M and L individually is not important; what matters is only $L_s = L + M$. In the following, the model of a synchronous generator consisting of (4.3), (4.4), (4.6), and (4.7) will be used to operate an inverter as a synchronverter.

4.2 Implementation of a Synchronverter

In this section, the details of how to implement an inverter as a synchronverter is described. A synchronverter consists of a power part and an electronic part. The power part is a simple DC–AC converter (inverter) used to convert DC power into three-phase AC, as shown in Figure 4.2, assuming that there is an independently controlled neutral line. The electronic part is an electronic controller that runs a program in a processor to control the switches shown in Figure 4.2. The core of the electronic part is the mathematical model of a synchronous generator shown in Figure 4.3. These two parts interact via the signals e and i, in addition to v and v_g that are used for controlling the synchronverter.

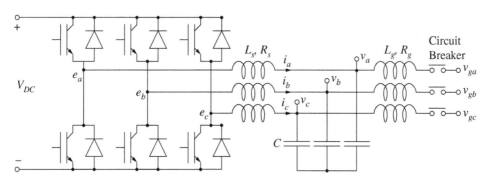

Figure 4.2 The power part of a synchronverter is a basic inverter.

Figure 4.3 The electronic part of a synchronverter without control.

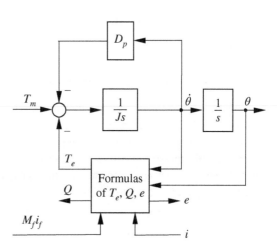

4.2.1 The Power Part

This part consists of three phase legs and a three-phase LC filter, which is used to suppress the switching noise. If the inverter is to be connected to the grid, then three more inductors and a circuit breaker can be adopted to interface with the grid.

Assume that the field (rotor) winding of the synchronverter is fed by an adjustable DC current source i_f instead of a voltage source v_f. In this case, the terminal voltage v_f varies, but this does not matter. As long as i_f is constant, the generated voltage from (4.4) is

$$e = \dot{\theta} M_f i_f \widetilde{\sin} \, \theta. \tag{4.8}$$

The terminal voltages $v = \begin{bmatrix} v_a & v_b & v_c \end{bmatrix}^T$ given in (4.3) should be the capacitor voltages, as shown in Figure 4.2. The inductance L_s and resistance R_s of the inductor can be chosen to represent the stator impedance of a synchronous generator. The switches in the inverter are operated so that the average values of e_a, e_b and e_c over a switching period should be equal to e given in (4.8), which can be achieved by the usual pulse width modulation (PWM) technique. Also shown in Figure 4.2 are three interfacing inductors L_g (with series resistance R_g) and a circuit breaker to facilitate the synchronization/connection with the grid.

4.2.2 The Electronic Part

Define the generated real power P and reactive power Q (as seen from the inverter legs) as

$$P = \langle i, \, e \rangle \text{ and } Q = -\langle i, \, e_g \rangle,$$

where e_g is the ghost signal of e, i.e.

$$e_g = \dot{\theta} M_f i_f \widetilde{\sin}(\theta + \frac{\pi}{2}) = \dot{\theta} M_f i_f \widetilde{\cos} \, \theta.$$

Then, the real power and reactive power are, respectively,

$$P = \dot{\theta} M_f i_f \langle i, \, \widetilde{\sin} \, \theta \rangle,$$
$$Q = -\dot{\theta} M_f i_f \langle i, \, \widetilde{\cos} \, \theta \rangle. \tag{4.9}$$

Note that if $i = i_0 \widetilde{\sin} \, \varphi$ (as would be the case in the sinusoidal steady state), then

$$P = \dot{\theta} M_f i_f \langle i, \, \widetilde{\sin} \, \theta \rangle = \frac{3}{2} \dot{\theta} M_f i_f i_0 \cos(\theta - \varphi),$$

$$Q = -\dot{\theta} M_f i_f \langle i, \, \widetilde{\cos} \, \theta \rangle = \frac{3}{2} \dot{\theta} M_f i_f i_0 \sin(\theta - \varphi).$$

These coincide with the conventional definitions for real power and reactive power, usually expressed in the *dq* coordinates. When the voltage and current are in phase, i.e. when $\theta - \varphi = 0$, the product of the RMS values of the voltage and current gives the real power P. When the voltage and current are $\frac{\pi}{2}$ rad out of phase, this product gives the reactive power Q. Positive Q corresponds to an inductive load. The above formulas for P and Q are used when regulating the real and reactive power of an SG.

The equation (4.6) can be written as

$$\ddot{\theta} = \frac{1}{J}(T_m - T_e - D_p \dot{\theta}),$$

where the input is the mechanical torque T_m, while the electromagnetic torque T_e depends on i and θ, according to (4.7). This equation, together with (4.7), (4.8) and (4.9), are implemented as the core of the electronic part of a synchronverter shown in Figure 4.3. Thus, the state variables of the synchronverter are i (which are actual currents), θ and $\dot{\theta}$ (which are a virtual angle and a virtual angular speed). The control inputs of the synchronverter are T_m and $M_f i_f$. In order to operate the synchronverter in a useful way, a controller should be added to generate the signals T_m and $M_f i_f$ so that the system stability is maintained and the desired values of real and reactive power are followed. The significance of Q will be discussed in the next section.

Note that the real power and the reactive power are calculated with e and i here. Instead, the voltage v can be measured and adopted to calculate the real power and reactive power.

4.3 Operation of a Synchronverter

4.3.1 Regulation of Real Power and Frequency Droop Control

For synchronous generators, the rotor speed is maintained by the prime mover and it is known that the damping factor D_p is due to mechanical friction. An important mechanism for SGs to share load evenly is to vary the real power it delivers according to the grid frequency, a function called "frequency droop". When the real power demand increases, the speed of the SGs drops due to increased T_e in (4.6). The speed regulation system of the prime mover then increases the mechanical power, e.g. by widening the throttle valve of an engine, so that a new power balance is achieved. This mechanism can be implemented in a synchronverter by comparing the virtual angular speed $\dot{\theta}$ with the angular frequency reference $\dot{\theta}_r$, e.g. the nominal angular speed $\dot{\theta}_n$, before feeding it into the damping block D_p; see the upper part of Figure 4.4. As a result, the damping factor D_p actually behaves as the frequency droop coefficient, which is defined as the ratio of the required change of torque ΔT to the change of speed (frequency) $\Delta \dot{\theta}$. That is,

$$D_p = \frac{\Delta T}{\Delta \dot{\theta}} = \frac{\Delta T}{T_{mn}} \frac{\dot{\theta}_n}{\Delta \dot{\theta}} \frac{T_{mn}}{\dot{\theta}_n},$$

where T_{mn} is the nominal mechanical torque. Note that in some references, e.g. (Sao and Lehn 2005), D_p is defined as $\frac{\Delta \dot{\theta}}{\Delta T}$. For example, it can be set for the torque (power) to change 100% for a frequency change of 5%. The active torque T_m can be obtained from the set point of real power P_{set} by dividing it with the nominal mechanical speed $\frac{\dot{\theta}_n}{p}$. This completes the feedback loop for real power; see the upper part of Figure 4.4. Because of the built-in frequency droop mechanism, a synchronverter automatically shares the load with other inverters of the same type and with SGs connected on the same bus. The power regulation loop is very simple, because no mechanical devices are involved and no extra measurements are needed for real power regulation (all the variables are available internally).

The regulation mechanism of the real power (torque) shown in the upper part of Figure 4.4 has a cascaded control structure, of which the inner loop is the frequency (speed) loop and the outer loop is the real power (torque) loop. The time constant of the frequency loop is $\tau_f = \frac{J}{D_p}$. In other words, J can be chosen as

$$J = D_p \tau_f.$$

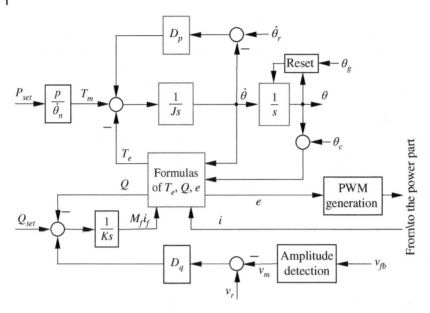

Figure 4.4 The electronic part of a synchronverter with the function of frequency and voltage control, and real and active power regulation.

Because there is no delay involved in the frequency (speed) loop, the time constant τ_f can be made much smaller than that of a physical synchronous generator. In order to make sure that the frequency loop has a quick response so that it can track the frequency reference quickly, τ_f should be made small. Hence, for a given frequency droop coefficient D_p, J should be made small. It is not necessary to have a large inertia like a physical synchronous generator, where a larger inertia means that more energy is stored mechanically. The energy storage function of a synchronverter can, and should, be decoupled from the inertia. This is the opposite of the approach proposed in (Driesen and Visscher 2008). The short term energy storage function can be implemented with a synchronverter using the same storage system, e.g., batteries, that is used for long term storage. Usually, a synchronverter would be operated in conjunction with a distributed power source and an energy storage unit in a certain arrangement.

4.3.2 Regulation of Reactive Power and Voltage Droop Control

The regulation of reactive power Q flowing out of the synchronverter can be realized similarly. Define the voltage droop coefficient D_q as the ratio of the required change of reactive power ΔQ to the change of voltage ΔV, i.e.

$$D_q = \frac{\Delta Q}{\Delta V} = \frac{\Delta Q}{Q_n} \frac{V_n}{\Delta V} \frac{Q_n}{V_n},$$

where Q_n is the nominal reactive power, which can be chosen as the nominal power, and V_n is the nominal amplitude of the terminal voltage v. Again, note that in much literature, e.g. (Sao and Lehn 2005), D_q is defined as $\frac{\Delta V}{\Delta Q}$. The regulation mechanism for the reactive power can be realized as shown in the lower part of Figure 4.4. The difference between the reference voltage v_r and the amplitude of the feedback voltage v_{fb} is amplified with the voltage droop coefficient D_q before adding to the difference between the set point Q_{set} and the reactive power Q, which is calculated according to (4.9). The resulting signal is then fed into an integrator with a gain $\frac{1}{K}$ to generate $M_f i_f$ (here, K is dual to the inertia J). Note that there is no need to measure reactive power Q because it is available internally.

Similarly, the regulation mechanism of the reactive power shown in the lower part of Figure 4.4 also has a cascaded control structure, if the effect of the LC filter is ignored or compensated (which means $v_{fb} \approx e$). The inner loop is the (amplitude) voltage loop and the outer loop is the reactive power loop. The time constant τ_v of the voltage loop is

$$\tau_v = \frac{K}{\dot{\theta}D_q} \approx \frac{K}{\dot{\theta}_n D_q}$$

as the variation of $\dot{\theta}$ is very small. Hence, K can be chosen as

$$K = \dot{\theta}_n D_q \tau_v.$$

The amplitude v_m of the voltage v can be obtained from the RMS values but it can also be obtained as follows. Assume that $v_a = v_{am} \sin\theta_a$, $v_b = v_{bm} \sin\theta_b$ and $v_c = v_{cm} \sin\theta_c$, then

$$v_a v_b + v_b v_c + v_c v_a$$
$$= v_{am} v_{bm} \sin\theta_a \sin\theta_b + v_{bm} v_{cm} \sin\theta_b \sin\theta_c + v_{cm} v_{am} \sin\theta_c \sin\theta_a$$
$$= \frac{v_{am} v_{bm}}{2} \cos(\theta_a - \theta_b) + \frac{v_{bm} v_{cm}}{2} \cos(\theta_b - \theta_c) + \frac{v_{cm} v_{am}}{2} \cos(\theta_c - \theta_a)$$
$$- \frac{v_{am} v_{bm}}{2} \cos(\theta_a + \theta_b) - \frac{v_{bm} v_{cm}}{2} \cos(\theta_b + \theta_c) - \frac{v_{cm} v_{am}}{2} \cos(\theta_c + \theta_a).$$

When the terminal voltages are balanced, i.e. when $v_{am} = v_{bm} = v_{cm} = v_m$ and $\theta_b = \theta_a - \frac{2\pi}{3} = \theta_c + \frac{2\pi}{3}$, then the last three terms in the above equality are balanced, having a doubled frequency. Hence,

$$v_a v_b + v_b v_c + v_c v_a = -\frac{3}{4} v_m^2,$$

and the amplitude v_m of the actual terminal voltage v can be obtained as

$$v_m = \frac{2}{\sqrt{3}} \sqrt{-(v_a v_b + v_b v_c + v_c v_a)}. \tag{4.10}$$

In real-time implementation, a low-pass filter is needed to filter out the ripples at the doubled frequency as the terminal voltages may be unbalanced. This also applies to T_e and Q.

4.4 Simulation Results

The parameters of the inverter for carrying out the simulations are given in Table 4.1.

The inverter is connected to a three-phase 400 V 50 Hz grid via a circuit breaker and a step-up transformer. The frequency droop coefficient is chosen as $D_p = 0.2026$ so that the frequency drops 0.5% when the torque (power) increases 100%. The voltage droop coefficient is chosen as $D_q = 117.88$ so that the voltage drops 5% when the reactive power increases 100%. The time constant of the frequency loop is chosen as $\tau_f = 0.002$ s and that of

Table 4.1 Parameters of the synchronverter for simulations.

Parameters	Values	Parameters	Values
L_s	0.45 mH	L_g	0.45 mH
R_s	0.135 Ω	R_g	0.135 Ω
C	22 μF	Frequency	50 Hz
R (parallel to C)	1000 Ω	Voltage (line–line)	20.78 Vrms
Rated power	100 W	DC-link voltage	42 V

the voltage loop is chosen as $\tau_v = 0.02$ s. The simulations were carried out in Matlab® 7.4 with Simulink™. The solver used in the simulations is ode23tb with a relative tolerance 10^{-3} and a maximum step size of 0.2 ms. The synchronverter can send pre-set real power and reactive power to the grid (called the set mode) and can automatically change the real power and reactive power sent to the grid according to the grid frequency and voltage (called the droop mode).

The simulation was started at $t = 0$ to allow the PLL and synchronverter to start-up (in real applications, these two can be started separately). The dynamics in the first half second is omitted. The circuit breaker was turned on at $t = 1$ s; the real power $P_{set} = 80$ W was applied at $t = 2$ s and the reactive power $Q_{set} = 60$ Var was applied at $t = 3$ s. The droop mechanism was enabled at $t = 4$ s and then the grid voltage decreased by 5% at $t = 5$ s.

4.4.1 Under Different Grid Frequencies

The system responses when the grid frequency was 50 Hz are shown in the left column of Figure 4.5. The synchronverter tracked the grid frequency very well all the time. The voltage difference between v and v_g before any power was applied was very small and the synchronization process was very quick. There was no problem turning the circuit breaker on at $t = 1$ s; there was not much transient response after this event either. The synchronverter responded to both the real power command at $t = 2$ s and the reactive command at $t = 3$ s very quickly and settled down in less than 10 cycles without any error. The coupling effect between the real power and the reactive power is reasonably small. When the droop mechanism was enabled at $t = 4$ s, there was not much change to the real power output as the frequency was not changed but the reactive power dropped by about 53 Var, about 50% of the power rating, because the local terminal voltage v was about 2.5% higher than the nominal value. When the grid voltage dropped by 5% at $t = 5$ s, the local terminal voltage dropped to just below the nominal value. The reactive power output then increased to just above the set point 60 Var.

The same simulation was repeated but with a grid frequency of 49.95 Hz, that is 0.1% lower than the nominal one. The system responses are also shown in the left column of Figure 4.5 for comparison. The synchronverter followed the grid frequency very well. When the synchronverter worked at the set mode, i.e. before $t = 4$ s, the real power and reactive power both responded to the set point exactly. After the droop mechanism was enabled at

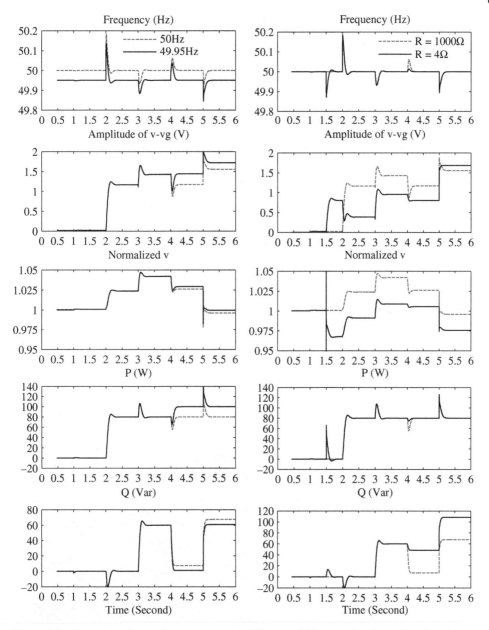

Figure 4.5 Operation of a synchronverter under different grid frequencies (left column) and different load conditions (right column).

$t = 4$ s, the synchronverter increased the real power output by 20 W, that is 20% of the rated power, corresponding to 0.1% drop of the frequency. This did not cause much extra change to the reactive power output, just slight adjustment corresponding to the slight change of the local voltage v.

4.4.2 Under Different Load Conditions

The same simulation with the nominal grid frequency was repeated but with the local load changed from $R = 1000\ \Omega$ to $R = 4\ \Omega$ at $t = 1.5$ s. The system responses are shown in the right column of Figure 4.5. The local voltage v dropped immediately, with some short big transients, after the load R was changed to $4\ \Omega$ at $t = 1.5$ s and the voltage difference between v and v_g increased because of the load current drawn from the grid. There was some transient power drawn from the synchronverter. When the power commands were applied, the synchronverter responded as if there was no load connected. The local voltage v recovered a bit because of the power delivered by the synchronverter. After the droop mechanism was enabled at $t = 4$ s, the synchronverter continued supplying the same real power to the local load but slightly less reactive power because the local voltage is slightly higher than the nominal value. When the grid voltage dropped at $t = 5$ s by 5%, the local voltage dropped immediately to about 2.5% below the nominal value, which caused the synchronverter to send an extra 50 Var of reactive power on top of the command. The responses when $R = 1000\ \Omega$ was not changed are also shown in the right column of Figure 4.5 for comparison.

4.5 Experimental Results

Experimental results from the microgrid shown in Figure 4.6 are described in this section. The system consists of synchronverters VSG and VSG2, and a three-phase local resistive load ($R_A = R_B = R_C = 45\ \Omega$). Each synchronverter has its own circuit breaker and the local bus is connected to the grid through another circuit breaker. The parameters of VSG and VSG2 are given in Tables 4.2 and 4.3, respectively. The synchronverters are connected

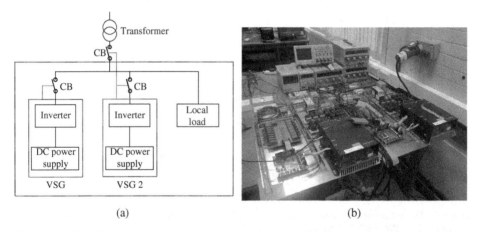

(a)   (b)

Figure 4.6 Experimental setup with two synchronverters. (a) System structure. (b) System photo.

Table 4.2 Parameters of VSG.

Parameter	Value	Rating	Parameter	Value	Rating
L_s	1.64 mH	7.2 A	R_s	0.308 Ω	7.2 A
L_g	0.82 mH	7.2 A	R_g	0.154 Ω	7.2 A
C_f	10 μF	160 V	f_{sw}	6 kHz	—

Table 4.3 Parameters of VSG2.

Parameter	Value	Rating	Parameter	Value	Rating
L_s	4.4 mH	4 A	R_s	0.988 Ω	4 A
L_g	2.2 mH	4 A	R_g	0.494 Ω	4 A
C_f	10 μF	160 V	f_{sw}	6 kHz	—

to the three-phase 400 V/50 Hz grid via a three-phase 190 V/400 V step-up transformer. The sampling and the switching frequencies of both VSG and VSG2 are 6 kHz. The DC-bus voltage of both synchronverters is 450 V with a 680 μF 800 V capacitor on the DC bus.

4.5.1 Grid-connected Set Mode

In this case, the synchronverter VSG was operated in the set mode to inject the set power to the grid, with the output currents i_A, i_B and i_C shown in Figures 4.7 and 4.8. Figure 4.7 shows the three-phase currents at the power output 2.25 kW. Figure 4.8 shows the output currents and the total harmonic distortion (THD) of phase-A current under different set power sent to the grid. When the power sent to the grid is low, the THD is high because of the harmonics in the grid voltage. The THD is improved when the power sent to the grid is increased. When the power reaches 1000 W, the THD is 4.03%, which meets most requirements.

4.5.2 Grid-connected Droop Mode

In this case, the droop mechanism was enabled. The real power reference value was set to $P_{set} = 1$ kW, with the results shown in Figure 4.9. The synchronverter automatically changed the real power according to the changing grid frequency. When the grid frequency increases, the real power decreases; when the grid frequency decreases, the real power increases. The nearly symmetric behavior shows that the primary frequency response is very fast.

4.5.3 Grid-connected Parallel Operation

In this case, both VSG and VSG2 were connected to the grid. The local load with $R_A = R_B = R_C = 45 \, \Omega$ was also connected, consuming over 800 W. Figure 4.10 shows the currents of

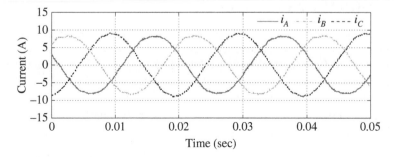

Figure 4.7 Experimental results in the set mode: output currents with 2.25 kW real power.

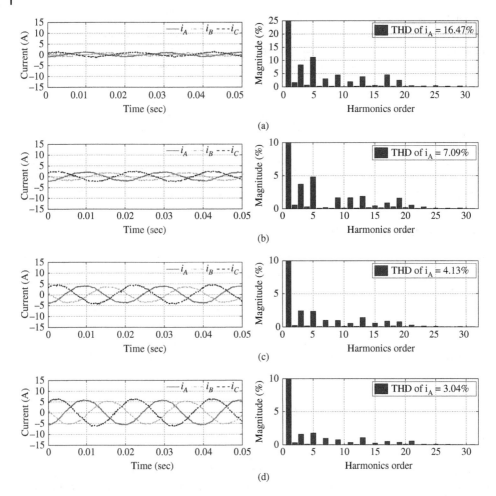

Figure 4.8 Experimental results in the set mode: output currents (left column) and the THD of phase-A current (right column) under different real powers. (a) 250 W. (b) 500 W. (c) 1000 W. (d) 1500 W.

VSG, VSG2, and the grid when VSG was set to send 1500 W and VSG2 was set to send 500 W, both in the set mode. Because the load is over 800 W, the power sent to the grid is about 1200 W.

4.5.4 Seamless Transfer of the Operation Mode

In this experiment, the synchronverter VSG was changed from the grid-connected mode to the stand-alone mode and then back. The resistive local load $R_A = R_B = R_C = 45\ \Omega$ was connected. The synchronverter was initially operated in the grid-connected set mode with $P_{set} = 750$ W. The experiment was carried out in the following sequence of actions, with the real power P and the reactive power Q of the VSG shown in Figure 4.11:

(1) The droop mechanism was enabled roughly at $t = 45$ s. Because the grid frequency was below 50 Hz, the real power P increased accordingly.

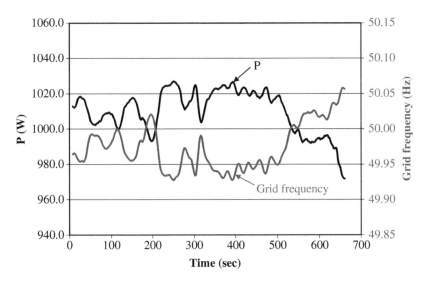

Figure 4.9 Experimental results in the droop mode: primary frequency response.

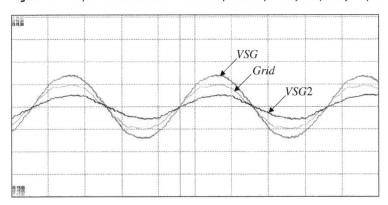

Figure 4.10 Experimental results: the currents of the grid, VSG, and VSG2 under the parallel operation of VSG and VSG2 with a local resistive load.

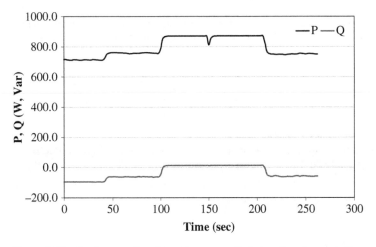

Figure 4.11 Real power P and reactive power Q during the change in the operation mode.

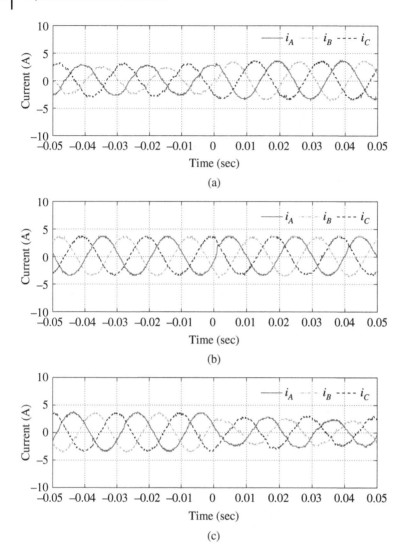

Figure 4.12 Transient responses of the synchronverter. (a) Transfer from grid-connected mode to standalone mode. (b) Re-synchronization. (c) Transfer from standalone mode to grid-connected mode.

(2) At $t = 100$ s, the inverter was disconnected from the grid, with the detailed response shown in Figure 4.12(a). There was no noticeable transients in the inverter voltage and the seamless disconnection from the grid was achieved. The synchronverter VSG took over the full local resistive load without any disruption.

(3) The re-synchronization impulse was activated, roughly at $t = 150$ s, for the synchronverter VSG to synchronize with the grid. As shown in Figure 4.11, there was a visible dip in the real power P, which is mainly due to the numerical calculation because the change in the currents shown in Figure 4.12(b) is very small.

(4) At $t = 210$ s, the inverter was connected back to the grid, with the detailed response shown in Figure 4.12(c). As shown in Figure 4.11, the real power and reactive power returned to the values before the VSG was disconnected from the grid smoothly.

In summary, the synchronverter is able to achieve seamless transfer of the operation mode from standalone to grid-connected and vice versa.

4.6 Summary

Based on (Hornik and Zhong 2011b; Zhong and Weiss 2011, 2009), the 1G VSM technology, i.e. the synchronverter technology, is described in detail. A synchronverter has the mathematical model of a synchronous machine embedded in the controller directly so it is able to behave like a conventional synchronous machine and provides a primary frequency response. It is also equipped with droop control to achieve power regulation and load sharing. Both simulation and experimental results are provided.

5

Synchronverter Based Loads

In next-generation smart grids, most loads will have PWM rectifiers at the front end for integration with the grid. In this chapter, a control strategy is presented to operate a PWM rectifier to mimic a synchronous motor. Two options are presented: one is to directly control the power exchanged with the grid and the other is to control the DC-bus voltage. The reactive power can be controlled as well, e.g., to obtain the unity power factor. Both simulation and experimental results are provided.

5.1 Introduction

Many three-phase rectifiers use a diode bridge circuit and a bulk storage capacitor. This has the advantage of being simple, robust and low cost. However, it allows only unidirectional power flow. Any energy returned from the load must be dissipated on power resistors controlled by a chopper connected across the DC link, which reduces the overall system efficiency and causes thermal management problems. Moreover, a diode bridge also causes low power factor and high THD in the input current. Voltage-source PWM rectifiers are increasingly used in home appliances, industrial DC power supplies, motor drives, etc. (Ericsen et al. 2006). PWM rectifiers are well known for their advantages compared to uncontrolled diode rectifiers, e.g. lower THD, high power factor, regenerative operation and active filtering (AF) capabilities (Cichowlas et al. 2005; Rodriguez et al. 2005).

Various strategies are available to control PWM rectifiers. These strategies can achieve the same major goals, such as high power factor and near-sinusoidal current waveforms, but they are often based on different principles. For example, voltage oriented control (VOC), which uses closed-loop current control in the rotating reference frame, offers a good dynamic and static performance (Dixon and Ooi 1988; Duarte et al. 1999; Hansen et al. 2000; Zhou and Wang 2003). It normally consists of an inner-loop high bandwidth current controller and an outer-loop low bandwidth DC-bus voltage controller. The current controller regulates the input line current to track a reference current that is in phase with the input voltage to ensure the unity power factor, while the DC-bus voltage controller regulates the DC-bus voltage and generates a current reference for the current controller, indirectly regulating the power fed into the rectifier (Cecati et al. 2005; Lee and Lim 2002; Zhou et al. 2009). Another method, direct power control (DPC), adopts instantaneous real and reactive power control loops to control the power directly through hysteresis

Power Electronics-Enabled Autonomous Power Systems: Next Generation Smart Grids,
First Edition. Qing-Chang Zhong.
© 2020 John Wiley & Sons Ltd. Published 2020 by John Wiley & Sons Ltd.

control, without using inner current loops. No PWM block is needed because the states of the converter switches can be selected from a table based on the instantaneous errors of real and reactive power between the reference and measured values (Chen and Joos 2008; Escobar et al. 2003; Malinowski et al. 2004; Noguchi et al. 1998; Shan et al. 2010). This leads to a variable and, often high, switching frequency, which makes it difficult for implementation in industry.

In this chapter, a control strategy is presented to operate a PWM rectifier to mimic a synchronous motor, following the idea of operating inverters to mimic synchronous generators (Zhong and Weiss 2011) as discussed in the previous chapter. This allows the use of a PWM block with a fixed switching frequency to convert the voltage into PWM signals to drive the rectifier switches.

5.2 Modeling of a Synchronous Motor

In this chapter, the model of a synchronous generator discussed in the previous chapter is adopted because the model for a generator or a motor is more or less the same, apart from that the direction of the stator current is defined opposite, flowing into the machine, as shown in Figure 5.1. Assume that the field and the three identical stator windings of a synchronous motor are distributed in slots around the periphery of the uniform air gap. The stator windings can be regarded as concentrated coils having self-inductance L and mutual inductance $-M$ with $M > 0$ (the negative sign is due to the $\frac{2\pi}{3}$ rad phase angle).

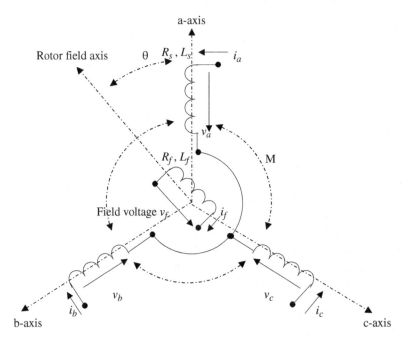

Figure 5.1 Structure of an idealized three-phase round-rotor synchronous motor.

Denote the flux vector and the current vector as

$$\Phi = \begin{bmatrix} \Phi_a \\ \Phi_b \\ \Phi_c \end{bmatrix}, \qquad i = \begin{bmatrix} i_a \\ i_b \\ i_c \end{bmatrix},$$

respectively, and

$$\widetilde{\cos}\theta = \begin{bmatrix} \cos\theta \\ \cos(\theta - \frac{2\pi}{3}) \\ \cos(\theta - \frac{4\pi}{3}) \end{bmatrix}, \qquad \widetilde{\sin}\theta = \begin{bmatrix} \sin\theta \\ \sin(\theta - \frac{2\pi}{3}) \\ \sin(\theta - \frac{4\pi}{3}) \end{bmatrix},$$

where θ is the rotor angle with respect to the phase A stator winding.

Assume that the resistance of the stator windings is R_s. Then the phase terminal voltages $v = [v_a \ v_b \ v_c]^T$ can be written as

$$v = e + R_s i + L_s \frac{di}{dt}, \tag{5.1}$$

where $L_s = L + M$ and $e = [e_a \ e_b \ e_c]^T$ is the back emf, often called the synchronous internal voltage, due to the rotor movement. It is given by

$$e = M_f i_f \dot{\theta} \widetilde{\sin}\theta, \tag{5.2}$$

where M_f is the maximum mutual inductance between the stator and rotor windings and i_f is the field current, which is often assumed constant or slowly time-varying.

The motor also satisfies

$$\ddot{\theta} = \frac{1}{J}(T_e - T_m - D_p\dot{\theta}), \tag{5.3}$$

where J is the moment of inertia of all the parts rotating with the rotor, D_p is the damping factor, T_m is the mechanical torque, and T_e is the electromagnetic torque given by

$$T_e = M_f i_f \langle i, \ \widetilde{\sin}\theta \rangle, \tag{5.4}$$

where $\langle \cdot, \ \cdot \rangle$ denotes the conventional inner product.

The real power and reactive power at the terminal can be calculated as

$$P = v_a i_a + v_b i_b + v_c i_c, \tag{5.5}$$

$$Q = \frac{1}{\sqrt{3}}[(v_b - v_c)i_a + (v_c - v_a)i_b + (v_a - v_b)i_c]. \tag{5.6}$$

Putting all the above together, the model of a synchronous motor can be described as shown in Figure 5.2.

5.3 Operation of a PWM Rectifier as a VSM

A three-phase PWM bridge rectifier is shown in Figure 5.3. The inductor L_s with its equivalent series resistance R_s, which are connected in each phase between the rectifier and the source to reduce the current ripples and to boost the voltage, can be treated as the stator winding of a three-phase synchronous machine.

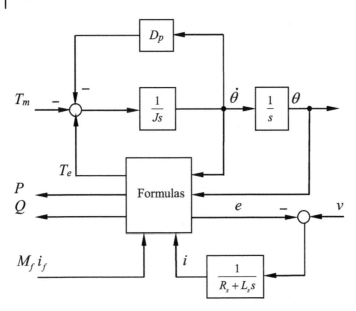

Figure 5.2   The model of a synchronous motor.

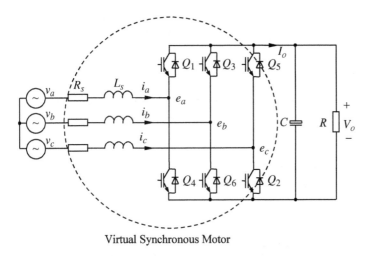

Virtual Synchronous Motor

Figure 5.3   PWM rectifier treated as a virtual synchronous motor.

The idea of operating a PWM rectifier as a VSM is to generate the voltage e according to the mathematical model of a synchronous motor developed in the previous section, or in other words, to embed the mathematical model into the controller. The key is how to generate the mechanical torque T_m and how to generate the field excitation current i_f.

5.3.1   Controlling the Power

As shown in Figure 5.4, the (virtual) mechanical torque T_m is generated by a PI controller that regulates the real power P. The reactive power Q can be controlled to track its reference

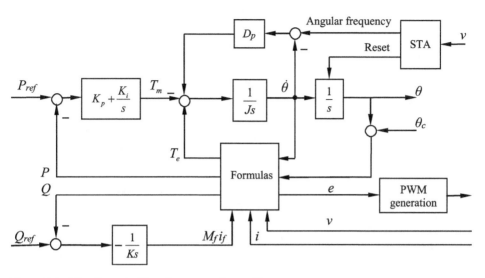

Figure 5.4 Directly controlling the power of a rectifier.

Q_{ref}. In order to obtain the unity power factor, Q_{ref} can be set as 0. The tracking error between the reference value Q_{ref} and reactive power Q is fed into an integrator with a gain $-\frac{1}{K}$ to regulate the field excitation $M_f i_f$. Here, the negative sign is due to the fact that the positive direction of the current is flowing into the rectifier. Here, the sinusoid tracking algorithm (STA) (Ziarani and Konrad 2004; Ziarani et al. 2003) is adopted to obtain the grid frequency and to reset the integrator so that the voltage e can be synchronized with the voltage v. An STA is a variant of phase-locked loops (PLL) and its details can be found in, e.g., (Ziarani and Konrad 2004; Ziarani et al. 2003) and (Zhong and Hornik 2013).

The real power channel shown in Figure 5.4 has three cascaded control loops. The inner loop is the frequency regulation loop with the feedback gain D_p, the middle loop is a torque loop with the feedback coming from the electromagnetic torque T_e and the outer loop is the real power loop with the feedback coming from the real power P. The first two loops are part of the model of a synchronous motor.

Before the PWM pulses are enabled, the rectifier is operated in the uncontrolled mode because of the parasitic diodes. This allows the control signal e to have the same frequency and phase as those of the terminal voltage e, which helps reduce the inrush current. This can be done via feeding the difference between the grid frequency obtained from the STA and the rectifier frequency θ to the D_p block and resetting the phase when the phase of voltage is 0.

5.3.2 Controlling the DC-bus Voltage

An important application of PWM rectifiers is to provide a constant DC-bus voltage. Instead of controlling the real power P, it is possible to control the DC-bus voltage. As shown in Figure 5.5, the outer-loop controller regulating the real power in Figure 5.4 is replaced with a DC-bus voltage controller. The error between the measured voltage V_o and its reference V_{ref} is fed into a PI controller to generate the virtual mechanical torque T_m.

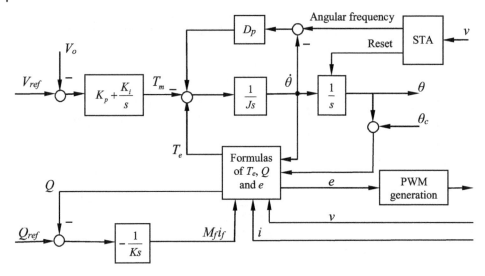

Figure 5.5 Controlling the DC-bus voltage of a rectifier.

5.4 Simulation Results

Simulations were carried out in Matlab7.6/Simulink/SimpowerSystems, with the solver ode23t with a relative tolerance of 10^{-3} and a maximum step size of 10 μs. The parameters of the rectifier simulated are given in Table 5.1. The frequency regulation coefficient is chosen as $D_p = 0.2026$. The rectifier was connected to the grid via a step-up transformer to facilitate the comparison with experimental results from a low voltage setup in the lab environment.

5.4.1 Controlling the Power

The parameters of the PI controller were chosen as $K_p = 0.05$ and $K_i = 0.5$. The simulation was started at $t = 0$. An STA was used for the initial synchronization during the first 0.5 s. During this period, the PWM rectifier worked as an uncontrolled rectifier supplying a load R of 50 Ω. This allowed the internal frequency to synchronize with the frequency of the

Table 5.1 Parameters of the rectifier under simulation.

Parameters	Values	Parameters	Values
L_s	0.45 mH	R_s	0.135 Ω
Nominal voltage	12 Vrms	Switch frequency	10 kHz
J	0.0405 kg m^2	C	5000 μF
Pole pairs p	1	K	740

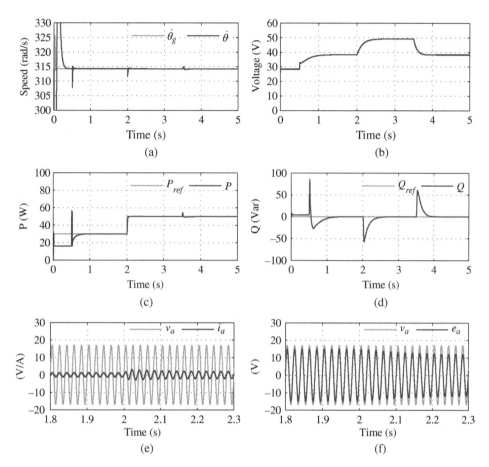

Figure 5.6 Simulation results when controlling the power. (a) Grid and internal frequencies. (b) DC-bus voltage. (c) Real power. (d) Reactive power. (e) Voltage v_a and current i_a. (f) Voltages v_a and e_a.

grid, as can be seen from Figure 5.6(a). At $t = 0.5$ s, the PWM pulses were enabled, with the reactive power reference Q_{ref} set at zero and the real power reference P_{ref} set at 30 W. As can be seen from Figure 5.6, the real power was regulated to 30 W and the reactive power was regulated to 0 after some dynamics. The DC-bus voltage increased accordingly. At $t = 2$ s, P_{ref} was changed to 50 W. The real power quickly increased to 50 W. There was some coupling effect in the reactive power, which is reasonable. The DC-bus voltage increased further because of the increased power drawn. Note that, as can be seen from Figure 5.6(e) and (f), the current drawn from the grid indeed increased and the generated voltage e_a decreased. Moreover, the current i was in phase with the voltage v_a, resulting in the unity power factor. At $t = 3.5$ s, the load was changed to $R = 30$ Ω. There was very small dynamics in the real power although the dynamics in the reactive power was more visible. The DC-bus voltage decreased accordingly. During the whole process, the internal frequency θ tracked the grid frequency $\dot{\theta}_g$ well, except some expected transients.

5.4.2 Controlling the DC-bus Voltage

The parameters of the PI controller were chosen as $K_p = 0.02$ and $K_i = 0.2$. An STA was used for the initial synchronization during the first 0.5 s. During this period, the PWM rectifier worked as an uncontrolled rectifier supplying a load R of 50 Ω. This allowed the internal frequency to synchronize with the frequency of the grid, as can be seen from Figure 5.7(a). At $t = 0.5$ s, the PWM pulses were enabled, with the reactive power reference Q_{ref} set at zero and the DC-bus voltage reference V_{ref} set at 40V. The DC-bus voltage increased and settled at the reference value, with some slight overshoot. The reactive power was regulated to zero after some transients. At $t = 2$ s, the output voltage reference was changed to $V_{ref} = 50$ V. The DC-bus voltage increased and settled down, again with some overshoot. The reactive power went through some transients but settled down at zero again. Because of the increased DC-bus voltage, the real power increased. Indeed, as can be seen from Figure 5.7(e) and (f), the current drawn from the grid increased and

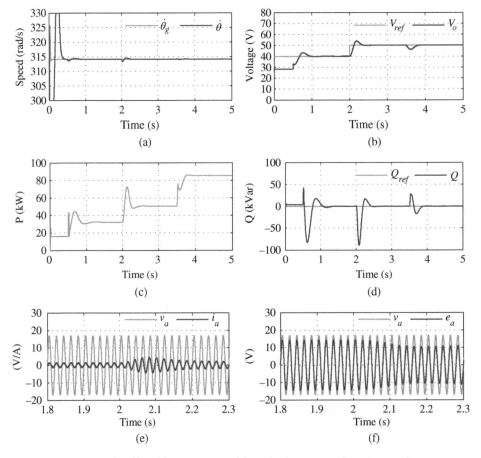

Figure 5.7 Simulation results when controlling the DC-bus voltage. (a) Grid and internal frequencies. (b) DC-bus voltage. (c) Real power. (d) Reactive power. (e) Voltage v_a and current i_a. (f) Voltages v_a and e_a.

the generated voltage e_a decreased. Moreover, the current i_a was in phase with the voltage v_a, resulting in the unity power factor. At $t = 3.5$ s, the load was changed to $R = 30\ \Omega$. However, the DC-bus voltage was regulated well after some transients. The real power drawn by the rectifier increased further. The reactive power is still regulated to be zero. During the whole process, the internal frequency $\dot{\theta}$ synchronized with the grid frequency $\dot{\theta}_g$ all the time, except for some expected transients.

5.5 Experimental Results

Experiments were carried out on a test rig with the system parameters roughly the same as those in Table 5.1 and the control parameters the same as those in the simulations. The experimental setup consists of a rectifier board, a three-phase grid interface inductor, a board consisting of voltage and current sensors, a 230 V/12 V step-down transformer, a dSPACE DS1104 R&D controller board with ControlDesk software and MATLAB Simulink/SimPower software package. The sampling frequency of the controller was 5 kHz and the switching frequency was 10 kHz. In order to minimize the inrush current, a resistor, which could be bypassed by a relay, was connected in series with the inductor.

5.5.1 Controlling the Power

The experimental results are shown in Figure 5.8.

The system was started with all IGBTs off to synchronize the rectifier with the grid. The DC-bus voltage was built up at around 26 V through the parasitic diodes. The real power and reactive power were zero because there was no load on the DC bus. Roughly at $t = 3$ s, the relay was turned on to bypass the current-limiting resistor. The real power increased a little bit to charge the DC-bus capacitor and then went back to zero. The DC-bus voltage increased to 28 V. Roughly at $t = 5$ s, a load $R = 50\ \Omega$ was connected to the DC bus. The real power increased to 16 W. Because of the load effect, the DC-bus voltage dropped to 26 V. Roughly at $t = 10$ s, the IGBTs were enabled with $P_{ref} = 40$ W and without the Q loop. The real power quickly increased to the reference value 40 W and the DC-bus voltage increased to 41 V gradually. Since the Q loop was not enabled, the reactive power was about -75 Var. Roughly at $t = 20$ s, the Q loop was enabled. The reactive power was regulated to zero to obtain unity power factor. Roughly at $t = 30$ s, P_{ref} was changed to 60 W. The real power increased quickly. The DC-bus voltage increased accordingly. The reactive power was maintained well around zero after some transients. As can be seen from Figures 5.8(e) and (f), the current i_a was in phase with e_a and the terminal voltage v_a was larger than e_a, indicating the real power flowing from the grid to the rectifier. Note that the THD of the current was low. Roughly at $t = 41$ s, the load was changed to $R = 30\ \Omega$. Both the real power and the reactive power was maintained well. The DC-bus voltage decreased accordingly.

As can be seen from Figure 5.8, the internal frequency $\dot{\theta}$ synchronized with the grid frequency $\dot{\theta}_g$ all the time, except for some expected transients.

5.5.2 Controlling the DC-bus Voltage

The experimental results are shown in Figure 5.9.

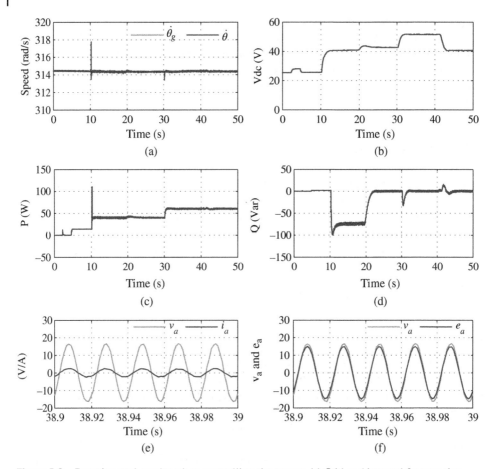

Figure 5.8 Experimental results when controlling the power. (a) Grid and internal frequencies. (b) DC-bus voltage. (c) Real power. (d) Reactive power. (e) Voltage v_a and current i_a. (f) Voltages v_a and e_a.

The system was started with all IGBTs off to synchronize the rectifier with the grid. Roughly at $t = 2$ s, the relay was turned on to bypass the current-limiting resistor. Roughly at $t = 4$ s, a load $R = 50\,\Omega$ was connected to the DC bus. The system responded similarly to the previous case. Then, roughly at $t = 10$ s, the IGBTs were enabled with $V_{\text{ref}} = 40$ V and without the Q loop. The DC-bus voltage increased to 40 V and the real power increased to about 40W. Since the Q loop was not enabled, the reactive power was about -70 Var. Roughly at $t = 20$ s, the Q loop was enabled. The reactive power was regulated to zero to obtain unity power factor. The DC-bus voltage was maintained well. Roughly at $t = 30$ s, V_{ref} was changed to 50 V. The DC-bus voltage responded properly and the real power increased accordingly. The reactive power was maintained well around zero after some transients. As can be seen from Figures 5.9(e) and (f), the current i_a was in phase with e_a and the terminal voltage v_a was larger than e_a, indicating the real power flowing from the grid to the rectifier. Note again that the THD of the current was low. Roughly at $t = 42$ s, the load was changed to $R = 30\,\Omega$. The DC-bus voltage was maintained well at 50 V. The real

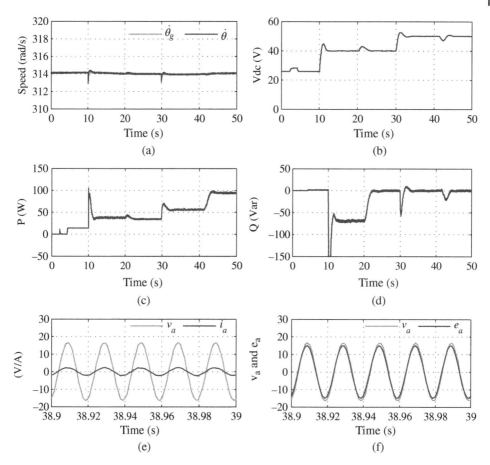

Figure 5.9 Experimental results when controlling the DC-bus voltage. (a) Grid and internal frequencies. (b) DC-bus voltage. (c) Real power. (d) Reactive power. (e) Voltage v_a and current i_a. (f) Voltages v_a and e_a.

power increased because of the heavier load while the reactive power was still maintained around zero.

Again, as can be seen from Figure 5.9, the internal frequency $\dot{\theta}$ synchronized with the grid frequency $\dot{\theta}_g$ all the time, except for some expected transients.

5.6 Summary

Based on (Ma et al. 2012), a PWM rectifier is operated as a VSM, by embedding the mathematical model of a synchronous machine in the controller. Depending on the requirement of the applications, two options are presented. One is to directly control the real power drawn by the rectifier and the other is to regulate the DC-bus voltage. The reactive power can be regulated as well, e.g., to obtain the unity power factor. Both simulation and experimental results are provided.

6

Control of Permanent Magnet Synchronous Generator (PMSG) Based Wind Turbines

PMSG wind turbines with full-scale back-to-back PWM converters are being widely adopted in industry. In this chapter, a control strategy based on the synchronverter technology is presented for back-to-back PWM converters. Both converters are operated as synchronverters, which are mathematically equivalent to the conventional synchronous generators. In contrast to common practice, the rotor-side converter (RSC) is responsible for maintaining the DC link voltage and the grid-side converter (GSC) is responsible for injecting the maximum power available to the grid. Since both converters are operated using the synchronverter technology, the whole wind power system is able to support the grid frequency and voltage when there is a grid fault, making it friendly to the grid. Extensive real-time digital simulation results are presented to verify the effectiveness of the presented method under normal and abnormal grid conditions.

6.1 Introduction

The large-scale utilization of renewable energy has been regarded as a major means to address sustainability and environmental issues (Blaabjerg and Ma 2013; DOE 2015; Yaramasu et al. 2015; Zhong and Hornik 2013). Many countries have set strategic plans to utilize renewable energy and make a transition to a low-carbon economy. Wind energy has become the fastest growing source of clean and renewable energy utilized in the world. The US Department of Energy (DOE) has targeted an increase in wind-generated electricity in the US to 20% in 2030 and 35% in 2050 (DOE 2015). As a result, the next-generation smart grid will see more and more wind power generation systems (WPGSs).

WPGSs have variable- or fixed-speed ones. Variable-speed ones are more attractive than fixed-speed ones because of the increased energy yield. Nowadays, the most commonly used topologies for variable-speed WPGSs (Blaabjerg and Ma 2013; Yaramasu et al. 2015) adopt doubly-fed induction generators (DFIG), often called type 3 systems, and permanent magnet synchronous generators (PMSGs), often called type 4 systems. A PMSG wind turbine is connected to the grid through full-scale back-to-back voltage-source converters, which consist of a rotor-side converter (RSC) and a grid-side converter (GSC) (Baroudi et al. 2005; Hong et al. 2013; Seixas et al. 2014). Because it is able to provide unity power factor, low current THD, and ancillary services such as low voltage ride through capability, it has become an industrial trend to adopt this topology.

Power Electronics-Enabled Autonomous Power Systems: Next Generation Smart Grids,
First Edition. Qing-Chang Zhong.
© 2020 John Wiley & Sons Ltd. Published 2020 by John Wiley & Sons Ltd.

Many control strategies have been developed for back-to-back PWM converters in wind power applications. Normally, the maximum power extraction from the wind is achieved by the RSC and the DC-bus voltage is controlled by the GSC (Bueno et al. 2008; Cardenas and Pena 2004; Konstantopoulos and Alexandridis 2014; Pena et al. 1996a,b; Portillo et al. 2006; Singh et al. 2011; Teodorescu and Blaabjerg 2004). In this case, when a grid fault occurs, the DC-bus voltage would increase because of the continuous generation of electricity by the wind turbine. As a fail-safe mechanism against equipment damage, wind turbines were conventionally designed to automatically disconnect themselves from the utility grid under these extreme conditions (Bollen 2000; McGranaghan et al. 1993). However, wind turbines are now required to ride through grid faults under some conditions. Alternatively, another option is for the RSC to control the DC-bus voltage and for the GSC to control the power fed to the grid (Hansen and Michalke 2009; Kim et al. 2010; Yuan et al. 2009). In both cases, the control principle is dominated by the well known vector control (Blaabjerg and Ma 2013; Li et al. 2012; Shariatpanah et al. 2013; Yaramasu et al. 2015), which converts quantities in the three-phase natural reference frame into quantities in the two-axis *dq* reference frame with controllers designed for the two axes, in a way similar to controlling DC machines (Bose, 2001, 2009). The vector control has some known drawbacks. For example, it is sensitive to parameters and uncertainties (Chen et al. 2016; Li et al. 2014; Zhong et al. 2015), which makes it difficult to capture the maximum available power when wind changes. The uncertainties could also lead to stability and reliability concerns (Li et al. 2012). Wind turbines are large flexible structures operating in noisy environments and the aerodynamic loads on the turbines are highly nonlinear (Johnson et al. 2004; Leith and Leithead 1997). The time constant of the armature winding limits the performance (Zhong et al. 1997) and the coupled current dynamics in the *dq* reference frame brings difficulties in current regulation as well (Mohamed and Lee 2006; Sneyers et al. 1985). Another problem with the vector control is the inherent difficulties in dealing with asymmetrical scenarios. For example, asymmetrical grid fault conditions have become new challenges for WPGSs (Yaramasu et al. 2015). In recent years, the conventional vector control has been improved by the feed-forward compensation (Mohamed and Lee 2006; Morimoto et al. 1994), the hysteresis band PWM strategy (Rebeiro and Uddin 2012) to deal with nonlinear effects, the direct-current vector control (Li et al. 2012) to enhance stability, reliability and efficiency, and the direct torque control to achieve fast response (Haque et al. 2014; Zhang et al. 2014; Zhong et al. 1997).

In this chapter, the synchronverter technology will be applied to control the back-to-back converters in PMSG wind turbines, by operating the RSC as a virtual synchronous motor and the GSC as a virtual synchronous generator. Since the RSC receives power from the PMSG at the AC side and injects it to the DC bus, its main task is to regulate the DC-bus voltage at the desired level while achieving unity power factor at the rotor side to minimize losses. Since the GSC is responsible for interacting with the grid, it is controlled to inject real and reactive power into the grid. Its main task is to achieve the maximum power point tracking (MPPT) while regulating the reactive power. As a result, the whole system behaves like a generator–motor–generator system, making PMSG wind turbines friendly with the grid. The synchronverter technology is independent of the machines parameters, offering significant advantages over vector control.

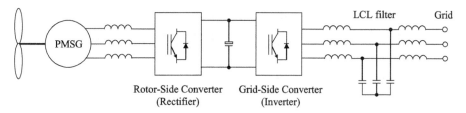

Figure 6.1 Integration of a PMSG wind turbine into the grid through back-to-back converters.

6.2 PMSG Based Wind Turbines

A PMSG adopts a permanent magnet to generate the magnetic field needed for electricity generation. There is no need to provide an external power supply for excitation or to have a controllable/accessible rotor circuit. This simplifies the system structure and reduces maintenance costs. PMSGs have higher efficiency and power density than induction generators (Li et al. 2012; Shariatpanah et al. 2013; Zhang et al. 2014). A PMSG based wind turbine operates at a synchronous speed and the frequency of the generated electricity is proportional to the rotor speed, which could eliminate the need for a mechanical speed sensor to measure the rotor speed.

For wind power applications, PMSGs are often equipped with full-scale RSCs and GSCs for grid connection, as shown in Figure 6.1. This allows the system to operate under a wide range of speeds. Moreover, the full-scale back-to-back converters make it possible to capture the maximum available energy in the wind and to provide high power quality with ancillary services (Seixas et al. 2014; Zhong et al. 2015). Because the grid and the PMSG are decoupled by the common DC bus, this system has great controllability over the full power and excellent capability of interacting with the grid (Yaramasu et al. 2015).

6.3 Control of the Rotor-Side Converter

As mentioned before, the main task of the RSC is to regulate the DC-bus voltage. This can be done via operating the RSC as a virtual synchronous motor, with the controller shown in Figure 6.2.

Denote the output voltage and current of the wind-turbine generator as $v_r = [v_{ra} \ v_{rb} \ v_{rc}]^T$ and $i_r = [i_{ra} \ i_{rb} \ i_{rc}]^T$, respectively. The controller shown in Figure 6.2 includes the model of a synchronous motor developed in the previous chapter, with subscript r added to relevant variables to differentiate them from the grid-side ones. There are

$$v_r = e_m + R_s i_r + L_s \frac{di_r}{dt},$$

and

$$\ddot{\theta}_m = \frac{1}{J_m}(T_{me} - T_{mm} - D_{mp}\dot{\theta}_m),$$

where the mechanical torque T_{mm} is the virtual mechanical torque generated by the PI controller to regulate the DC-bus voltage.

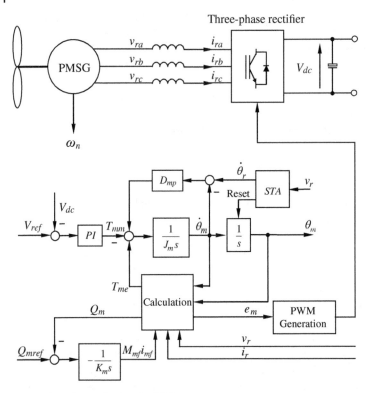

Figure 6.2 Controller for the RSC.

In order to operate the RSC as a virtual synchronous motor properly, the frequency $\dot{\theta}_{\mathrm{m}}$ inside the rectifier should be the same or very close to the frequency $\dot{\theta}_{\mathrm{r}}$ of the PMSG voltage v_{r} and the terminal voltage should be synchronized with v_{r}. Before the rectifier is connected to v_{r}, its phase θ_{m} should be kept the same as that of v_{r} in order to minimize the settling time following the connection. The traditional PLL in the dq frame is not fast enough for this purpose (Ziarani et al. 2003). Here, the STA (Ziarani and Konrad 2004) is adopted to quickly provide the frequency and the phase information of the voltage signal.

The control strategy involves two channels: one channel to regulate the real power and the other to regulate the reactive power, as discussed in the previous chapter. The real power channel has three cascaded control loops. The inner loop is the frequency regulation loop (with the feedback gain D_{mp}), the middle loop is a torque loop (with the feedback coming from the current i_{r} via the electromagnetic torque T_{me}) and the outer loop is the DC-bus voltage loop (with the feedback coming from the DC voltage V_{dc}). The first two loops are part of the model of a virtual synchronous motor.

When the RSC is connected to v_{r}, its virtual internal frequency should track the frequency of v_{r}. This can be done via feeding the difference between the frequency $\dot{\theta}_{\mathrm{r}}$ obtained from the STA and the rectifier frequency $\dot{\theta}_{\mathrm{m}}$ to the D_{mp} block. In this case, the constant D_{mp} represents the (virtual) mechanical friction coefficient of the virtual motor. The frequency droop coefficient is defined as $D_{\mathrm{mp}} = -\frac{\Delta T}{\Delta \dot{\theta}_{\mathrm{m}}}$, where ΔT denotes the amount of change in the torque that leads to the change $\Delta \dot{\theta}_{\mathrm{m}}$ in the frequency.

The reactive power Q_m of the RSC can be controlled to track the reference Q_{mref}. The tracking error between the reference value Q_{mref} and reactive power Q_m is fed into an integrator with a gain $-\frac{1}{K}$ to regulate the field excitation $M_f i_f$ so that the voltage e_m is changed accordingly. In order to obtain the unity power factor, Q_{mref} can be set at 0.

6.4 Control of the Grid-Side Converter

As discussed above, the control objective for the GSC is to inject the maximum power extracted from wind to the grid. Normally, variable-speed wind generation systems are more attractive than fixed-speed systems due to the high productivity. However, this does not necessarily mean that the actual efficiency is always high in variable-speed wind generation systems. It depends on the control algorithm used to extract the output power from the wind turbine and other factors. In a variable-speed wind generation system, the wind turbine can be operated at its maximum power operating point for various wind speeds by maintaining the tip speed ratio to the value that maximizes its aerodynamic efficiency. In order to achieve this ratio, the permanent magnet synchronous generator load line should be matched very closely to the maximum power line of the wind turbine. There are many MPPT algorithms available; see, e.g., (Femia et al. 2005; Hong et al. 2013; Koutroulis and Kalaitzakis 2006; Linus and Damodharan 2015). Since this is not a focus of this chapter, a simple algorithm is discussed below.

In a variable-speed wind turbine, the power P_w that can be obtained from the wind is a cubic function of the wind speed v_w, i.e.

$$P_w = \frac{1}{2}\rho\pi R^2 v_w^3 C_p, \tag{6.1}$$

where ρ is the air density, R is the turbine radius and $C_p = f(\lambda, \beta)$ is the power coefficient that depends on the tip speed ratio λ of the wind turbine and the angle of blades β. The tip speed ratio is defined as

$$\lambda = \frac{\omega_n R}{v_w},$$

where ω_n is the rotor speed of the turbine. Consequently, for different wind speeds, the optimal values of the tip speed and the pitch angle need to be followed in order to achieve the maximum C_p and therefore the maximum power output. The maximum power of the wind turbine can be calculated as

$$P_m = \frac{1}{2}\rho\pi R^2 C_{pm}(\frac{R\omega_n}{\lambda_{opt}})^3 = K_{opt}\omega_n^3,$$

where $K_{opt} = \frac{1}{2}\rho\pi C_{pm}(\frac{R^5}{\lambda_{opt}^3})$. Taking the losses into account, the grid power reference can be chosen slightly smaller than P_m. In this chapter, this is chosen as

$$P_{ref} = 0.95 K_{opt}\omega_n^3. \tag{6.2}$$

The coefficient, 0.95, is an estimated value after considering losses, which may be changed for different systems. Taking into account the converter and the PMSG ratings, the coefficient can be estimated by considering a wide range of wind speeds to cover the widest

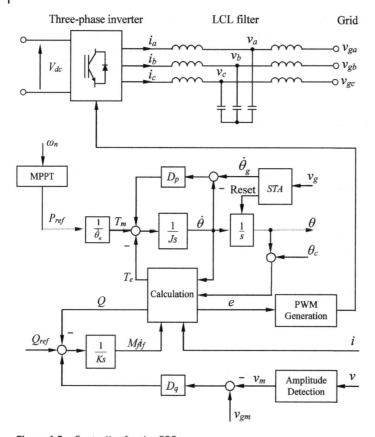

Figure 6.3 Controller for the GSC.

possible scenarios. This reference power can be passed to the GSC so that the right amount of power is fed into the grid.

The GSC control strategy is shown in Figure 6.3, where the inductor current, the capacitor voltage and the grid voltage are denoted as $i = [i_a \ i_b \ i_c]^T$, $v = [v_a \ v_b \ v_c]^T$ and $v_g = [v_{ga} \ v_{gb} \ v_{gc}]^T$, respectively. This is very similar to the synchronverter presented in Chapter 4, with the real power reference set according to (6.2). Because of the inherent capability of the synchronverter, this strategy is able to feed the real power to the grid while regulating the reactive power. As a result, the PMSG wind turbine can provide support to the grid frequency and voltage in the same way as a conventional synchronous generator.

6.5 Real-time Simulation Results

Real-time simulations were carried out to verify the control strategy described above. The back-to-back converter was implemented using the universal bridge with IGBTs as switches and the switching frequency was set to 5 kHz. The other parameters of the system used in the simulations are given in Table 6.1.

Table 6.1 Parameters of a PMSG wind turbine system.

Parameter	Value
Air density	$1.2\,\text{kg}\,\text{m}^{-3}$
Rotor radius	$1.2\,\text{m}$
Inertia of the PMSG	$0.032\,\text{kg}\,\text{m}^2$
$M_f i_f$ of the PMSG	$0.5\,\text{V}\,\text{s}$
Friction factor of the PMSG	$0\,\text{N}\,\text{m}\,\text{s}$
Pole pairs of the PMSG	4
Winding inductance of the PMSG	$4\,\text{mH}$
Winding resistance of the PMSG	$0.135\,\Omega$
DC-bus voltage	$400\,\text{V}$
DC-bus capacitance	$5000\,\mu\text{F}$
Inductance on the inverter side	$1.1\,\text{mH}$
Inductance on the grid side	$1.1\,\text{mH}$
Capacitance of the LCL filter	$22\,\mu\text{F}$
Grid phase voltage (RMS)	$110\,\text{V}$

The power coefficient C_p is calculated by using the following formula (Sun et al. 2003):

$$C_p(\lambda, \beta) = 0.22(\frac{116}{\beta} - 0.4 - 5)e^{-\frac{12.5}{\beta}}$$

with

$$\beta = \frac{1}{\frac{1}{\lambda + 0.08\lambda} - \frac{0.035}{\lambda^3 + 1}}.$$

For the simulated wind turbine, the optimal tip speed ratio λ_{opt} is 6.8 and the maximum power coefficient C_{pm} is 0.419.

6.5.1 Under Normal Grid Conditions

The simulation was carried out according to the following sequence of actions:

(1) Start the system, but keep all the IGBTs off with an initial wind speed of $12\,\text{m}\,\text{s}^{-1}$ to establish the DC-bus voltage first (with $R_{dc} = 150\,\Omega$).

(2) Start operating the IGBTs for both converters at $t = 0.4\,\text{s}$ with $V_{ref} = 400\,\text{V}$ and $Q_{ref} = 0$ for the RSC to regulate the DC-bus voltage (with $R_{dc} = 150\,\Omega$) and with $P_{ref} = 0$ and $Q_{ref} = 0$ for the GSC to synchronize with the grid.

(3) Turn the circuit breaker on to connect the GSC to the grid at $t = 0.8\,\text{s}$.

(4) Set $P_{ref} = 0.95 K_{opt}\omega_n^3$ at $t = 3\,\text{s}$ and disconnect the DC load $R_{dc} = 150\,\Omega$ to send the maximum power to the grid.

(5) Change the wind speed to $15\text{m}\,\text{s}^{-1}$ at $t = 6\,\text{s}$.

It is worth noting that, in the real system, the DC load R_{dc} is often referred to as a DC-bus chopper, which can be activated when the DC-bus voltage exceeds a certain level, which

Table 6.2 GSC control parameters.

Parameter	Value	Parameter	Value
J	0.0122	K	2430.1
D_p	6.0793	D_q	386.7615

is chosen according to the converter ratings and is usually around 5% or 10% above the nominal DC-bus voltage (Chaudhary et al. 2009).

6.5.1.1 GSC Performance

The parameters for the GSC controller are shown in Table 6.2. The real-time simulation results for the dynamic response are shown in Figure 6.4(a). The connection of the system to the grid was very smooth without noticeable dynamics and the synchronverter frequency was able to synchronize with the grid frequency at all times. The power coefficient C_p was kept almost equal to its maximum value after the system was requested to send the maximum power to the grid at $t = 3$ s and the real power P accurately tracked the reference power P_{ref}. Therefore, the maximum power point tracking was achieved by the grid-side converter, even after the wind speed changed at $t = 6$ s. The phase a grid current and grid voltage around the wind speed change are shown in Figure 6.4(b). The current increased smoothly, with unity power factor achieved for the complete wind power system.

6.5.1.2 RSC Performance

The parameters for the RSC controller are given in Table 6.3. The dynamic response of the RSC is shown in Figure 6.5(a). The DC-bus voltage was maintained well at 400 V after the system was connected to the grid. The RSC was able to track the generator frequency, which is proportional to the rotor speed ω_n. During the wind change at $t = 6$ s, the responses of the RSC voltage and current are shown in Figure 6.5(b). The current drawn by the RSC, i.e. the generator current, is in phase with the voltage, achieving unity power factor.

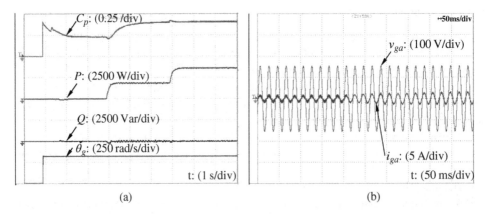

(a)   (b)

Figure 6.4 Dynamic response of the GSC. (a) Full simulation process. (b) Voltage and current around the wind change.

Table 6.3 RSC control parameters.

Parameter	Value	Parameter	Value
J_m	0.0122	K_p	0.8
D_{mp}	6.0793	K_i	2
K_m	2430.1		

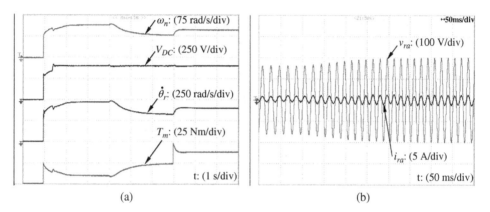

(a) (b)

Figure 6.5 Dynamic response of the RSC. (a) Full simulation process. (b) Voltage and current around the wind change.

6.5.2 Under Grid Faults

In order to verify the fault ride-through capability of the control strategy, two real-time simulations were carried out, assuming 50% grid voltage drop and 1% grid frequency drop, respectively. Each fault occurred at 6 s and lasted for 0.1 s, respectively. The sequence of actions were the same as the ones in the previous experiments except that the fifth action was replaced with the grid faults at $t = 6$ s. The system parameters remained unchanged, apart from that the impedance of the feeder was considered explicitly as 2.2 mH and 0.2 Ω.

The simulations were started with all IGBTs turned OFF, a DC-bus load of 150 Ω, and a wind speed of 12 m s^{-1}. The DC-bus voltage was established by the RSC acting as a diode bridge. At $t = 0.4$ s, the IGBTs of both converters was enabled for the RSC to regulate the DC-bus voltage with $V_{ref} = 400$ V and $Q_{ref} = 0$ and for the GSC to synchronize with the grid with $P_{ref} = 0$ and $Q_{ref} = 0$. The DC-bus voltage was quickly established, with a slight short drop in the rotor speed ω_n. At $t = 0.8$ s, the circuit breaker was turned ON to connect the GSC to the grid (still with $P_{ref} = 0$ and $Q_{ref} = 0$). Because the GSC was already synchronized with the grid, there was not much transient. At $t = 3$ s, the DC-bus chopper was disconnected and the GSC P_{ref} was set according to the MPPT algorithm. The power sent to the grid and the current reached the steady state quickly and the DC-bus voltage was maintained well. As expected, it took longer for the mechanical quantities, such as the power coefficient C_p, the torque T_m, and the rotor speed ω_n, to reach the steady state. At $t = 6$ s, one of the two grid faults was applied for 0.1 s. When the grid voltage dropped by 50%, as shown in Figure 6.6(a), the real power sent to the grid increased initially before it dropped while the current increased, which means the current became more reactive.

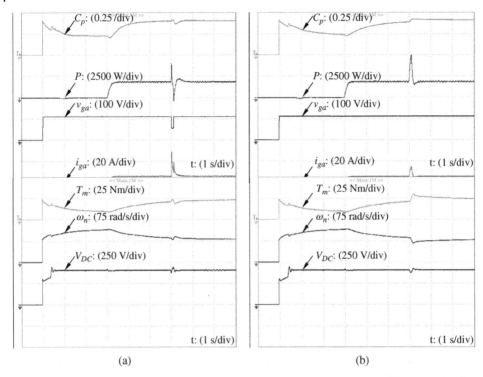

(a)   (b)

Figure 6.6 Real-time simulation results with a grid fault appearing at $t = 6$ s for 0.1 s. (a) When the grid voltage drops by 50%. (b) When the grid frequency drops by 1%.

There was a drop in the rotor speed and the DC-bus voltage in order to release more energy to support the grid. After the voltage recovered, the system returned back to normal. When the grid frequency dropped by 1%, as shown in Figure 6.6(b), the real power sent to the grid increased to support the grid and the current increased as well, which means the current stayed active. There was a big drop in the rotor speed and the DC-bus voltage in order to release more energy to support the grid. After the frequency recovered, the real power and the current returned back to normal quickly but it took a little longer for the mechanical quantities to return back to normal. Indeed, as demonstrated, the WPGS could support the grid well when there were grid faults. It is worth noting that the DC-bus voltage was maintained well during the faults.

6.6 Summary

Based on (Zhong et al. 2015), the synchronverter technology is applied to operate a PMSG wind turbine as a virtual synchronous generator on the grid side and as a virtual synchronous motor on the rotor side. The whole system becomes a (virtual) generator–motor–generator system. It is able to provide grid support while maintaining the DC-bus voltage. The unity power factor can be achieved for both sides, reducing the overall system loss. Real-time simulation results verified the control strategy under both normal and abnormal grid conditions, demonstrating excellent fault ride-through capability.

7

Synchronverter Based AC Ward Leonard Drive Systems

In this chapter, the VSM concept is applied to control the speed of an AC machine in four quadrants, via powering the AC machine with a VSM that generates a variable-voltage-variable-frequency supply. This is a natural and mathematical, but not physical, extension of the conventional Ward Leonard drive systems for DC machines to AC machines. If the rectifier providing the DC bus for the AC drive is controlled as a VSM according to Chapter 5, then an AC drive is equivalent to a motor–generator–motor system. This facilitates the analysis of AC drives and the introduction of some special functions. Control strategies with and without a speed sensor are presented, together with experimental results.

7.1 Introduction

Motors consume the majority of electricity, of which 50–70% is consumed by asynchronous electric motors and 3–10% by synchronous electric motors[1] . Variable speed drives (VSDs), often equipped with inverters, are widely used nowadays to save energy, increase productivity and improve quality in many applications, such as home appliances, robots, pumps, fans, automotives, railways, industrial processes, and, recently, renewable energy. AC motors are the main driving force in industry because of their small size, reliability, low cost and low maintenance (Boldea 2008; Bose 1993; Kazmierkowski et al. 2011; Pacas 2011). Due to the advancement of power electronics, digital signal processing (DSP), etc., the technology of VSDs for AC motors has matured and AC drives have replaced DC drives in many application areas. There are mainly three approaches developed for AC drives (Boldea 2008; Bose 2001; Kazmierkowski et al. 2011):

1. V/f control. The idea is to generate a variable-voltage-variable-frequency sinusoidal power supply from a constant DC power source. The control variables are voltage and frequency while maintaining their ratio constant to provide (almost) constant flux. It is widely used in open-loop drives, where the requirement of performance, e.g. speed accuracy and response, is not high and/or the controller needs to be simple (Sun et al. 2009). This is also called scalar control because only the amplitude of the voltage is

---

1 http://encyclopedia2.thefreedictionary.com/Power+System+Load

Power Electronics-Enabled Autonomous Power Systems: Next Generation Smart Grids,
First Edition. Qing-Chang Zhong.
© 2020 John Wiley & Sons Ltd. Published 2020 by John Wiley & Sons Ltd.

controlled. It is possible to add the feedback of speed, torque and/or flux to improve the performance (Ancuti et al. 2010; Suetake et al. 2011).

2. Vector control. The idea is to control AC motors in a way similar to controlling separately excited DC motors, after introducing some transformations. The three-phase currents are converted into d, q current components i_d and i_q, which correspond to the field and armature currents of DC motors, respectively. If i_d is oriented (aligned) in the direction of the rotor flux and i_q is perpendicular to it, then the control of i_d and i_q is decoupled, as in the case of DC motors. The frequency is not directly controlled, as in scalar control, but indirectly controlled; the torque is controlled indirectly via controlling the current. The advantage of vector control is that it provides good performance that is similar to DC drives. The drawbacks of vector control are: (i) the flux estimation and field orientation are dependent on motor parameters, which change in reality, e.g. with temperature; (ii) the controller is very complicated; and (iii) the inverter is often current controlled via a hysteresis-band PWM, which makes the system analysis difficult (Bascetta et al. 2010; Gulez et al. 2008; Jain and Ranganathan 2011; Mengoni et al. 2008; Shao et al. 2009). A lot of patches have been developed for vector control to improve the performance; see e.g. (Al-Nabi et al. 2012; Chatterjee 2011; Hatua et al. 2012; Leidhold 2011; Liu and Li 2012; Mohseni et al. 2011; Patel et al. 2011; Pellegrino et al. 2009; Trentin et al. 2009).

3. Direct torque (and flux) control. The torque (and stator flux) is (are) directly controlled via selecting appropriate inverter voltage space vectors through a look-up table but the frequency is indirectly controlled (Depenbrock 1988; Takahashi and Noguchi 1986; Zaid et al. 2010; Zheng et al. 2011). It uses hysteresis based control, which generates flux and torque ripples, and the switching frequency is not constant. It also needs motor parameters to estimate the torque (and stator flux) (Khoucha et al. 2010; Shyu et al. 2010). Again, the hysteresis-based control makes system analysis very difficult.

These schemes have been further advanced for a long period with the development of related technologies in control theory and microelectronics. They are suitable for different applications because of their different characteristics (Boldea 2008; Kazmierkowski et al. 2011; Rodriguez et al. 2012). The vector control and direct torque (and flux) control provide very good performance but the control algorithms involve several transformations and are very complicated. What is worse is that look-up tables are used in the direct torque (and flux) control, which makes the analytical analysis of the system very difficult. The high order of the resulting complete system also means that the system stability is difficult to guarantee. V/f control is simple but the performance needs to be improved. Hence, a simple high-performance AC drive that facilitates the system analysis is desirable.

From the viewpoint of control system design, the AC motor is simply the load to an inverter. The main control objective of a drive is to regulate the speed and the torque to obtain fast and good response. The change of motor parameters (including the load) should not impose a major problem to the system. Such an attempt is made in this chapter, following the concept of operating inverters to mimic synchronous generators (Zhong and Hornik 2013; Zhong and Weiss 2009, 2011) and motivated by the conventional Ward Leonard drive systems (WLDSs). The physical interpretation of this is that the AC motor is powered by a synchronous generator (SG) driven by a variable-speed prime mover. The synchronous generator and the prime mover is then replaced by an inverter that behaves as an SG. The torque and speed of the AC motor is then controlled via controlling the torque and frequency of the

Table 7.1 Comparison of different control strategies for AC VSDs.

Control type	Frequency control	Torque control	Flux control
V/f control	Direct	None	None
Vector control	Indirect	Indirect	Direct
Direct torque control	Indirect	Direct	Direct
AC WLDS	Direct	Direct	Open-loop

SG. The resulting control scheme is very simple as it does not involve vector transformations nor the estimation of flux. No complicated concepts, e.g. vector control and field orientation, are needed and the scheme is very easy to understand. This also unifies the drive for synchronous machines (SMs) and induction machines (IMs). In this scheme, the attention of how to design AC drives has shifted from motor-oriented to inverter-oriented. This has led to an extremely simple controller. The presented AC drive can be regarded as powered by a synchronous generator while a vector-controlled AC drive is powered by a DC generator with some transformations. Another important advantage is that the complete system can be described by the analytic mathematical models of generators and motors, which facilitates the analysis of the whole system. A comparison of these different VSDs is given in Table 7.1.

7.2 Ward Leonard Drive Systems

Induction motors, particularly those of the squirrel-cage type, have been the principal workhorse for a long time. However, until the beginning of the 1970s, they had been operated in the constant-voltage-constant-frequency (CVCF) uncontrolled mode, which is still very common nowadays. VSDs were dominated by DC motors in the Ward Leonard arrangement. Ward Leonard drive systems, also known as Ward Leonard control, were widely used DC motor speed control systems introduced by Harry Ward Leonard in 1891. A Ward Leonard drive system, as shown in Figure 7.1, consists of a motor (prime mover) and a generator with shafts coupled together. The motor, which turns at a constant speed, may be AC or DC powered. The generator is a DC generator, with field windings and armature windings. The field windings are supplied with a variable DC source to produce a variable output voltage in the armature windings, which then supplies a second DC motor to drive the load.

   A natural analogy is to replace the DC generator with a synchronous generator and the DC motor with an AC machine (an induction motor or a synchronous motor); see Figure 7.2(a). This configuration is called an AC Ward Leonard drive system (Zhong 2010a,b). The prime mover in a DC WLDS maintains a constant speed and the flux of the generator is variable; the prime mover in an AC WLDS needs to have a variable speed (so that the frequency of the output can be varied) and the flux of the generator is constant. The output of the generator (voltage) in a DC WLDS is varied via controlling the field

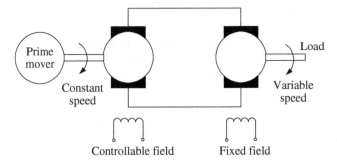

Figure 7.1 Conventional (DC) Ward Leonard drive system.

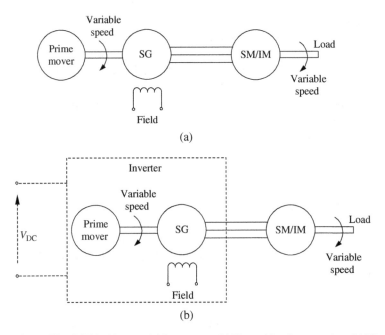

Figure 7.2 AC Ward Leonard drive system. (a) Natural implementation. (b) Virtual implementation.

voltage while the output of the generator (voltage and frequency) in an AC WLDS is varied via controlling the speed of the prime mover. If the speed of the prime mover could be varied, it could have been used to drive the load directly. Hence, there is no need to have a physical AC WLDS. However, instead of having a physical synchronous generator that is driven by a variable-speed prime mover, an inverter that captures the main dynamics of the physical system (the synchronous generator, the variable-speed prime-mover, and its controller), as shown in Figure 7.2(b), can be used to power the motor, following the synchronverter technology of operating inverters to mimic synchronous generators (Zhong and Hornik 2013; Zhong and Weiss 2009, 2011). Ideally, if the motor has the same pole number as the generator and there is no loss, the torque of the generator would be the same as the torque of the motor. Hence, the torque of the motor could be controlled via controlling the torque entering the synchronous generator.

7.3   Model of a Synchronous Generator

The model of synchronous generators is very well documented and can be found in Chapter 4. Here, some changes are made to the model developed in Chapter 4, assuming that the flux established in the stator by the field windings is sinusoidal and that the stator winding resistance and inductance are zero. The saturation effect of the machine is introduced by limiting the voltage, as shown in Figure 7.3. When the field current I_f is constant, the generated voltage $e = \begin{bmatrix} e_a & e_b & e_c \end{bmatrix}^T$ is

$$e = \lambda \dot{\theta} \widetilde{\sin}\ \theta \tag{7.1}$$

where θ is the electrical rotor position (hence $\dot{\theta}$ is the electrical angular speed), λ is the amplitude of the mutual flux linkage between the stator winding and the rotor winding, and $\widetilde{\sin}\ \theta$ is the vector $\begin{bmatrix} \sin\ \theta & \sin(\theta - \frac{2\pi}{3}) & \sin(\theta - \frac{4\pi}{3}) \end{bmatrix}^T$. In fact, λ is also the ratio of the generated voltage (amplitude) to the speed (angular) and $\lambda = M_f I_f$, where M_f is the maximum mutual inductance between the stator winding and the rotor winding. The mechanical dynamics of the machine is governed by

$$J\ddot{\theta} = T_\mathrm{m} - T_\mathrm{e} - D_\mathrm{p}\dot{\theta},$$

where J is the moment of inertia of all parts rotating with the rotor, D_p is the damping factor, T_m is the mechanical torque applied to the synchronous generator by the prime mover and T_e is the electromagnetic torque given by

$$T_\mathrm{e} = p\lambda \langle i,\ \widetilde{\sin}\ \theta \rangle. \tag{7.2}$$

Here, p is the number of pole pairs per phase, $i = \begin{bmatrix} i_a & i_b & i_c \end{bmatrix}^T$ is the state current vector, and $\langle \cdot,\ \cdot \rangle$ denotes the conventional inner product. It is worth noting that if $i = I_0 \widetilde{\sin}\ \varphi$ then

$$T_\mathrm{e} = p\lambda I_0 \langle \widetilde{\sin}\ \varphi,\ \widetilde{\sin}\ \theta \rangle = \frac{3}{2} p\lambda I_0\ \cos(\theta - \varphi),$$

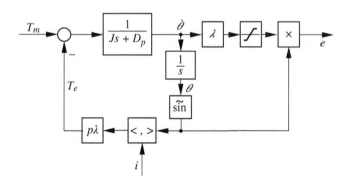

Figure 7.3   Mathematical model of a synchronous generator.

which is a constant DC value. This is a very important property, from which a simple control strategy can be designed to regulate the speed of the AC machine.

7.4 Control Scheme with a Speed Sensor

7.4.1 Control Structure

As explained before, the idea of the AC WLDS is to power the AC motor with a synchronous generator, driven by a variable-speed prime mover that is implemented via an inverter. Hence, the focus of the control system is to control the generator instead of the motor. The mechanical torque T_m applied to the generator can be easily generated by a speed controller (governor), e.g. a PI controller, which compares the actual speed $\dot{\theta}_f$ with the reference speed $\dot{\theta}_r$. If the motor is synchronous, then the actual speed can be directly taken from the generator without a speed sensor as the motor runs at the synchronous speed $\dot{\theta}$. If the motor is inductive, then the actual speed (mechanical) can be measured from the motor and it should be converted to the electrical speed via multiplying it with the number of pole pairs p. Usually this involves in a low-pass filter to reduce the measurement noise. Another aspect that could be easily taken into account is the voltage drop on the stator winding of the motor, particularly, when the speed is low. It can be compensated for via a feed-forward path containing the stator winding resistance R_s from current i to the generated voltage e. Thus, the resulting complete controller consists of a synchronous generator model, a speed measurement unit, a speed controller and a current feed-forward controller, as shown in Figure 7.4. In order to speed up the system response and to minimize the number of tuning parameters, the inertia of the generator can be chosen to be $J = 0$ (i.e. zero inertia). This also reduces the system order by one, which helps improve system stability.

It is worth noting that the λ of the generator is always kept constant. When the speed of the generator exceeds the rated speed, the generated voltage is bounded by the rated voltage so that the insulation of the motor is not damaged (the voltage boost due to the

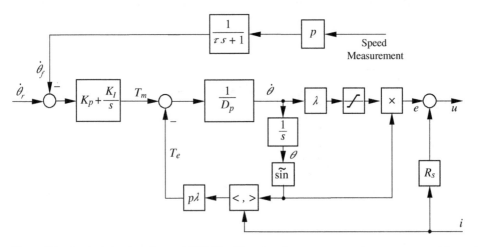

Figure 7.4 Control structure for an AC WLDS with a speed sensor.

current feed-forward path should not exceed the margin allowed, which is normally the case). It should be pointed out that p, λ, and R_s should be chosen the same as those of the motor.

The output u of the controller, which is the sum of the generated voltage and the compensated voltage drop on the motor stator winding, can be passed though a three-phase inverter, after appropriate scaling according to the DC-link voltage, to power the AC motor. The switches in the inverter are operated so that the average values of the inverter output over a switching period should be equal to u, which can be achieved by many known pulse width modulation (PWM) techniques. Because of the inherent low-pass filtering effect of the motor, there is no need to connect an LC filter to improve current quality.

7.4.2 System Analysis and Parameter Selection

In order to simplify the analysis, assume that the speed feedback is taken from $\dot\theta$, i.e., $\dot\theta_f = \frac{1}{\tau s + 1}\dot\theta$. The torque T_e can be regarded as a disturbance to simplify the analysis. The transfer function from the speed reference $\dot\theta_r$ to the speed $\dot\theta$, assuming zero load, is then

$$H_{\dot\theta}(s) = \frac{(K_{\mathrm{P}}s + K_{\mathrm{I}})(\tau s + 1)}{D_{\mathrm{p}}\tau s^2 + (D_{\mathrm{p}} + K_{\mathrm{P}})s + K_{\mathrm{I}}}.$$

Hence, if there is a step change in the reference speed $\dot\theta_r$, there is a step change in the speed $\dot\theta$ by $\frac{K_{\mathrm{P}}}{D_{\mathrm{p}}}\dot\theta_r$. This is normally regarded as aggressive. In order to avoid this, K_{P} can be chosen as 0. As a result,

$$H_{\dot\theta}(s) = \frac{K_{\mathrm{I}}(\tau s + 1)}{D_{\mathrm{p}}\tau s^2 + D_{\mathrm{p}}s + K_{\mathrm{I}}}.$$

This is a second order system with the poles at

$$s_{1,2} = \frac{-1 \pm \sqrt{1 - 4\tau K_{\mathrm{I}}/D_{\mathrm{p}}}}{2\tau}.$$

If K_{I} is chosen as

$$K_{\mathrm{I}} = \frac{D_{\mathrm{p}}}{4\tau},$$

then the two poles are $s_{1,2} = -\frac{1}{2\tau}$ and

$$H_{\dot\theta}(s) = \frac{\tau s + 1}{(2\tau s + 1)^2}.$$

This leaves enough margin for the controller to cope with uncertainties and parameter variations. The transfer function from torque $-T_e$ as a disturbance to the speed $\dot\theta$ is

$$H_{\mathrm{T}}(s) = \frac{4\tau s(\tau s + 1)}{D_{\mathrm{p}}(2\tau s + 1)^2},$$

which means that any step change in the load does not cause a static error in the speed $\dot\theta$. If the torque has a step change T_e, then the speed jumps by $\frac{1}{D_{\mathrm{p}}}T_e$ and recovers. Hence, in order to reduce the impact of the load change on the speed, D_{p} should not be too small.

The speed response is directly related to the time constant of the low-pass filter used in the speed measurement unit. The smaller the time constant, the faster the system response.

The above analysis is approximate because the loop involving the motor that affects T_e is not fully considered and the speed feedback is not exactly taken from the motor. However, it does offer some insightful understanding to the system and, in principle, reflects the system dynamics as can be seen from the experimental results in the next section. It is worth noting that, although the above design leads to a non-oscillatory response, the closed-loop system in real implementation could be oscillatory because of the reasons mentioned above.

The four-quadrant operation of AC machines comes automatically with the presented AC WLDS. There is no need to add any extra effort or device; the change in the sign of the speed reference changes the direction of the motor rotation. A positive frequency (speed) reference leads to a positive speed and a negative frequency (speed) reference leads to a negative speed. A change in the frequency from negative to positive, or from positive to negative, leads to the reversal of the motor rotation.

In summary, the speed response is determined by the time constant τ of the speed measurement unit and the torque response is determined by D_p. The parameters of the controller can be chosen as follows:

(1) Choose p, λ and R_s the same as, or close to, those of the motor and choose $J = 0$.

(2) Determine the time constant τ to meet the requirement of the speed response (and the measurement noise) and D_p to meet the requirement of the torque response.

(3) Choose $K_P = 0$ and $K_I = \frac{D_p}{4\tau}$.

7.5 Control Scheme without a Speed Sensor

7.5.1 Control Structure

If the motor is synchronous, then there is no need to have a speed sensor because the speed of a synchronous motor converges to the speed $\dot{\theta}$ of the generator, which is internally available in the controller. Even for induction motors, $\dot{\theta}$ is the synchronous speed and can be used to reflect the actual motor speed (with the difference being the slip). In this case, D_p can be chosen as 0. The slip of an induction motor can be compensated for to some extent as well. It is well known that the speed drop $\dot{\theta}_s$ is in proportion to the torque over a wide speed range, i.e.

$$\dot{\theta}_s = K_T T_e.$$

This can be obtained from the torque-speed characteristics of the motor. For synchronous motors, $K_T = 0$. This load (torque) effect can be compensated via adding $K_T T_e$ to the speed reference $\dot{\theta}_r$. The resulting speed–sensorless control scheme for AC machines is shown in Figure 7.5. It consists of the model of a synchronous generator, a speed controller, a load-effect compensator and a current feed-forward controller, which is a feed-forward path containing the stator winding resistance R_s from current i to the generated voltage e. This scheme is applicable for both synchronous (with $K_T = 0$) and induction motors. For a synchronous motor, it provides zero-static-error speed control; for an induction motor, there is normally a small static error depending on the compensation accuracy of the load (torque)

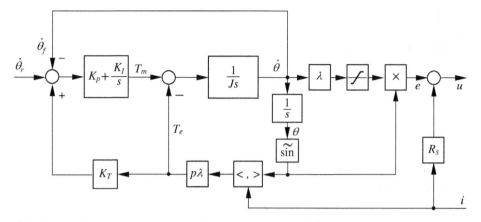

Figure 7.5 Control structure for an AC WLDS without a speed sensor.

effect. The accuracy can be improved via using a two-dimensional table to determine K_T according to the torque-speed characteristics of the motor, taking into account both the synchronous speed and the torque.

7.5.2 System Analysis and Parameter Selection

In order to simplify the exposition below, consider the case when the motor is synchronous, i.e. $K_T = 0$. The transfer function from the speed reference $\dot{\theta}_r$ to the speed $\dot{\theta}$, assuming zero load, is

$$H_{\dot{\theta}}(s) = \frac{K_P s + K_I}{Js^2 + K_P s + K_I}$$

and the transfer function from the torque $-T_e$ (regarded as a disturbance) to the speed $\dot{\theta}$ is

$$H_T(s) = \frac{s}{Js^2 + K_P s + K_I}.$$

It is a second-order system with the poles at

$$s_{1,2} = \frac{-1 \pm \sqrt{1 - 4K_I J/K_P^2}}{2J/K_P}.$$

Increasing K_I tends to make the system response oscillatory. Define $\tau = \frac{J}{K_P}$ or choose

$$K_P = \frac{J}{\tau}$$

and

$$K_I = \frac{K_P}{4\tau} = \frac{J}{4\tau^2},$$

then the two poles are $s_{1,2} = -\frac{1}{2\tau}$. Under this set of parameters,

$$H_{\dot{\theta}}(s) = \frac{4\tau s + 1}{(2\tau s + 1)^2},$$

$$H_T(s) = \frac{4\tau^2 s}{J(2\tau s + 1)^2}.$$

The speed response can be tuned by changing τ and the torque response can be tuned by changing J. In summary, the control parameters can be chosen as follows:

(1) Choose p, λ, and R_s the same as, or close to, those of the motor and choose $D_p = 0$.

(2) Determine the time constant τ to meet the requirement of the speed response and J to meet the requirement of the torque response.

(3) Choose $K_P = \frac{J}{\tau}$ and $K_I = \frac{K_P}{4\tau}$.

(4) Choose K_T according to the torque-speed characteristics of the motor ($K_T = 0$ for a synchronous motor).

7.6 Experimental Results

The AC WLDS strategy was verified on an experimental system shown in Figure 7.6. The system consists of an inverter, a board consisting of current sensors, a dSPACE DS1104 R&D controller board equipped with ControlDesk software, and an induction motor. The motor parameters are given in Table 7.2. According to the parameters, it can be found that $\lambda = 0.0305$ and $K_T = 86.82$. The inverter has the capability to generate PWM voltages from a constant $42V$ DC voltage source and the motor is equipped with a bi-directional encoder with 1000 lines for speed measurement.

Figure 7.6 An experimental AC drive.

Table 7.2 Parameters of the motor.

Parameter	Value	Parameter	Value
R_s	$0.17\,\Omega$	Rated frequency	128 Hz
p	2	Rated speed	3621 rpm
Rated voltage (line-to-line)	30 VRMS	Rated torque	0.528 N m

7.6.1 Case 1: With a Speed Sensor for Feedback

The control parameters were chosen as $\tau = 0.1$ s and $D_p = 0.08$, which results in $K_I = 0.2$. Many experiments were carried out to test the performance of the system and some of the results are shown here.

7.6.1.1 Reversal from a High Speed without a Load

The reference speed was changed from -3600 rpm to 3600 rpm at around $t = 0.6$ s. The responses (speed, torque, current and voltage) are shown in Figure 7.7. The motor quickly reversed from -3600 rpm to 3600 rpm and settled down in about 1.2 s. There was a very short period of over-current around 70%; the voltage dropped when the reversal was started and then gradually built up after the reversal. The phase sequence of the currents were changed at around 0.85 s, which corresponds to the change of the rotating direction of the magnetic field, to enable the reversal.

7.6.1.2 Reversal from a High Speed with a Load

The reference speed was changed from -1800 rpm to 1800 rpm at around $t = 1$ s. The responses are shown in Figure 7.8. The motor quickly reversed from -1800 rpm to 1800 rpm in about 1.5 s, which is slightly longer than the case without a load. There was about 11% overshoot in the speed and the over current increased to about 150%.

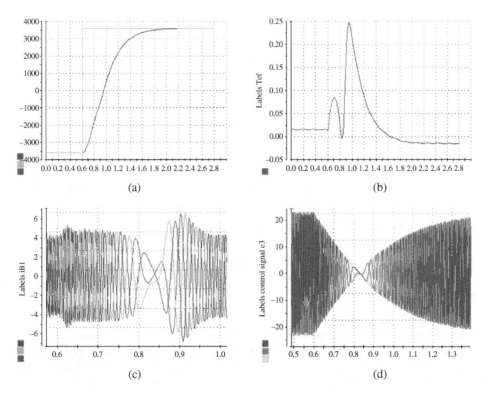

Figure 7.7 Reversal from a high speed without a load. (a) Speed. (b) Torque of the generator. (c) Current. (d) Voltage.

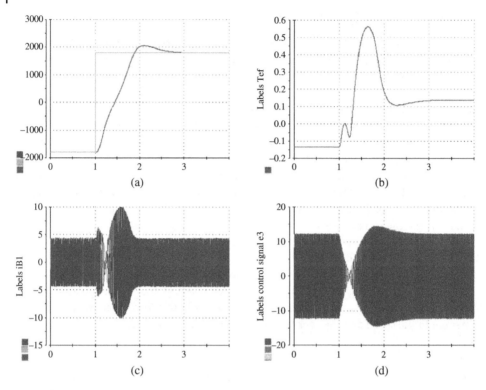

Figure 7.8 Reversal from a high speed with a load. (a) Speed. (b) Torque of the generator. (c) Current. (d) Voltage.

7.6.1.3 Reversal from a Low Speed without a Load

The reference speed was changed from −150 rpm to 150 rpm at around $t =0.6$ s. The responses are shown in Figure 7.9. It took about 1.5 s to complete the reversal and settle down. The over-current was only about 15% and the speed overshoot was about 16%. Note that the sequence of the three phase currents/voltages changed at around $t = 0.8$ s.

7.6.1.4 Reversal from a Low Speed with a Load

The reference speed was changed from −300 rpm to 300 rpm at around $t =2.2$ s. The responses are shown in Figure 7.10. The motor quickly reversed from −300 rpm to 300 rpm in about 2 s. There was a noticeable stop in the middle of the reversal process. The over current was about 50%.

7.6.1.5 Reversal at an Extremely Low Speed without a Load

The reference speed was changed from −4.5 rpm to 4.5 rpm at around $t =5$ s. The responses are shown in Figure 7.11. It took about 15 s to complete the reversal and settle down due to the extremely low speed. There were some ripples in the measured speed, which was owing to the error in the measurement unit (the motor actually rotated smoothly). The motor speed dropped to 0 quickly and remained at a standstill for about 12 s, during which time

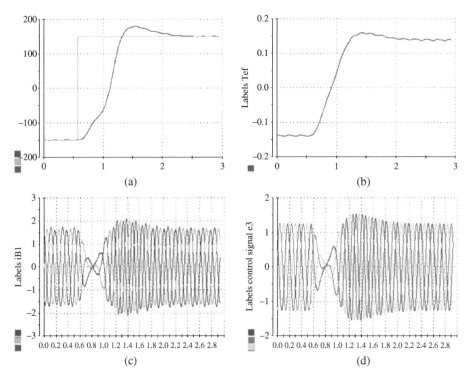

Figure 7.9 Reversal from a low speed without a load. (a) Speed. (b) Torque of the generator. (c) Current. (d) Voltage.

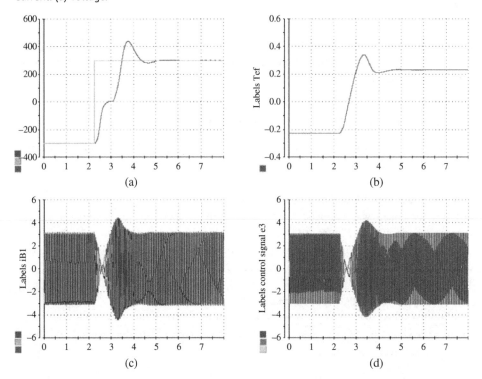

Figure 7.10 Reversal from a low speed with a load. (a) Speed. (b) Torque of the generator. (c) Current. (d) Voltage.

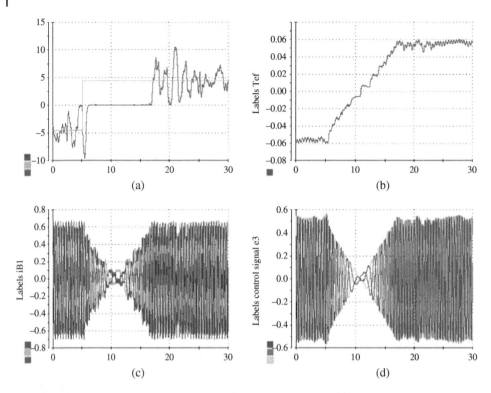

Figure 7.11 Reversal at an extremely low speed without a load. (a) Speed. (b) Torque of the generator. (c) Current. (d) Voltage.

the torque increased almost linearly, before the torque was accumulated high enough to start the motor. Once the current gradually increased to a level that was enough to generate the required torque, the motor started rotating.

7.6.2 Case 2: Without a Speed Sensor for Feedback

The control parameters were chosen as $\tau = 0.1$ s and $J = 0.08$, which resulted in $K_P = 0.8$ and $K_I = 2$. Some of the results are shown here.

7.6.2.1 Reversal from a High Speed without a Load
The reference speed was changed from -3600 rpm to 3600 rpm at around 0.7 s. The responses are shown in Figure 7.12. The motor quickly reversed from -3600 rpm to 3600 rpm in about 1.3 s. There was a very short period of over 100% over-current and the speed overshoot was about 14%. The static error was almost zero.

7.6.2.2 Reversal from a High Speed with a Load
The reference speed was changed from -1800 rpm to 1800 rpm at around $t = 0.4$ s. The responses are shown in Figure 7.13. It took about 1.6 s to complete the reversal and there was about 200% over-current.

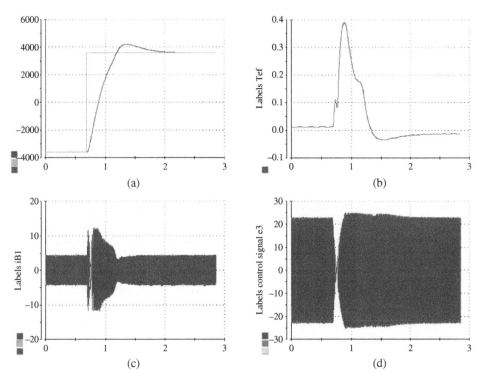

Figure 7.12 Reversal from a high speed without a load (without a speed sensor). (a) Speed. (b) Torque of the generator. (c) Current. (d) Voltage.

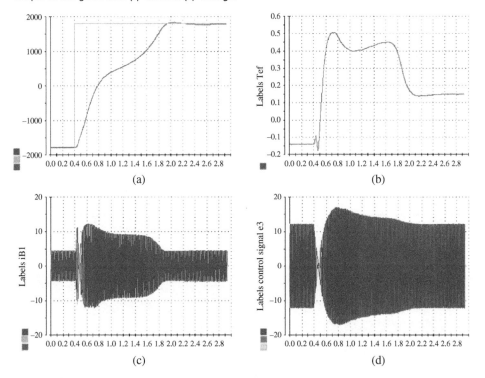

Figure 7.13 Reversal from a high speed with a load (without a speed sensor). (a) Speed. (b) Torque of the generator. (c) Current. (d) Voltage.

7.7   Summary

Based on (Zhong 2013a), the concept of the conventional DC Ward Leonard drive system is extended to AC machines. Instead of implementing it physically, it is implemented mathematically and an inverter is controlled to replicate the dynamics of the physical set of a variable-speed prime mover and a synchronous generator. This leads to the smooth operation of an AC machine in four quadrants. The change of the sign of the frequency set-point leads to the change of the phase sequence of the current and, furthermore, the change of the rotational direction. Two control schemes, one with a speed sensor and the other without a speed sensor, are presented for the system. Extensive experimental results validated the feasibility.

While it is vital for some applications to achieve the fastest possible torque response, this may have been aggressively pushed for some applications where the response speed is not critical. For the former case, it is often achieved via controlling the current provided by the inverter, but for the latter case, the speed regulation can be achieved by controlling the voltage provided by the inverter. In other words, AC drives can be classified as current-controlled AC drives and voltage-controlled AC drives. Vector control and direct torque control belong to the current-controlled AC drives while V/f control and the AC WLDS presented in this chapter belong to the voltage-controlled AC drives. Similarly to AC dives, inverters in smart grid integration can be current-controlled or voltage-controlled as well; see (Zhong and Hornik 2013) for detailed and systematic treatment of controlling inverters for renewable energy and smart grid integration.

8

Synchronverter without a Dedicated Synchronization Unit

A synchronverter is an inverter that mimics synchronous generators. Similar to other grid-connected inverters, the original synchronverter discussed in Chapter 4 also includes a dedicated synchronization unit, e.g. a phase-locked loop (PLL), to provide the phase, frequency and amplitude of the grid voltage as references. In this chapter, a radical step is taken to improve the synchronverter as a self-synchronized synchronverter by removing the dedicated synchronization unit. It can automatically synchronize itself with the grid before connection and maintain synchronization with the grid after connection. This considerably improves the performance, and reduces the complexity and computational burden of the controller. All the functions of the original synchronverter, such as frequency and voltage regulation, real power and reactive power control, are maintained. Both simulation and experimental results are presented. Experimental results have shown that the presented control strategy can improve the performance of frequency tracking by more than 65%, the performance of real power control by 83% and the performance of reactive power control by about 70%.

8.1 Introduction

More and more renewable energy sources are being connected to power systems, often via DC/AC converters (also called inverters). The most important and basic requirement for such applications is to keep inverters synchronized with the grid before and after being connected to the grid so that (1) an inverter can be connected to the grid and (2) the inverter can feed the right amount of power to the grid even when the grid voltage changes its frequency, phase, and amplitude (Blaabjerg et al. 2006; Rodriguez et al. 2007a; Shinnaka 2008; Thacker et al. 2011; Wildi 2005; Zhang et al. 2010; Zhang et al. 2011b; Zhong and Hornik 2013). It has been the norm (Svensson 2001) to adopt a synchronization unit, e.g. a PLL and its variants (Ciobotaru et al. 2006; Karimi-Ghartemani and Iravani 2001, 2002; Karimi-Ghartemani and Ziarani 2004; Shinnaka 2008; Yuan et al. 2002; Ziarani and Konrad 2004), to make sure that the inverter is synchronized with the grid. This practically adds an outer-loop controller (the synchronization unit) to the inverter controller.

Power Electronics-Enabled Autonomous Power Systems: Next Generation Smart Grids,
First Edition. Qing-Chang Zhong.
© 2020 John Wiley & Sons Ltd. Published 2020 by John Wiley & Sons Ltd.

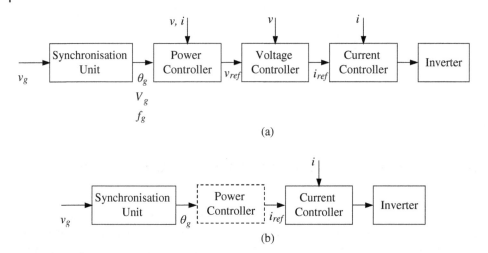

Figure 8.1 Typical control structures for a grid-connected inverter. (a) When controlled as a voltage supply. (b) When controlled as a current supply.

A grid-connected inverter can be controlled as a voltage supply or a current supply. When it is controlled as a voltage supply, the typical control structure is shown in Figure 8.1(a). It consists of a synchronization unit to synchronize with the grid, a power loop to regulate the real power and reactive power exchanged with the grid, a voltage loop to regulate the output voltage and a current loop to control the current. The synchronization unit often needs to provide the frequency and the amplitude, in addition to the phase, of the fundamental component of the grid voltage as the references for the power controller (Shinnaka 2011). The negative impact of a synchronization unit on control performance is well known (Harnefors et al. 2007; Jovcic et al. 2003). Moreover, because PLLs are inherently nonlinear and so are the inverter controller and the power system, it is extremely difficult and time-consuming to tune the PLL parameters to obtain satisfactory performance.

A slow synchronization unit could directly affect control performance and degrade system stability but a complex synchronization unit, on the other hand, is often computationally intensive, which adds significant burden to the controller. Hence, the synchronization needs to be done quickly and accurately in order to maintain synchronism, which makes the design of the controller and the synchronization unit very challenging because the synchronization unit is often not fast enough with acceptable accuracy, and it also takes time for the power and voltage controllers to track the references provided by the synchronization unit. When a grid-connected inverter is controlled as a current supply, the output voltage is maintained by the grid and the inverter only regulates the current exchanged with the grid. In this case, the typical control structure is shown in Figure 8.1(b). Normally, it does not have a voltage controller; the power controller is often a simple static one as well, which does not require much effort for tuning. Moreover, the synchronization unit is often required to provide the phase of the grid only. Some simple synchronization methods can be adopted and no extra effort is needed to design the synchronization unit. Because of the simplified control structure and the reduced demand on the synchronization unit, it is well known that it is much easier to

Figure 8.2 A compact controller that integrates synchronization and voltage/frequency regulation together for a grid-connected inverter.

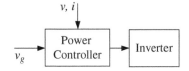

control a grid-connected inverter as a current supply than to control it as a voltage supply (Kazmierkowski and Malesani 1998). However, when an inverter is controlled as a current supply, it causes undesirable problems. For example, the controller needs to be changed when the inverter is disconnected from the grid to operate in the standalone mode or when the grid is weak because it does not have the capability of regulating the voltage. A current-controlled inverter may also continue injecting currents into the grid when there is a fault on the grid, which might cause excessively high voltage. Moreover, it is difficult for a current-controlled inverter to take part in the regulation of the grid frequency and voltage, which is a must when the penetration of renewable energy exceeds a certain level. Hence, the industry is increasingly demanding for voltage-controlled inverters and the challenge of achieving fast synchronization cannot be circumvented. A lot of research activity has been done in recent years to increase the speed and accuracy of synchronization (da Silva et al. 2010; Dong et al. 2011; Escobar et al. 2017; Guo et al. 2011; Karimi-Ghartemani et al. 2011; Santos Filho et al. 2008; Wang and Li 2011) but it is still not fast enough to obtain adequate accuracy. In this chapter, a radical step is taken to remove the synchronization unit after embedding the synchronization function into the power controller.

A synchronverter includes the mathematical model of a synchronous machine and behaves as a synchronous generator to provide a voltage supply. Its controller is in principle a power controller with integrated capability of voltage and frequency regulation so it is able to achieve real power control, reactive power control, frequency regulation, and voltage regulation. Because of the embedded mathematical model, a utility company is able to control a synchronverter in the same way as controlling synchronous generators, which considerably facilitates the grid connection of renewable energy and smart grid integration. Since a synchronous machine is inherently able to synchronize with the grid, it should be possible to integrate the synchronization function into the power controller and make a synchronverter to synchronize with the grid without a dedicated synchronization unit. This leads to a compact control structure shown in Figure 8.2. A current controller is not included but can be easily added for over-current protection, if needed.

8.2 Interaction of a Synchronous Generator (SG) with an Infinite Bus

The per-phase model of an SG, or a synchronverter, connected to an infinite bus is shown in Figure 8.3. The generated real power P and reactive power Q are (Singh et al. 2009; Wildi 2005)

$$P = \frac{3V_g E}{2X_s} \sin(\theta - \theta_g),$$

(8.1)

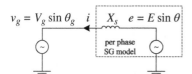

$v_g = V_g \sin\theta_g$ i X_s $e = E\sin\theta$

per phase SG model

Figure 8.3 The per-phase model of an SG connected to an infinite bus.

and

$$Q = \frac{3V_g}{2X_s}[E\cos(\theta - \theta_g) - V_g], \tag{8.2}$$

where V_g is the amplitude of the infinite bus voltage; E is the induced voltage amplitude of the SG, which can be regulated by controlling the exciting current/voltage, or $M_f i_f$ in the case of a synchronverter; θ_g and θ are the phases of the grid voltage and of the SG, respectively; and X_s is the synchronous reactance of the SG. The phase difference

$$\delta = \theta - \theta_g$$

is often called the power angle, which is regulated by controlling the driving torque of the machine. The factor $\frac{1}{2}$ in (8.1) and (8.2) is because V_g and E are amplitude values instead of RMS values.

The voltage V_g and the corresponding phase θ_g of the infinite bus can be used as the references for E and θ to generate the preferred P and Q according to (8.1) and (8.2). When the driving torque T_m is increased, δ increases and the real power delivered to the grid increases until the electrical power is equal to the mechanical power supplied by the turbine. The maximum δ that the generator is able to synchronize with the grid is $\frac{\pi}{2}$ rad (Wildi 2005). If the mechanical power keeps increasing and results in a δ that is larger than $\frac{\pi}{2}$ rad, then the rotor of the SG accelerates and loses synchronization with the grid. This should be avoided.

The reactive power Q can be regulated by controlling E, according to (8.2). When θ and E are controlled to be

$$\begin{cases} E = V_g, \\ \theta = \theta_g, \end{cases} \tag{8.3}$$

there is no real power or reactive power exchanged between the connected SG and the grid. In other words, if P and Q are controlled to be zero then the condition (8.3) is satisfied and the generated voltage e is the same as the grid voltage v_g. This condition is not common in the normal operation of an SG, but when it is satisfied the SG can be connected to or disconnected from the grid without causing large transient dynamics. This can be used to synchronize a synchronverter with the grid before connection.

8.3 Controller for a Self-synchronized Synchronverter

The controller for a self-synchronized synchronverter is shown in Figure 8.4, after making some necessary changes to the core of the synchronverter controller shown in Figure 4.4. It is able to be connected to the grid safely and to operate without the need of a dedicated synchronization unit. There are two major changes made: (i) a virtual current i_s generated from the voltage error between e and v_g is added, and the current fed into the controller

Figure 8.4 The controller for a self-synchronized synchronverter.

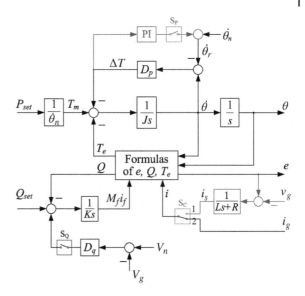

can be either i_s or the grid current i_g; (ii) a PI controller is added to regulate the output ΔT of the frequency droop block D_p to be zero and to generate the reference frequency $\dot{\theta}_r$ for the original synchronverter. In order to facilitate the operation of the self-synchronized synchronverter, three switches S_C, S_P, and S_Q are added to change the operation mode. When switch S_C is thrown at position 1 (with S_P turned ON and S_Q turned OFF), the synchronverter is operated under the set mode. If P_{set} and Q_{set} are both 0, then the operation mode is called the self-synchronization mode and the synchronverter is able to synchronize with the grid. When it is synchronized with the grid, the circuit breaker in the power part can be turned on to connect the synchronverter to the grid. When switch S_C is thrown at position 2, the synchronverter can be operated in four different modes. All the possible operation modes are shown in Table 8.1. In order to safeguard the operation in the self-synchronization mode, S_P can be turned ON and S_Q can be turned OFF automatically whenever switch S_C is thrown at position 1. In this chapter, only the characteristics that

Table 8.1 Operation modes of a self-synchronized synchronverter.

Switch S_C	Switch S_P	Switch S_Q	Mode
1	ON	ON	N/A
1	ON	OFF	Self-synchronization
1	OFF	ON	N/A
1	OFF	OFF	N/A
2	ON	ON	P-mode, V_D-mode
2	ON	OFF	P-mode, Q-mode
2	OFF	ON	f_D-mode, V_D-mode
2	OFF	OFF	f_D-mode, Q-mode

are different from the original synchronverter are described. The details about the original synchronverter, including tuning of parameters, can be found in Chapter 4.

8.3.1 Operation after Connection to the Grid

As mentioned before, the power angle δ of a synchronverter can be controlled by the virtual mechanical torque T_m calculated from the power command P_{set} as

$$T_m = \frac{P_{set}}{\dot{\theta}} \approx \frac{P_{set}}{\dot{\theta}_n},$$

where $\dot{\theta}_n$ is the nominal grid frequency. When S_p is turned on, ΔT is controlled to be 0 in the steady state via the PI controller. Hence, T_e is the same as T_m and $\dot{\theta}$ is controlled to be

$$\dot{\theta} = \dot{\theta}_r = \dot{\theta}_n + \Delta\dot{\theta}, \tag{8.4}$$

where $\Delta\dot{\theta}$ is the output of the PI controller. The power angle δ settles down at a constant value that results in $P = P_{set}$. This operation mode is called the set mode. In order to differentiate the set mode for real power and reactive power, the set mode for the real power is called the P-mode and the set mode for the reactive power is called the Q-mode. If $P_{set} = 0$, then $\theta = \theta_g$, in addition to $\dot{\theta} = \dot{\theta}_r$. When the switch S_p is turned OFF, the PI controller is taken out of the loop and the synchronverter is operated in the frequency droop mode (in short, the f_D-mode, meaning that the real power P is not the same as P_{set} but deviated from P_{set})[1] with the frequency droop coefficient defined as

$$D_p = -\frac{\Delta T}{\Delta\dot{\theta}}, \tag{8.5}$$

where

$$\Delta\dot{\theta} = \dot{\theta} - \dot{\theta}_n \tag{8.6}$$

is the frequency deviation of the synchronverter from the nominal frequency. It is also the input to the frequency droop block D_p (because S_p is OFF). This recovers the synchronverter frequency as

$$\dot{\theta} = \dot{\theta}_n + \Delta\dot{\theta},$$

which is the same as (8.4) but with a different $\Delta\dot{\theta}$. Actually, in both cases, $\dot{\theta}$ converges to the grid frequency $\dot{\theta}_g$ when the power angle δ is less than $\frac{\pi}{2}$ rad, as will be shown below.

As discussed in Chapter 4, the time constant $\tau_f = J/D_p$ of the frequency loop is much smaller than the time constant $\tau_v \approx \frac{K}{\dot{\theta}D_q} \approx \frac{K}{\dot{\theta}_n D_q}$ of the voltage loop. Therefore, $M_f i_f$ can be assumed constant when considering the dynamics of the frequency loop. Moreover, according to (8.1), the real power delivered by the synchronverter (or an SG) is proportional to $\sin\delta$. As a result, the electromagnetic torque T_e is proportional to $\sin\delta$. For $\delta \in (-\frac{\pi}{2}, \frac{\pi}{2})$, T_e increases when the power angle δ increases and T_e decreases when the power angle δ decreases. If the grid frequency $\dot{\theta}_g$ decreases, then the power angle δ and the electromagnetic torque T_e increase. As a result, the input to the integrator block $\frac{1}{Js}$ in Figure 8.4

---

1 This means the PI controller is active only in the self-synchronization mode and the P-mode but not in the f_D-mode.

decreases and the synchronverter frequency $\dot{\theta}$ decreases. The process continues until $\dot{\theta} = \dot{\theta}_g$. If the grid frequency increases, then a similar process happens until $\dot{\theta} = \dot{\theta}_g$. Hence, the synchronverter frequency $\dot{\theta}$ automatically converges to $\dot{\theta}_g$ (when $\delta \in (-\frac{\pi}{2}, \frac{\pi}{2})$) and there is no need to have a synchronization unit to provide $\dot{\theta}_g$ as the reference frequency.

The reactive power control channel of the original synchronverter is preserved, with the added switch S_Q to turn ON/OFF the voltage droop function. When S_Q is OFF, $M_f i_f$ is generated from the tracking error between Q_{set} and Q by the integrator with the gain $1/K$. Therefore, the generated reactive power Q tracks the set point Q_{set} without any error in the steady state regardless of the voltage difference between V_n and V_g. This operation mode is the set mode for the reactive power, denoted the Q-mode. When the switch S_Q is ON, the voltage droop function is enabled and the voltage error $\Delta V = V_n - V_g$ is taken into account while generating $M_f i_f$. Hence, the reactive power Q does not track Q_{set} exactly but with a steady-state error $\Delta Q = Q_{set} - Q$ that is determined by the voltage error ΔV governed by the voltage droop coefficient

$$D_q = -\frac{\Delta Q}{\Delta V}.$$

This operation mode is the voltage droop mode denoted the V_D-mode, meaning that the reactive power Q is deviated from Q_{set}.

8.3.2 Synchronization before Connection to the Grid

Before the synchronverter is connected to the grid, the frequency of the generated voltage is required to be the same as that of the grid voltage v_g. Moreover, the amplitude E is also required to be equal to the amplitude V_g and the phase of e and v_g must be the same as well. For a conventional SG, a synchroscope is often used to measure the phase difference between e and v_g so that the mechanical torque is adjusted accordingly to synchronize the SG with the grid. For grid-connected inverters, PLLs are often adopted to measure the phase of the grid voltage so that the generated voltage is locked with the grid voltage.

As mentioned before, the controller shown in Figure 8.4 is able to operate the synchronverter under the set mode with $P_{set} = 0$ and $Q_{set} = 0$. As a result, the condition (8.3) can be satisfied when it is connected to the grid. However, the current i_g flowing through the grid inductor is 0 until the circuit breaker is turned on and, hence, no regulation process can happen. In order to mimic the process of connecting a physical machine to the grid, a virtual per-phase inductor $Ls + R$ is introduced to connect the synchronverter with the grid and the resulting current

$$i_s = \frac{1}{Ls + R}(e - v_g)$$

can be used to replace i_g for feedback so that T_e and Q can be calculated according to (4.7) and (4.9). This allows the synchronverter to operate in the P-mode for the real power with $P_{set} = 0$ and in the Q-mode for the reactive power with $Q_{set} = 0$ so that the generated voltage e is synchronized with the grid voltage v_g. The only difference is that the (virtual) current i_s, instead of the grid current i_g, is routed into the controller via the switch S_C thrown at position 1. Since the current i_s is not physical, the inductance L and resistance R of the virtual synchronous reactance X_s can be chosen within a wide range. Small values lead to a large transient current i_s to speed up the synchronization process before connection. However, too small L and R may cause oscillations in the frequency estimated. Normally, the L and R can be chosen slightly smaller than the corresponding values of L_s and R_s. Moreover, the ratio $\frac{R}{L}$ defines the cut-off frequency of the filter $\frac{1}{sL+R}$, which determines the capability of filtering out the harmonics in the voltage v_g.

When the virtual current i_s is driven to zero, the synchronverter is synchronized with the grid. Then, the circuit breaker can be turned on at any time to connect the synchronverter to the grid. When the circuit breaker is turned on, the switch S_C should be turned to position 2 so that the real current i_g is routed into the controller for normal operation. After the synchronverter is connected to the grid, the switches S_P and S_Q can be turned ON/OFF to achieve any operation mode shown in Table 8.1.

8.4 Simulation Results

The control strategy was verified with simulations carried out in MATLAB 7.9/Simulink/ Sim PowerSystems, with the parameters given in Table 8.2. The solver used was ode23 with a maximum step size of 0.1 ms. The inverter was connected to the grid via a step-up transformer, which allows an equivalent comparable experimental set-up at low voltage. Similar to the implementation in Chapter 4, $D_p = 0.2026$ was chosen so that the drop of 0.5% in the frequency (from the nominal frequency f_n) causes the torque (hence, the power) to increase by 100% (of the nominal power) and the voltage droop coefficient was chosen as $D_q = 117.88$ so that the drop of 5% in the voltage requires the increase of 100% reactive power. The time constants for the droop control loops were chosen to be $\tau_f = 0.002$ s and $\tau_v = 0.02$ s.

8.4.1 Normal Operation

The simulation was started at $t = 0$ s, with switch S_C at position 1, S_P turned ON, S_Q turned OFF and the circuit breaker turned OFF, i.e. in the self-synchronization mode with $P_{set} = 0$ and $Q_{set} = 0$. The grid voltage was set to be 2% higher than the nominal value. The synchronverter synchronized itself with the grid very quickly. At $t = 2.0$ s, the circuit breaker was turned ON and S_C was switched to position 2. The set point for the real power

Table 8.2 Parameters used in simulations and experiments.

Parameter	Value	Parameter	Value
L_s	0.45 mH	f_n	50 Hz
R_s	0.135 Ω	V_n	$12\sqrt{2}$ V
C	22 μF	DC-link voltage	42 V
Parasitic resistance of C	1000 Ω	L	0.2 mH
Rated power	100 VA	R	0.05 Ω
L_g	0.15 mH	K_p	0.5
R_g	0.045 Ω	K_i	20

P_{set} = 80 W was applied at t = 5 s and the set point for the reactive power Q_{set} = 60 Var was applied at t = 10 s. The grid frequency was stepped up to f = 50.1 Hz (i.e. increased by 0.2%) at t = 15 s. The f_D-mode was enabled at t = 20 s by turning S_P OFF and the V_D-mode was enabled at t = 25 s by turning S_Q ON. The grid frequency was changed back to 50 Hz at t = 30 s and the simulation was stopped at t = 35 s. The system responses are shown in Figure 8.5.

The synchronverter was operated in the self-synchronization mode at first to synchronize itself with the grid. The virtual current i_s was replaced with the real current i_g as soon as the synchronverter was connected to the grid. The transition between the two currents was very smooth, with P and Q well maintained at zero before and after connection. There was small dynamics at the moment of connection (mainly because of the filter). After the connection, regardless of the set points P_{set} and Q_{set} applied and the modes of operation, the synchronverter tracked the grid frequency very well without any problem, even when the frequency stepped up by 0.1 Hz at t = 15 s. Note again that no dedicated

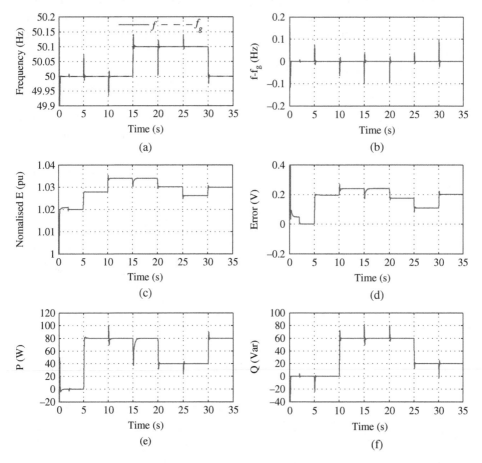

Figure 8.5 Simulation results: under normal operation. (a) Frequencies f and f_g. (b) Frequency tracking error $f - f_g$. (c) Amplitude E of the generated voltage e. (d) Amplitude of $v - v_g$. (e) Real power. (f) Reactive power.

synchronization unit was adopted in the controller and the frequency f is the internal frequency of the synchronverter. There are transient peaks in the frequency curve at each event, as shown in Figure 8.5(a) and Figure 8.5(b). This is expected, similar to the change of the speed of a synchronous machine when the operating condition is changed.

The operation modes of the original synchronverter were preserved well. The system under the set mode (i.e. the *P*-mode and the *Q*-mode) had reasonable responses with small overshoot and zero steady-state error. The frequency was also tracked well in the droop modes, but the real power and reactive power changed with the grid frequency, as expected. When the f_D-mode was enabled and the frequency was increased to $f_g = 50.1$ Hz, the real power dropped by 40W, i.e. $\frac{0.2\%}{0.5\%} = 40\%$ of the nominal value, because the frequency f_g at that time was 0.2% higher than f_n. The power quickly jumped back to the set point when f_g returned to the nominal frequency f_n. Similarly, when the V_D-mode was enabled, the reactive power dropped by 40 Var, i.e. $\frac{2\%}{5\%} = 40\%$ of the nominal value, because the grid voltage V_g was 2% higher than the nominal value V_n.

The voltages v and v_g and their difference around the connection time are shown in Figure 8.6, with phase *b* taken as an example. Before connection, the difference was about 100 mV peak-to-peak, which was very small, and there was no problem in connecting it to the grid.

In order to compare the performance with the original synchronverter, a three-phase PLL was adopted to provide the grid frequency. After a great deal of effort was made to tune the PLL parameters, the best performance achieved is shown in Figure 8.7. The overshoots of the frequencies were quite similar. The frequency of the original synchronverter reached 50.1 Hz quicker than the self-synchronized synchronverter but it had a long oscillatory

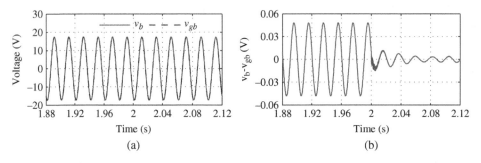

Figure 8.6 Simulation results: connection to the grid. (a) v_b and v_{gb}. (b) $v_b - v_{gb}$.

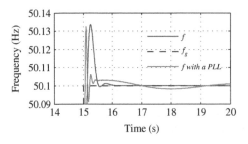

Figure 8.7 Comparison of the frequency responses of the self-synchronized synchronverter (*f*) and the original synchronverter with a PLL (*f* with a PLL).

tail and took much longer to settle down. The frequency of the self-synchronized synchronverter settled down within 1 s but the frequency of the original synchronverter had noticeable errors even after 5 s. Because the frequency in a real power system changes all the time, the long tail of errors could lead to large tracking error in the frequency.

In order to demonstrate the dynamic performance of the synchronverter, the responses when the grid frequency increased by 0.1 Hz at 15 s and returned to normal at 30 s are shown in Figure 8.8. Note that during the former case the PI controller to generate the reference frequency $\dot{\theta}_r$ was active (S_P was ON) but during the latter case the PI controller was not active (S_P was OFF). This led to different response speeds. Moreover, the currents did not change much in the former case because the synchronverter was operated in the set mode but the currents changed significantly in the latter case because the synchronverter was operated in the droop mode. There were no noticeable changes in the voltage amplitude in both cases.

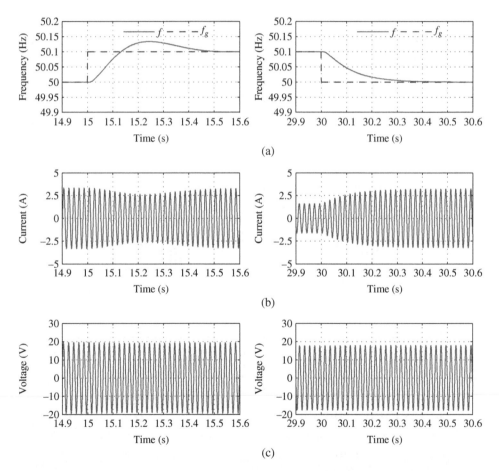

Figure 8.8 Dynamic performance when the grid frequency increased by 0.1 Hz at 15 s (left column) and returned to normal at 30 s (right column). (a) Synchronverter frequency f and the grid frequency f_g. (b) Output current of phase a. (c) Output voltage of phase a.

8.4.2 Operation under Grid Faults

In order to demonstrate the performance of the self-synchronized synchronverter under grid faults, two simulations were carried out. One is when the grid had a 50% voltage dip and the other is when the grid had a 1% frequency drop. Each fault happened at 36 s and lasted for 0.1 s, respectively. Everything remained the same as those under the normal operation, apart from that the impedance of the feeder was considered explicitly as 1.35 mH and 0.405 Ω.

The simulation results when the grid frequency dropped by 1% for 0.1 s are shown in the left column of Figure 8.9. The synchronverter frequency dropped and then recovered, following the grid frequency. The output current increased and then returned back to normal. It is worth noting that the current was not excessively high. There was no noticeable change in the amplitude of the output voltage.

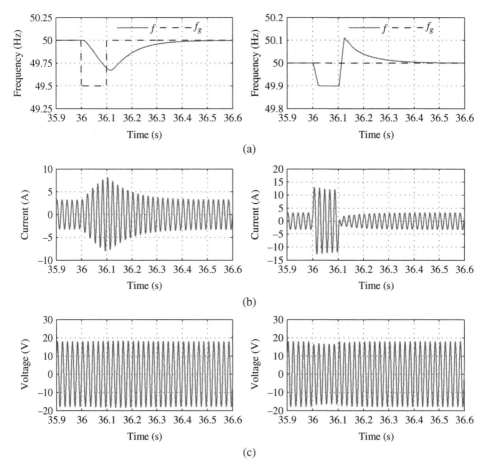

Figure 8.9 Simulation results under grid faults: when the frequency dropped by 1% (left column) and the voltage dropped by 50% (right column) at $t = 36$ s for 0.1 s. (a) Synchronverter frequency f and the grid frequency f_g. (b) Output current of phase a. (c) Output voltage of phase a.

The simulation results when the grid voltage dropped by 50% for 0.1 s are shown in the right column of Figure 8.9. When the grid fault happened, the output current increased immediately and when the fault was removed the output current decreased immediately and returned to normal. The maximum current is about 3.5 times of the normal current, which is acceptable because power inverters are often designed to cope with excessive over-currents for a short period. The frequency dropped by 0.1 Hz during the fault and then recovered after the fault was removed. The output voltage decreased accordingly but recovered very quickly. Although the frequency took about 0.2 s to return to normal the voltage and current returned to normal within 0.1 s after the fault was removed.

It is worth emphasizing that, during the two faults described above, the synchronverter demonstrated excellent fault ride-through capability although the PI controller introduced to provide the grid frequency was not active (because the synchronverter was working in the droop mode). This is due to the inherent self-synchronization mechanism of the controller based on the mathematical model of synchronous machines.

8.5 Experimental Results

The control strategy was also tested with an experimental synchronverter. The parameters used in the experimental set-up are roughly the same as those used in the simulations given in Table 8.2. The synchronverter was connected to a three-phase 400 V 50 Hz grid via a circuit breaker and a step-up transformer. The sampling frequency of the controller was 5 kHz and the switching frequency was 15 kHz.

The experiments were carried out according to the following sequence of actions:

(1) Starting the system with the IGBT switched off in the self-synchronization mode (S_C: 1; S_P: ON and S_Q: OFF) with $P_{set} = 0$ and $Q_{set} = 0$.
(2) Starting the IGBT, roughly at $t = 5$ s.
(3) Turning the circuit breaker on and switching S_C to position 2, roughly at $t = 10$ s.
(4) Applying $P_{set} = 60$ W, roughly at $t = 20$ s.
(5) Applying $Q_{set} = 20$ Var, roughly at $t = 30$ s.
(6) Turning S_P OFF to enable the f_D-mode, roughly at $t = 40$ s.
(7) Turning S_Q ON to enable the V_D-mode, roughly at $t = 50$ s.
(8) Stopping data acquisition at $t = 60$ s.

In order to demonstrate the frequency tracking performance of the self-synchronized synchronverter, a three-phase PLL block from Simulink/SimPowerSystems was used to obtain the grid frequency f_g for comparison. Note that the PLL was not used for control purposes. Many experiments were done, but only two typical cases are shown here.

8.5.1 Case 1: With the Grid Frequency Below 50 Hz

The responses of the system when the grid frequency was lower than 50 Hz are shown in the left column of Figure 8.10. The self-synchronized synchronverter tracked the grid frequency very well before the connection, with achieved peak–peak ripples of around 0.0053 Hz. However, the grid frequency obtained from the PLL had peak–peak ripples of around 0.035 Hz. If it was adopted to provide the reference for the power

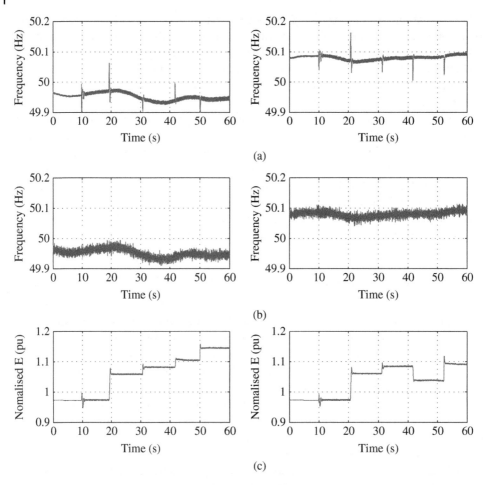

Figure 8.10 Experimental results: when the grid frequency was lower (left column) and higher (right column) than 50 Hz. (a) Synchronverter frequency f. (b) Grid frequency f_g from a three-phase PLL for comparison. (c) Amplitude E of the generated voltage e. (d) Real power at the terminal. (e) Reactive power at the terminal.

controller, the smallest achievable ripples would be larger than 0.035 Hz, which means the self-synchronized synchronverter has significantly improved the performance of frequency tracking, by almost six times. There are transient changes in the frequency shown in Figure 8.10(a). This is similar to the change in the speed of a synchronous machine when the operational condition is changed. There were small transient responses at the connection time, but they quickly settled down after the connection. The responses at all actions were very smooth and the frequency was tracked well, with achieved peak–peak ripples of around 0.011 Hz. The frequency ripples nearly doubled after connection but it was still three times better than that obtained from the PLL. In the set mode, the real and reactive power followed the reference set points. Note that the real power shown was the actual power sent to the grid from the synchronverter, which was a little less than the set point because there was a small loss on the internal

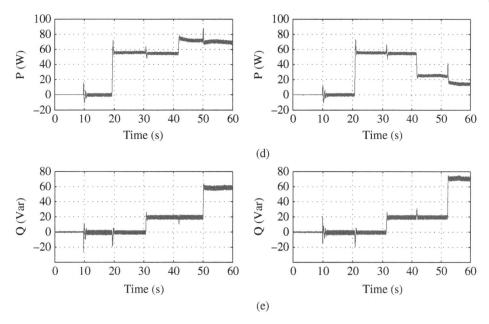

Figure 8.10 (Continued)

resistance of the inductors. When the f_D-mode was enabled at about 42 s, the real power immediately increased to the new steady state, which also changed with the grid frequency, as expected.

During the experiment, the grid voltage V_g was about 98% of the nominal voltage V_n as tracked by E before $t = 20$ s. The set point of Q was chosen to be as small as 20 Var so that in the V_D-mode the synchronverter would not generate excessive reactive power that could cause over-currents. The response of the reactive power generated was stable and smooth. When the V_D-mode was enabled at about $t \approx 50$ s, the reactive power increased by about 40 Var, as expected because of the low V_g.

Some experimental results from the synchronverter with a PLL given in Chapter 4 when the grid frequency was lower than 50 Hz are shown in the left column of Figure 8.11 for comparison. Although the experiments were done slightly differently, it can be seen that the original synchronverter equipped with a PLL had caused frequency ripples of around 0.037 Hz, which means the performance of frequency tracking was been improved by 70% after removing the PLL. This leads to much smaller ripples in the real power and reactive power, from around 21.3 W and 19.6 Var obtained from the original synchronverter with a PLL to around 3.7 W and 6.1 Var obtained from the self-synchronized synchronverter, in addition to the considerably simplified controller. The performance of real power control and reactive power control was improved by 83% and 69%, respectively, which is very significant.

The phase b voltages and their difference around the connection time are shown in the left column of Figure 8.12. The voltage v_b was synchronized with the grid voltage v_{gb} with a small error before the connection.

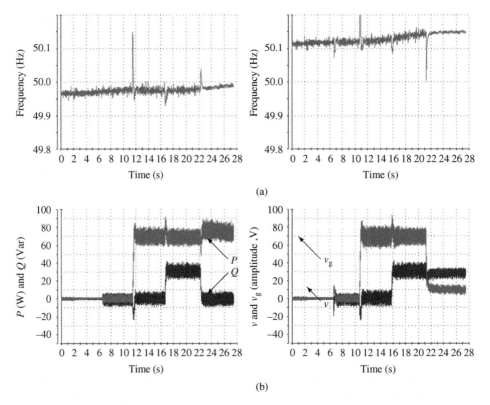

Figure 8.11 Experimental results of the original synchronverter: when the grid frequency was lower than 50 Hz (left column) and higher than 50 Hz (right column). (a) Synchronverter frequency. (b) P and Q.

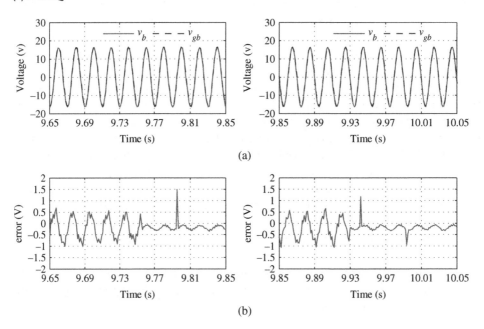

Figure 8.12 Voltages around the connection time: when the grid frequency was lower (left column) and higher (right column) than 50 Hz. (a) v_b and v_{gb}. (b) $v_b - v_{gb}$.

8.5.2   Case 2: With the Grid Frequency Above 50 Hz

The results when the grid frequency was higher than 50 Hz are shown in the right column of Figure 8.10. There was not much difference in the responses before the droop modes were enabled compared to the previous case. The frequency was tracked well and the set points for the real power and reactive power were followed with smooth transition and small overshoots. When the f_D-mode was enabled, the generated active power reduced quickly to a new steady state, which also changed with the frequency. This was expected because the grid frequency f_g was higher than 50 Hz. The grid voltage level V_g in this case was slightly smaller than that in the previous case and was about 97.5% V_n. Therefore, when the V_D-mode was enabled at about 50 s, the reactive power increased by about 50 Var.

Some experimental results from the synchronverter with a PLL in (Zhong and Weiss 2011) when the grid frequency was higher than 50 Hz are shown in the right column of Figure 8.11 for comparison. Again, although the experiments were done slightly differently, it can be seen that the self-synchronized synchronverter offers much better performance of frequency tracking, which leads to much smaller ripples in the real power and reactive power, in addition to the considerably simplified controller. The self-synchronized synchronverter achieved frequency ripples of around 0.012 Hz but the original synchronverter equipped with a PLL had caused frequency ripples of around 0.035 Hz, which means the performance of frequency tracking was improved by 66%. This leads to much smaller ripples in the real power and reactive power, from around 24.6 W and 18.5 Var obtained from the original synchronverter with a PLL to around 3.9 W and 5.3 Var obtained from the self-synchronized synchronverter. The performance of real power control and reactive power control was improved by 84% and 71%, respectively, which is again very significant.

The phase b voltages and their difference around the connection time are shown in the right column of Figure 8.12. Again, the voltage v_b was synchronized with the grid voltage v_{gb} with a small error before the connection.

8.6   Benefits of Removing the Synchronization Unit

Both simulations and experiments have demonstrated that it is possible to operate a grid-connected inverter without a dedicated synchronization unit. The synchronization function can be embedded into the controller itself. It has been shown by the experimental results that the performance of frequency tracking can be improved by more than 65%, the performance of real power control by 83% and the performance of reactive power control by about 70%.

In order to demonstrate the impact of removing the synchronization unit on the complexity of the overall controller and the demand for the computational capability of the micro-controller adopted to implement the controller, the original synchronverter and the self-synchronized synchronverter, together with a three-phase inverter, a three-phase LCL filter connected to the grid was built in MATLAB 7.9/Simulink/SimPowerSystems and implemented with a xPC target[2] running on a Intel® Core™ i7 3.2GHz CPU with

---

2 The details of the xPC target are: xPC version 4.2 (R2009b); Compiler Open Watcom 1.7; Fix-step 20 kHz sampling time; solver ode3.

Table 8.3 Impact on the complexity of the overall controller and the demand for the computational capability.

	Code size (bytes)	Average TET (μs)
Power part only	44112	1.64
With the original synchronverter	60753	4.66
With the self-synchronized synchronverter	48920	4.08
Net value of the original controller	16641	3.02
Net value of the self-synchronized controller	4808	2.44
Improvement	71.1%	19.2%

4GB DDR3 RAM. The code sizes and the average TET (target execution time) are shown in Table 8.3. The code size is reduced by 71.1% and the average target execution time is shortened by 19.2%. It is not a surprise that the code size has been reduced significantly because a three-phase PLL is a very complex system, including a $abc/dq0$ transformation, an automatic gain control block for each phase, a variable-frequency mean value block, a second-order filter and other simple blocks. There are in total 11 integrators, 7 variable delay blocks and 2 trigonometric function blocks. It is more complicated than the synchronverter itself. Hence, removing the dedicated synchronization unit also reduces the development costs and improves the software reliability.

8.7 Summary

Based on (Zhong et al. 2014), it was shown that there is no need to include a dedicated synchronization unit in a grid-tied converter. The synchronization function can be implemented by the controller itself. This leads to much improved performance, simplified control, reduced demand for computational power, reduced development cost and effort, and improved software reliability. It is able to synchronize itself with the grid before connection and to track the grid frequency automatically after connection. Moreover, it is able to operate in different modes as the original synchronverter, without having a dedicated synchronization unit to provide the grid frequency as the reference frequency. Both simulation and experimental results are presented. Experimental results have shown that the presented control strategy can improve the performance of frequency tracking by more than 65%, the performance of real power control by 83% and the performance of reactive power control by about 70%.

9

Synchronverter Based Loads without a Dedicated Synchronisation Unit

Following the removal of the dedicated synchronization unit from the original synchronverter in the previous chapter, the dedicated synchronization unit is removed from synchronverter based loads (rectifiers) in this chapter. Two controllers are presented: one is to directly control the power exchanged with the grid and the other is to control the DC-bus voltage. Both simulation and experimental results are provided.

9.1 Controlling the DC-bus Voltage

The controller to regulate the DC-bus voltage is shown in Figure 9.1. Compared to the controller with a dedicated synchronization unit shown in Figure 5.5, it is easy to see that three major changes have been made. First, a virtual current i_s generated from the error between the grid voltage v and the control signal e is introduced and the current fed into the controller can be either the virtual current i_s or the grid current i. Second, the dedicated synchronization unit is removed. Finally, a PI controller is added to generate the deviation of the reference frequency $\dot{\theta}_r$ from the rated frequency $\dot{\theta}_n$ for the VSM while making the error $\Delta\dot{\theta}$ between the reference frequency $\dot{\theta}_r$ and virtual frequency $\dot{\theta}$ zero.

Note that a voltage droop loop is not included in the control strategy but it can be easily added if the rectifier is required to provide voltage support automatically.

9.1.1 Self-synchronization

It is well known that a three-phase PWM rectifier consists of six switches with anti-paralleled diodes. When all the switches are off, the three-phase PWM rectifier works as an uncontrolled rectifier with six diodes. When turning on the switches, the control signal e must be synchronized with the grid voltage v. Otherwise, large transient currents may appear because of the large error between the control signal e and the grid voltage v, which might damage the converter. Synchronizing the control signal e with the grid voltage v means both signals have the same frequency, the same amplitude E, and the same phase. In other words, e should be the same as v (Zhong and Hornik 2013).

During the uncontrolled mode, the V_o-loop should be disabled with $T_m = 0$ (because V_o is not controllable). The initial field excitation $M_f i_f$ can be set as $\frac{E}{\dot{\theta}_n}$, where $\dot{\theta}_n$ is the nominal

Power Electronics-Enabled Autonomous Power Systems: Next Generation Smart Grids,
First Edition. Qing-Chang Zhong.

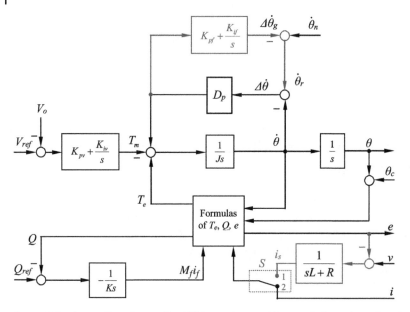

Figure 9.1 Controlling the rectifier DC-bus voltage without a dedicated synchronization unit.

grid frequency. However, the current i flowing through the line impedance from the grid to the rectifier cannot be adopted for control because the current i cannot be changed by the controller when the rectifier is operated in the uncontrolled mode. In order to solve this problem, the virtual current

$$i_s = \frac{1}{sL + R}(v - e) \tag{9.1}$$

generated by passing the voltage error between v and e through the virtual inductor $sL + R$ is introduced to replace i for feedback so that T_e and Q can be calculated. Since i_s is not physical, the values of L and R can be chosen small to speed up the synchronization process. More details about their selection can be found in the previous chapter.

In the steady state, there is $\Delta\dot\theta = 0$ because of the PI controller that generates $\Delta\dot\theta_g$, which forces $T_e = T_m = 0$. At the same time, $Q = Q_{ref} = 0$ because of the integrator that generates $M_f i_f$. As a result, $i_s = 0$ and the control signal e becomes the same as the grid voltage v according to (9.1). In other words, the converter is synchronized with the grid. After synchronization, the power switches can be operated in the PWM-controlled mode.

9.1.2 Normal Operation

When the rectifier is operated in the PWM-controlled mode, the switch S is turned to position 2 so that the physical current i is fed into the controller. Moreover, the V_o-loop is enabled. The operation of the Q-loop is the same as that in the case with a dedicated synchronization unit. The DC-bus voltage V_o-loop is also very similar. The major difference lies in the generation of the reference frequency $\dot\theta_r$, which is the grid frequency $\dot\theta_g$ in the case with a dedicated synchronization unit. It will be shown below that the controller shown in Figure 9.1 actually makes $\dot{\theta}_r = \dot{\theta}_g$ automatically.

It is well known that the phase difference

$$\delta = \theta_g - \theta, \tag{9.2}$$

called the power angle, is in proportion to the real power when δ is small. If the grid frequency $\dot{\theta}_g$ decreases, then the power angle δ decreases. As a result, the real power decreases, which forces the DC-bus voltage of the rectifier to decrease. Because of the regulation of the DC-bus voltage PI controller ($K_{pv} + \frac{K_{iv}}{S}$), the virtual mechanical torque T_m increases. As a result, the input to the integrator block $\frac{1}{Js}$ in Figure 9.1 decreases (note the negative sign of T_m) and the frequency $\dot{\theta}$ decreases. This process continues till $\dot{\theta} = \dot{\theta}_g$, i.e., when the steady state is reached (assuming that the system is stable). At the same time, $\Delta\theta$ increases, which forces $\Delta\dot{\theta}_g$ to increase as well through the regulation of the frequency PI controller ($K_{pf} + \frac{K_{if}}{S}$). The frequency reference $\dot{\theta}_r = \dot{\theta}_n - \Delta\dot{\theta}_g$ decreases till $\dot{\theta}_r = \dot{\theta}$. As a result,

$$\dot{\theta} = \dot{\theta}_r = \dot{\theta}_n - \Delta\dot{\theta}_g = \dot{\theta}_g.$$

If the grid frequency θ_g increases, then a similar process happens until $\dot{\theta} = \dot{\theta}_g$. Therefore, the controller shown in Figure 9.1 is able to automatically make $\dot{\theta}_r = \dot{\theta}_g$ and make $\dot{\theta}$ track $\dot{\theta}_g$ automatically, without a dedicated synchronization unit to provide $\dot{\theta}_g$.

9.2 Controlling the Power

As discussed in Chapter 5, the DC-bus voltage controller can be replaced with a real power controller to generate the virtual torque T_m. The resulting control strategy is shown in Figure 9.2. The error between the real power reference P_{ref} and the real power P is fed into the real power PI controller ($K_{pp} + \frac{K_{ip}}{S}$) to generate the virtual torque T_m. Here, the real power P can be obtained with e and i or with v and i. The difference is minor because of the different points of measurement. When the real power P decreases, the error to the PI controller increases. As a result, the torque T_m increases so more power is to be drawn from the grid, leading to a power balance. The operation of the rest of the controller is the same as the one in the previous section.

Because $\Delta\theta$ is controlled to be zero, the controller shown in Figure 9.2 operates in the set mode. The PI controller can be reset after connecting the rectifier to the grid. If so, the rectifier works in the droop mode. Also, a voltage droop loop can be easily added if the rectifier is required to provide voltage support automatically.

9.3 Simulation Results

The control strategies were tested in Matlab7.6/Simulink/SimpowerSystems. The solver used in the simulations was ode23t with a relative tolerance of 10^{-3} and a maximum step size of 10 μs. The parameters of the rectifier used in the simulation are given in Table 9.1. The rectifier was connected to the grid via a set-up transformer.

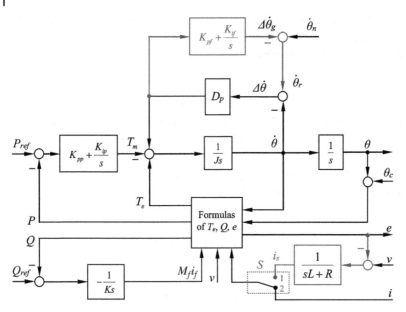

Figure 9.2 Controlling the rectifier power without a dedicated synchronization unit.

Table 9.1 Parameters of the rectifier.

Parameter	Value	Parameter	Value
L_s	0.45 mH	R_s	0.135 Ω
C	5000 μF	Switching frequency	10 KHz
Input voltage	12 Vrms	Grid frequency	50 Hz

Table 9.2 Parameters for controlling the DC-bus voltage.

Parameter	Value	Parameter	Value
K_{pf}	0.5	K_{if}	7
K_{pv}	0.01	K_{iv}	0.1
J	0.0002 kg m^2	K	370

9.3.1 Controlling the DC-bus Voltage

The parameters of the controller used in the simulation are given in Table 9.2. The responses are shown in Figure 9.3. The grid frequency $\dot{\theta}_g$ was obtained through a PLL and shown in Figure 9.3(a) for comparison. Initially, it took some time for the PLL to derive $\dot{\theta}_g$.

The simulation was started at $t = 0$s, with switch S at position 1. The rectifier worked in the uncontrolled mode with the load $R = 50$ Ω. The V_o-loop was disabled so that the rectifier could be synchronized with the grid. The DC-bus voltage is about

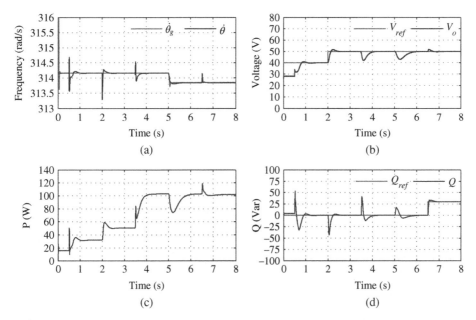

Figure 9.3 Simulation results when controlling the DC bus voltage. (a) Frequencies. (b) DC bus voltage. (c) Real power. (d) Reactive power.

$12 \times 2.34 = 28$ V and the power is about $28^2/50 = 15.7$ W. The frequency was synchronized with the grid. At $t = 0.5$s, the V_o-loop was enabled and the switch S was turned to position 2. The DC-bus voltage reference V_{ref} was set at 40 V and the reactive power reference Q_{ref} was set at zero to obtain the unity power factor. The DC-bus voltage increased to V_{ref} in about 0.3 s. There were some acceptable transients when the PWM was enabled because it was not able to produce the required voltage due to the low DC-bus voltage. The reactive power reached 0 in about 0.5 s. At $t = 2$ s, the DC-bus voltage reference was increased to $V_{ref} = 50$ V. Again, the DC-bus voltage quickly reached the reference with a small overshoot. The reactive power had a spike but returned to normal very quickly. At $t = 3.5$ s, the load was changed to $R = 25\ \Omega$. Because of the load increase, the DC-bus voltage dropped but quickly recovered. The real power increased. At $t = 5$ s, the grid voltage frequency decreased from 50 Hz to 49.95 Hz. The rectifier quickly followed the frequency change and maintained the DC-bus voltage. At $t = 6.5$ s, the reactive power reference was changed to $Q_{ref} = 30$ Var. Indeed, the reactive power Q changed accordingly. There were some dynamics in the DC-bus voltage, the real power, and the frequency but they all quickly returned to normal.

Figure 9.4(a) shows the grid voltage and the control signal in the controlled mode. It can be seen that e_a and v_a are almost the same. In other words, they are synchronized. Figure 9.4(b) shows the grid voltage and the control signal in the PWM-controlled mode. The control signal e_a is smaller than the grid voltage v_a, indicating the current flowing from the grid to the rectifier.

Figure 9.5 shows the grid voltage and the input current in the uncontrolled mode and the PWM-controlled mode with $Q_{ref} = 0$. The current is highly distorted in the uncontrolled

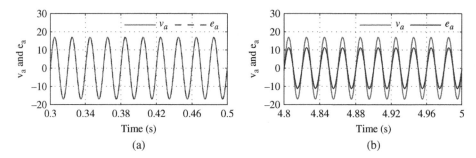

Figure 9.4 Grid voltage and control signal. (a) Uncontrolled mode. (b) PWM-controlled mode.

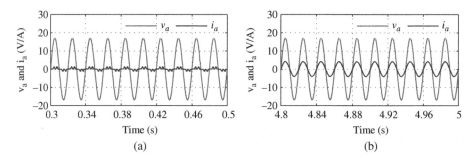

Figure 9.5 Grid voltage and input current. (a) Uncontrolled mode. (b) When Q_{ref}=0 Var.

mode but is nearly sinusoidal (with THD= 0.5%) and in phase with the line voltage when $Q_{ref} = 0$. Therefore, the unity power factor operation is achieved with high current quality.

9.3.2 Controlling the Power

The parameters of the controller used in the simulation are given in Table 9.3. The simulation results are shown in Figure 9.6. Again, the grid frequency $\dot{\theta}_g$ was obtained through a PLL and shown in Figure 9.6(a) for comparison.

The simulation was started at $t = 0$ s, with switch S at position 1. The rectifier worked in the uncontrolled mode with $P_{ref} = 0$ and $Q_{ref} = 0$. The load was $R = 50\,\Omega$. The frequency was quickly synchronized with the grid frequency. At $t = 0.5$ s, the switch S was turned to position 2 to operate the rectifier in the PWM-controlled mode with $P_{ref} = 50$ W and $Q_{ref} = 0$. The real power P quickly increased and settled down at the reference value after some transients. There were also some transients in the reactive power before settling

Table 9.3 Parameters for controlling the the power.

Parameter	Value	Parameter	Value
K_{pf}	0.5	K_{if}	7
K_{pp}	0.001	K_{ip}	0.1
J	0.0002 kg m²	K	370

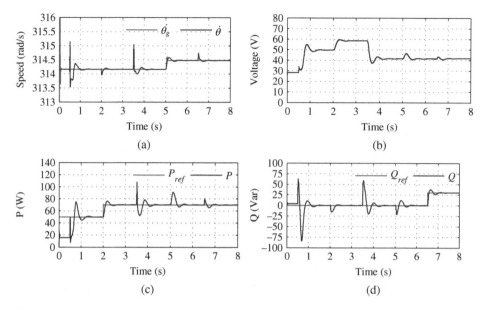

Figure 9.6 Simulation results when controlling the real power. (a) Frequencies. (b) DC-bus voltage. (c) Real power. (d) Reactive power.

down to 0. These transients can be reduced by putting the reference value through a low-pass filter and further tuning the control parameters. At $t = 2$ s, the real power reference was increased to $P_{ref} = 70$W. The real power response was very reasonable. There was some coupling effects in the reactive power but the reactive power quickly returned to 0. The DC-bus voltage increased due to the increase of the power reference. There was some dynamics in the frequency but it quickly settled down. At $t = 3.5$ s, the load was changed to $R = 25\ \Omega$. The DC-bus voltage dropped accordingly due to the increase in the load. The real power was maintained after some dynamics. Again, the frequency and the reactive power were maintained, after some transients. At $t = 5$ s, the grid voltage frequency increased from 50 Hz to 50.05 Hz. The real power maintained at its reference value after some dynamics. The frequency quickly followed the change of the grid frequency. The DC-bus voltage reacted upon the change but returned to normal; so did the reactive power. At $t = 6.5$ s, the reactive power reference Q_{ref} was changed to 30 Var. Indeed, the reactive power increased. There were some transients in the frequency, the DC-bus voltage and the real power but they all returned to normal quickly.

9.4 Experimental Results

The parameters of the experimental system are roughly the same as those given in Table 9.1. The control parameters were not changed either. The experimental setup consists of a rectifier board, a three-phase grid interface inductor, a board consisting of voltage and current transducers, a step-up transformer, a dSPACE DS1104 R&D controller board with ControlDesk software and MATLAB Simulink/SimPower software package. A resistor in parallel with a relay was inserted into the input to reduce the inrush current. The sampling frequency of the controller is 5kHz and the switching frequency is 10 kHz.

9.4.1 Controlling the DC-bus Voltage

The experiment was carried out according to the following sequence of actions:

(1) Start the system, with all the IGBTs OFF, the V_o-loop and the Q-loop disabled, and the switch S at position 1.

(2) Turn the relay ON roughly at $t = 3$ s for the system to work as an uncontrolled rectifier without a load.

(3) Connect the load $R = 50\,\Omega$ roughly at $t = 5$ s.

(4) start operating the IGBTs roughly at $t = 10$ s, turn switch S to position 2, enable the V_o-loop with $V_{ref} = 40$ V.

(5) Enable the Q-loop with $Q_{ref} = 0$ roughly at $t = 20$ s.

(6) Change V_{ref} to 50 V roughly at $t = 31$ s.

(7) Change the load to $R = 30\,\Omega$ roughly at $t = 42$ s.

Figure 9.7 shows the responses of the system. The frequency $\dot{\theta}$ tracked the grid frequency $\dot{\theta}_g$ very well all the time. Before $t = 3$ s, the DC-bus voltage was established at about 26 V. The real power and reactive power were zero because there was no load on the DC bus. When the relay was turned ON at $t = 3$ s, the real power increased a little bit to charge the DC-bus capacitor and then went back to zero. The DC-bus voltage went up to 28 V. When the load $R = 50\,\Omega$ was connected to the DC bus at $t = 5$ s, the DC-bus voltage dropped to 26 V because of the load effect, making the real power about 16 W. Up to this point, the system worked as an uncontrolled rectifier, with the control signal synchronized with the grid voltage, and was ready for connection.

When the control strategy was enabled at $t = 10$ s, the DC-bus voltage increased to the reference value 40 V and the real power increased to around 37 W. Since the Q-loop was not enabled, the reactive power was about 70 Var. After enabling the Q-loop at $t = 20$ s, the reactive power was regulated around zero to obtain the unity power factor. The real power dropped a little bit to 33 W because of the significantly reduced losses following the current reduction by the reactive power compensation. Roughly, the current was reduced by $\frac{\sqrt{70^2+37^2}}{33} = 2.4$ times, which means the conduction loss was reduced by nearly 6 times. The DC-bus voltage was maintained at the reference value after some transients. After that, the DC-bus voltage increased to 50 V quickly due to the reference change. The real power increased as well but the reactive power was maintained at around 0. When the load changed from $50\,\Omega$ to $30\,\Omega$ at $t = 42$ s, the real power went up to about 90 W accordingly, but the DC-bus voltage and the reactive power were maintained well. The phase a voltage and current in Figure 9.7 showed that the current was highly distorted during the uncontrolled mode but very clean during the PWM-controlled mode.

9.4.2 Controlling the Power

The experiment was carried out according to the following sequence of actions:

(1) Start the system, with all the IGBTs OFF, the P-loop and the Q-loop disabled, and the switch S at position 1.

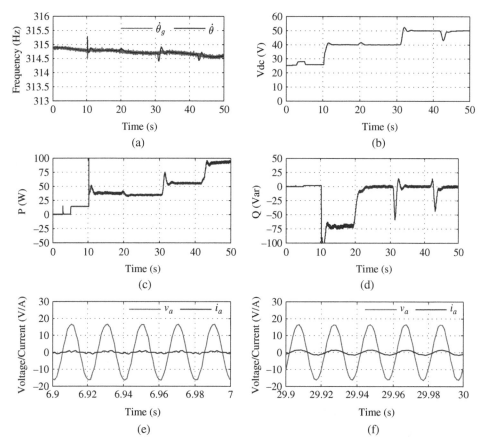

Figure 9.7 Experiment results: controlling the DC-bus voltage. (a) Frequencies. (b) DC bus voltage. (c) Real power. (d) Reactive power. (e) v_a and i_a in the uncontrolled mode. (f) v_a and i_a in the PWM-controlled mode.

(2) Turn the relay ON roughly at $t = 2$ s for the system to work as an uncontrolled rectifier without a load.

(3) Connect the load $R = 50\ \Omega$ roughly at $t = 7$ s.

(4) Start operating the IGBTs roughly at $t = 10$ s, turn switch S to position 2, enable the P-loop with $P_{ref} = 40$ W.

(5) Enable the Q-loop roughly at $t = 20$ s.

(6) Change P_{ref} to 60 W roughly at $t = 30$ s.

(7) Change the load to $R = 30\ \Omega$ roughly at $t = 41$ s.

Figure 9.8 shows the responses of the system. From $t = 0$ to 10 s, the results are almost the same as the previous one. When the control strategy was enabled at $t = 10$ s, the real power quickly went up to the reference value 40 W and the DC-bus voltage increased to 42 V gradually because of the big capacitor on the DC-bus. Since the Q-loop was closed, the reactive power decreased to 70 Var. After turning the Q-loop on at $t = 20$ s, the reactive power remains at zero to obtain the unity power factor. In addition, the DC bus voltage increased to 43 V because of the reactive power compensation from 20 s to 30 s. After that, the real power

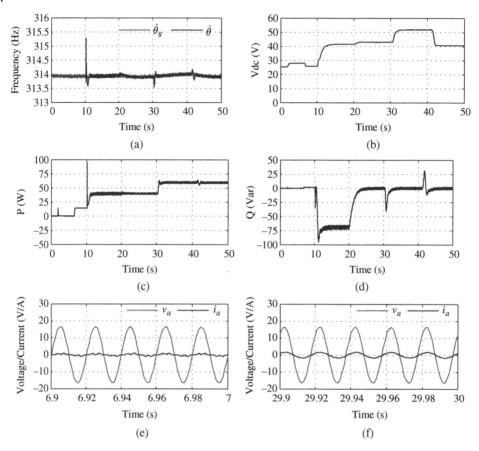

Figure 9.8 Experiment results: controlling the power. (a) Grid and internal frequencies. (b) DC-bus voltage. (c) Real power. (d) Reactive power. (e) v_a and i_a in the uncontrolled mode. (f) v_a and i_a in the PWM-controlled mode.

increased to 60 W quickly due to the reference changed. The DC-bus voltage increased as well but the reactive power was maintained well after some transients. When the load changed from 50 Ω to 30 Ω at $t = 41$ s, the real power and the reactive power were maintained well after some transients but the DC-bus voltage dropped to 41 V, as expected. The phase a voltage and current in Figure 9.8 show that the current was highly distorted during the uncontrolled mode but very clean during the PWM-controlled mode.

9.5 Summary

Based on (Zhong et al. 2012b), a PWM-controlled rectifier is operated to behave as a VSM without the need of a dedicated synchronization unit such as a PLL. Two control strategies are presented. One is to control the DC-bus voltage and the other is to control the real power directly. The reactive power can be regulated as well. Simulation and experimental results are provided to demonstrate the excellent performance of the control strategies.

10

Control of a DFIG Based Wind Turbine as a VSG (DFIG-VSG)

In Chapter 6, a control strategy is presented to operate a PMSG based wind turbine as a VSM. In this chapter, a control strategy is presented to operate a DFIG based wind turbine as a VSG, after revealing the analogy between differential gears and DFIGs. An electromechanical model is presented to represent a DFIG as a differential gear that links a rotor shaft driven by a prime mover (the wind turbine), a virtual stator shaft coupled with a virtual synchronous generator and a virtual slip shaft coupled with a virtual synchronous motor. Moreover, an AC drive, which consists of a rotor-side converter (RSC) and a grid-side converter (GSC), is adopted to regulate the speed of the slip virtual synchronous motor. Based on this, the concept of a DFIG-VSG is presented to operate the whole DFIG-converter system as one VSG with the stator shaft synchronously rotating at the grid frequency, even when the rotor shaft speed changes, through controlling the virtual slip shaft to synchronously rotate at the slip frequency. Following the concept of the AC Ward Leonard drive systems discussed in Chapter 7, the RSC is controlled as a virtual synchronous generator to generate an appropriate slip voltage, via regulating the real power and reactive power sent to the grid. In order to facilitate this, the GSC is controlled as a virtual synchronous motor to maintain a stable DC-bus voltage without drawing reactive power in the steady state. In other words, the slip virtual synchronous motor together with the AC drive consisting of the RSC and the GSC are operated as a virtual synchronous motor–generator–motor set when the rotor speed is below the synchronous speed and as a virtual synchronous generator–motor–generator set when the rotor speed is above the synchronous speed. A prominent feature is that there is no need to adopt a phase-locked loop (PLL), either on the grid side or on the rotor side. The system is not only able to provide the kinetic energy stored in the turbine/rotor shaft as inertia to support the grid frequency, but it is also able to provide reactive power to support the grid voltage. Simulation results from a 1.5 MW wind turbine-driven DFIG connected to a weak AC grid and experimental results from a prototype system are presented to demonstrate the feasibility and effectiveness of the strategy. The technology is patented.

10.1 Introduction

Nowadays, the most commonly used topologies for variable-speed WPGSs (Blaabjerg and Ma 2013; Yaramasu et al. 2015) adopt doubly-fed induction generators (DFIGs), often called

Power Electronics-Enabled Autonomous Power Systems: Next Generation Smart Grids,
First Edition. Qing-Chang Zhong.
© 2020 John Wiley & Sons Ltd. Published 2020 by John Wiley & Sons Ltd.

type 3 systems, and permanent magnet synchronous generators (PMSGs), often called type 4 systems. Because the stator windings of DFIGs are directly connected to the grid and only the slip power goes through the power electronic converters, the converter capacity needed is only a fraction of the rated power, which reduces system cost, size, weight, and power losses. Moreover, the stator windings are able to stand high over-currents because of the high thermal capacity of the machine. The voltage level can be high as well, which further reduces losses and cost. However, the current technologies do not offer full control over the total power, which leads to problems when interacting with the grid, in particular, when there are grid faults. As discussed in Chapter 6, the industry is being forced to move towards type 4 systems with PMSGs, which interact with the grid through a set of full-scale back-to-back converters. Such wind turbines are able to interact with the grid in a friendly manner and take part in the regulation of grid frequency and voltage with full control over the power, which is extremely important when the penetration of wind energy reaches a certain level. However, the need of full-scale back-to-back power converters increases cost, size, weight, and power losses. Moreover, power electronic converters do not have the thermal capacity of electric machines and cannot provide the short-circuit capacity needed to clear faults.

In this chapter, a hidden analogy between DFIGs and differential gear is revealed. The mechanics of a DFIG is then modeled as a virtual differential gear. Moreover, a DFIG based WPGS can be modeled and controlled as one fully-controllable VSG. This leads to a DFIG based WPGS that behaves like a VSG to have the advantages of a PMSG based WPGS, referred to as a DFIG-VSG. Different WPGSs are compared in Table 10.1.

As shown in Chapter 6, a PMSG based wind turbine can be controlled to behave like a VSM. Since a DFIG based wind turbine also adopts a set of back-to-back RSCs and GSCs, the same concept can be applied to DFIG based wind turbines as well. Indeed, attempts have been made with inertia effects elaborated to prove the advantages of applying such a method in weak grids (Wang et al. 2015a; Wang et al. 2015b; Zhao et al. 2015). Most DFIG based wind farms that have adopted the VSG method generally rely on additional energy storage to realize this (Alamri and Alamri 2009; Papaefthymiou et al. 2010), but the additional equipment increases the cost and reduces system reliability. The concept of synthetic inertia is also able to provide frequency regulation for DFIGs (Anaya-Lara et al. 2006; Guo et al. 2013; Morren et al. 2006b). However, the implementation of the synthetic inertia and frequency regulation requires the information of the grid frequency and the rate of change of frequency (ROCOF), which cannot avoid the use of a PLL (Zhu et al. 2011) and could lead to poor performance because of the noises introduced in calculating the ROCOF. As discussed in Chapter 8, the dedicated synchronization unit in a converter can be removed

Table 10.1 Comparison of different wind power generation systems.

	Thermal capacity	Voltage level	Converter power	Grid-friendly	Size	Cost	Efficiency	Controllability
DFIG based	**High**	**Can be high**	**Partial**	No	**Small**	**Low**	**High**	Low
PMSG based	Low	Limited	Full	**Yes**	Large	High	Low	**High**
DFIG-VSG	**High**	**Can be high**	**Partial**	Yes	**Small**	**Low**	**High**	**High**

to achieve self-synchronization (Nguyen et al. 2012; Zhong et al. 2012b; Zhong et al. 2014). The utilization of the VSG technique and the self-synchronization method to DFIG turbines would ultimately harmonize the relationship between wind turbines and power systems.

10.2 DFIG Based Wind Turbines

A DFIG has a stator winding and a rotor winding, which is accessible via slip rings and offers the possibility of controlling the rotor current. As a result, the operational speed of a DFIG can be varied in a controlled manner, making it suitable for variable-speed WPGSs. A typical DFIG based wind turbine WPGS is shown in Figure 10.1. The stator winding is directly connected to the grid while the rotor winding is connected to an AC drive, i.e. a set of back-to-back power electronic converters consisting of an RSC and a GSC sharing the same DC bus. The RSC can be controlled to inject a current with a variable frequency into the rotor circuit, making up the difference between the mechanical speed of the rotor and the electrical speed of the grid. Hence, the rotor speed can be varied in a wide range via controlling the frequency of the rotor current.

The stator always feeds real power to the grid but the real power in the rotor circuit can flow bidirectionally, from the grid to the rotor or from the rotor to the grid, depending on the rotor speed ω_r. Ignoring the losses, the power processed by the rotor circuit is

$$P_{gs} = s \cdot P_s, \tag{10.1}$$

where P_s is the stator power and s is the slip defined as the ratio between the slip speed

$$\omega_{rs} = \omega_s - \omega_r, \tag{10.2}$$

and the synchronous speed ω_s, i.e. $s = \frac{\omega_{rs}}{\omega_s}$. The power sent to the grid is

$$P_g = P_s - P_{gs} = (1 - s)P_s. \tag{10.3}$$

The choice of the rated power for the rotor circuit is a trade-off between cost and the desired speed range. A DFIG is often operated with a speed range of ±30% around ω_s. Hence, the majority of the power flows through the stator circuit and only a fractional power flows through the rotor circuit. As a result, a sufficient range of operational speed can be achieved at a reasonably low cost. Moreover, the converters can provide reactive power if needed.

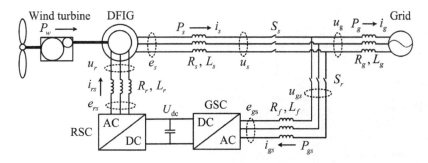

Figure 10.1 Typical configuration of a turbine-driven DFIG connected to the grid.

Although a DFIG offers a sufficient range of operational speeds and many other advantages, with the current technologies, it is very sensitive to voltage disturbances, especially voltage sags (Guo et al. 2012). Abrupt voltage drops at the terminals often cause large voltage disturbances on the rotor, which may exceed the voltage rating of the RSC, making the rotor current uncontrollable and even damaging the RSC. These problems are not intrinsic but mainly due to improper control strategies applied.

10.3 Differential Gears and Ancient Chinese South-pointing Chariots

The concept of differential gears is very fundamental in mechanical engineering. A differential gear is a mechanical device that consists of some gears and three input/output shafts. Any of the three shafts can serve as either input or output as long as there is one input and one output at any given moment. The three rotational degrees of freedom are subject to one gear constraint

$$k_1\omega_1 + k_2\omega_2 + k_3\omega_3 = 0, \tag{10.4}$$

where k_i $(i = 1, 2, 3)$ is a constant and ω_i $(i = 1, 2, 3)$ is the speed, so there are only two independent degrees of freedom. The main function of a differential gear is to generate a speed that is the sum or the difference of the other two. Basically, it provides the implementation of addition with mechanical devices and hence is very fundamental.

Differential gears are crucial for automobiles and other wheeled vehicles. It allows the outer wheel to rotate faster than the inner wheel during a turn so that the outer wheel can rotate farther. Basically, it splits the input power of the drive shaft to drive the two wheels. An increase in the speed of one wheel is balanced by a decrease in the speed of the other.

Figure 10.2 A model of an ancient Chinese south-pointing chariot (WIKIpedia 2018).

Actually, the usage of differential gears can be traced back to the ancient Chinese south-pointing chariots in the first millennium BCE (Needham 1986). Figure 10.2 illustrates a model. A south-pointing chariot provides the cardinal direction as a non-magnetic, mechanized compass. Also, differential gears have been used to build analogue computers and other devices. See (WIKIpedia 2018) for an overview about differential gears and (Moule 1924) for the interesting history of the south-pointing chariot.

10.4 Analogy between a DFIG and Differential Gears

It is well recognized that the power flow of a DFIG has two channels, one through the stator windings and the other through the rotor windings, often coupled with a back-to-back converter, as shown in Figure 10.1. For the power flow channel through the back-to-back converter, the GSC can be operated to behave like a virtual synchronous motor (denoted as a GS-VSM) with the inertia J_{gs} and the angular speed ω_{gs}, as demonstrated in (Zhong et al. 2012b). Conceptually, the RSC can also be operated as a virtual synchronous generator (denoted as a RS-VSG) to provide the voltage for the rotor windings, but this does not provide insightful understanding about the system. The power flow channel through the stator windings needs to be understood better. The relationship between the stator, the rotor, and the slip needs to be clarified in order to clearly understand the operation of a DFIG as a VSG.

Since the rotor speed ω_r, the slip speed/frequency ω_{rs}, and the stator synchronous speed/ frequency ω_s satisfy (10.2), which is a special case of (10.4), there is a fundamental analogy between a DFIG and differential gears. The mechanics of a DFIG can then be modeled as a differential gear, which has a physical rotor shaft coupled with the prime mover (the turbine rotor in the case of a wind turbine), a virtual stator shaft rotating at the synchronous speed ω_s, a virtual slip shaft rotating synchronously with the slip frequency ω_{rs}, as illustrated in Figure 10.3. The virtual stator shaft works with the stator windings to form a (stator) virtual synchronous machine, which is operated as a generator, while the virtual slip shaft works together with the rotor windings to form another (slip) virtual synchronous machine, which can be operated as a motor or a generator according to the rotor speed ω_r. The rotor shaft is an input shaft and the stator shaft is an output

Figure 10.3 A differential gear that illustrates the mechanics of a DFIG, where the figure of the differential gear is modified from (Shetty 2013).

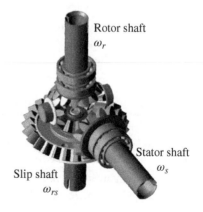

Rotor shaft
ω_r

Stator shaft
ω_s

Slip shaft
ω_{rs}

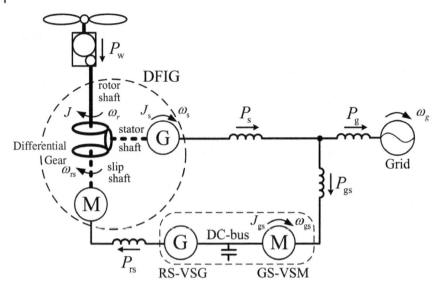

Figure 10.4 The electromechanical model of a DFIG connected to the grid.

shaft while the slip shaft can be either an input or output shaft. For the varying rotor speed ω_r, it is always possible to control the slip shaft speed ω_{rs} so that the stator shaft rotates synchronously with the grid frequency. Without loss of generality, ω_s and ω_r are assumed to be positive but ω_{rs} can be positive or negative, depending on whether the rotor speed ω_r is lower or higher than the speed ω_s. As a result, the equivalent electromechanical model of the system in Figure 10.1 in terms of virtual synchronous machines is obtained, as shown in Figure 10.4. When $\omega_r \geq \omega_s$, the slip shaft changes the direction of rotation to act as an output shaft and the power flow in the rotor channel changes direction. Note that, because of the energy conservation law, the torques on the three shafts are the same. If there is no torque on any of the three shafts, then there is no torque on all of them. The back-to-back converter plays the role of a variable frequency drive to control the speed of the slip synchronous motor. The GSC can be operated as a virtual synchronous motor, denoted as a GS-VSM while the RSC can be operated as a virtual synchronous generator, denoted as a RS-VSG. As a result, a DFIG based WPGS is modeled as a turbine driven differential gear that has one shaft coupled with a stator virtual synchronous generator and another shaft coupled with a slip virtual synchronous generator–motor–generator (or motor–generator–motor) set. This provides insightful understanding about DFIG based WPGSs, paving the way to control a DFIG based WPGS as one VSG.

10.5 Control of a Grid-side Converter

Since the overall objective is to control a DFIG based WPGS as a VSG, the net real power P_g and reactive power Q_g exchanged with the grid should be regulated according to the frequency dynamics and voltage dynamics of the grid via controlling the VSG. Since the stator windings are connected to the grid directly, it is desirable for the majority of real power

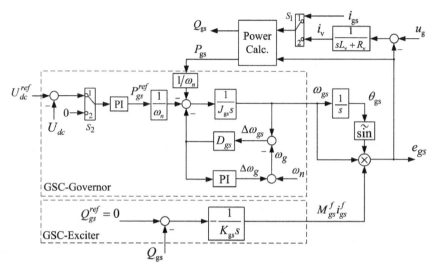

Figure 10.5 Controller to operate the GSC as a GS-VSM.

and reactive power to go through the stator windings while the GSC is only responsible for maintaining the DC-bus voltage to facilitate the control of the RSC. In other words, the back-to-back GSC-RSC converters should be a local channel inside the system.

10.5.1 DC-bus Voltage Control

The controller for the GS-VSM is shown in Figure 10.5. Since this is in principle to operate a rectifier as a virtual synchronous machine, the DC-bus voltage control strategy presented in Chapter 9 can be adopted. The DC-bus voltage U_{dc} is maintained at the reference voltage U_{dc}^{ref} through a voltage PI controller to generate the real power reference P_{gs}^{ref} for the GS-VSM. The built-in swing equation of the GS-VSM is

$$J_{gs} \cdot \frac{d\omega_{gs}}{dt} = T_{gs} - T_{gs}^{ref} - D_{gs}(\omega_{gs} - \omega_g),$$

where J_{gs} is the virtual inertia of the GS-VSM, $T_{gs} = \frac{P_{gs}}{\omega_n}$ is the electromagnetic torque calculated from the real power P_{gs}, $T_{gs}^{ref} = P_{gs}^{ref}/\omega_n$ is the corresponding load torque generated by the voltage PI controller that regulates the DC-bus voltage U_{dc}, and D_{gs} is the virtual friction/damping coefficient. It also includes a PI controller to generate the difference between the rated frequency ω_n and the grid frequency ω_g so that the speed/frequency ω_{gs} of the GS-VSM can synchronize with the grid frequency ω_g. Hence, there is no need to have a dedicated synchronization unit, e.g. a PLL.

10.5.2 Unity Power Factor Control

The controller also includes a GSC exciter consisting of a PI controller that regulates the reactive power Q_{gs} to track the reference reactive power Q_{gs}^{ref} and generate the virtual field excitation $M_{gs}^f i_{gs}^f$. The reference reactive power Q_{gs}^{ref} can be set at zero so that the rotor channel does not draw any reactive power from the grid. This helps reduce the capacity (and cost) of the converter. The back-EMF e_{gs} of the GS-VSM is then

$$e_{gs} = M_{gs}^f i_{gs}^f \omega_{gs} \widetilde{\sin}\theta_{gs}, \tag{10.5}$$

which can be converted into PWM pulses to drive the power electronic switches of the GSC. Here,

$$\widetilde{\sin}\theta_{gs} = \left[\sin\theta_{gs} \ \sin\left(\theta_{gs} - \frac{2\pi}{3}\right) \ \sin\left(\theta_{gs} + \frac{2\pi}{3}\right) \right]^T$$

represents the three-phase sinusoidal vector. Note that the terminal voltage u_{gs} satisfies

$$u_{gs} = R_f i_{gs} + L_f \frac{di_{gs}}{dt} + e_{gs}, \tag{10.6}$$

where R_f and L_f are the resistance and inductance of the RL filter of the GSC. It is the same as the grid voltage u_g once the rotor circuit breaker S_r is turned ON.

Note that, as shown in Figure 10.5, the real power P_{gs} and reactive power Q_{gs} are calculated according to the generated back-EMF e_{gs} and the sampled current i_{gs}. This helps reduce the number of voltage sensors needed and, hence, the cost.

As shown in Figure 10.5, the controller that regulates the virtual frequency ω_{gs} is similar to the function of a governor. The governor has three cascaded loops, including the inner frequency loop, the middle torque loop and the outer DC-bus voltage loop. The first two loops perform the function of a VSM while the third loop regulates the DC-bus voltage to generate the desired real power reference P_{gs}^{ref} for the VSM.

10.5.3 Self-synchronization

In order to minimize the inrush current when turning on the circuit breaker S_r, a virtual impedance $sL_v + R_v$ is introduced to generate a virtual current according to the voltage difference between e_{gs} and u_g while operating the GSC in the uncontrolled mode with the mode switches S_1 and S_2 at position 2. The virtual current

$$i_v = \frac{u_g - e_{gs}}{L_v s + R_v}, \tag{10.7}$$

instead of the current i_{gs}, is used to calculate the real power P_g and reactive power Q_g. As a result, the voltage e_{gs} is regulated to be the same as the grid voltage u_g. Once e_{gs} is synchronized with u_g, it can be sent out to the switches after PWM conversion while turning S_r ON and the mode switches S_1 and S_2 to position 1 to start normal operation.

The guidelines for tuning the parameters can be found in Chapter 9 or (Zhong et al. 2014).

10.6 Control of the Rotor-Side Converter

The DFIG stator VSG cannot be controlled directly. Because of (10.2), the virtual stator VSG can be controlled through the RSC. The control structure is shown in Figure 10.6. The key is that the RSC is controlled according to the real power and reactive power exchanged with the grid, i.e. the voltage and current at the grid side, instead of the real power and reactive power (or the voltage and current) at the rotor terminals. As a result, the whole system behaves as a VSG, providing frequency and voltage support to the grid.

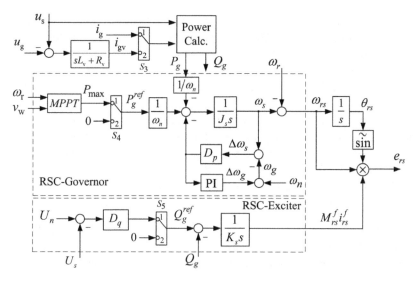

Figure 10.6 Controller to operate the RSC as a RS-VSG.

10.6.1 Frequency Control

According to the electromechanical model shown in Figure 10.4, the real power exchanged with the grid is regulated according to the swing equation of the stator VSG designed as

$$J_s \cdot \frac{d\omega_s}{dt} = T_g^{ref} - T_g - D_p(\omega_s - \omega_g),$$

where J_s is the virtual inertia of the stator shaft, $T_g^{ref} = P_g^{ref}/\omega_n$ is the mechanical torque applied to the stator shaft, $T_g = P_g/\omega_n$ is the electromagnetic torque and D_p is the frequency droop coefficient or the virtual friction coefficient. Note that the real power P_g is obtained by measuring the grid current i_g and the stator terminal voltage u_s, which is the same as the grid voltage u_g when the stator circuit breaker S_s is ON. Hence, this reflects the net real power exchanged with the grid, i.e. the total power extracted by the wind turbine (less losses). In order to maximize the yield, it is very important to extract the maximum power from the wind. There are many MPPT algorithms available. Since this is not the focus of this chapter, it is assumed that the maximum power P_{max} corresponding to the wind speed v_w is obtained through MPPT as the real power reference P_g^{ref}.

Since the stator shaft speed ω_s cannot be directly controlled, the slip shaft speed/frequency ω_{rs} obtained according to (10.2) with the stator frequency ω_s and the rotor shaft speed ω_r can be adopted to generate the phase angle θ_{rs} of the control signal e_{rs} for the RSC. In other words, the speed/frequency of the RS-VSG is indirectly controlled through the stator VSG via regulating the real power sent to the grid.

10.6.2 Voltage Control

Similarly, the voltage amplitude of the RS-VSG can be controlled through regulating the reactive power exchanged with the grid. The field excitation $M_{rs}^f i_{rs}^f$ of the RS-VSG is generated through an integrator or a PI controller that regulates the reactive power Q_g sent to the grid to its reference value Q_g^{ref}. Moreover, a voltage droop controller is added through the droop coefficient D_q so that the RS-VSG can regulate the RMS value of the terminal voltage u_s around its nominal value U_n. Note that the stator terminal voltage u_s, instead of the rotor terminal voltage, is used here. Hence, the reactive power reflects the net reactive power exchanged with the grid. This does not only reduce the number of voltage sensors needed and hence the cost but also facilitates the control design. Otherwise, it would have been difficult to determine the reference values for the voltages, currents, and power of the rotor windings because of the varying operational condition. The rotor currents and voltages are only intermediate variables and there is no need to measure them for the purpose of control. This is a distinctive feature of the control strategy.

Because of (10.2), the slip shaft speed $\omega_{rs} = \omega_s - \omega_r$ can be integrated to obtain the slip shaft angle θ_{rs}. As a result, the control signal for the RSC can be formed as

$$e_{rs} = M_{rs}^f i_{rs}^f \omega_{rs} \widetilde{\sin}\theta_{rs}, \tag{10.8}$$

according to the dynamics of synchronous machines (Zhong and Weiss 2011). This can be converted into PWM pulses to drive the RSC so that the rotor winding voltage

$$u_r = -R_r i_{rs} - L_r \frac{di_{rs}}{dt} + e_{rs},$$

where R_r and L_r are the rotor resistance and leakage inductance, can regulate the speed of the slip synchronous motor as ω_{rs}.

It is worth noting that the voltage of the stator VSG is indirectly controlled through the RSC-VSG via regulating the reactive power sent to the grid.

10.6.3 Self-synchronization

The connection of the stator windings to the grid also needs some care. The stator voltage u_s needs to be synchronized with the grid voltage u_g before the stator circuit breaker S_s is turned ON. A PI controller is included to generate the difference between the rated frequency ω_n and the grid frequency ω_g so that the virtual stator shaft rotating at ω_s can synchronize with the grid frequency ω_g, without using a dedicated synchronization unit.

As shown in Figure 10.6, the self-synchronization technique introduced in (Zhong et al. 2014) is adopted to achieve synchronization without the need of a dedicated synchronization unit. In order to facilitate this, three mode switches S_3, S_4, and S_5 are introduced to operate the controller in the normal operational mode (at position 1) or in the self-synchronization mode (at position 2). When it is in the self-synchronization mode, the reference real power P_g^{ref} and reactive power Q_g^{ref} are all set at zero. Moreover, a virtual impedance $L_v s + R_v$ is introduced to generate a virtual grid current

$$i_{gv} = \frac{u_s - u_g}{L_v s + R_v}, \tag{10.9}$$

according to the voltage difference between the stator voltage u_s and the grid voltage u_g. This virtual current, instead of the grid current i_g, is adopted to calculate the real power P_g and reactive power Q_g. Hence, before turning S_s ON, the controller regulates the stator voltage u_s until it is the same as u_g, i.e. synchronized, by regulating the real power and the reactive power to zero. Once it is synchronized, the mode switches S_3, S_4, and S_5 can be turned to position 1 and the stator circuit breaker S_s can be turned ON to start the normal operation.

10.7   Regulation of System Frequency and Voltage

The controller shown in Figure 10.6 regulates $\Delta\omega_s$ to zero in the steady state, which means $\omega_s = \omega_g$. In other words, the system works in the set mode with $P_g = P_g^{\text{ref}}$ in the steady state and the reference real power P_g^{ref} is injected to the grid even if the grid frequency ω_g deviates from the nominal frequency ω_n by $\Delta\omega_g$. It is possible for the wind turbine to take part in the regulation of system frequency by disabling the PI controller that regulates $\Delta\omega_g$, i.e. to make $\Delta\omega_g = 0$. In this case, the actual real power P_g sent to the grid in the steady state is no longer P_g^{ref} but

$$P_g = P_g^{\text{ref}} - D_p(\omega_g - \omega_n)\omega_n,$$

where D_p is the frequency droop coefficient defined as

$$D_p = -\frac{\Delta T}{\Delta\omega} = -\frac{\Delta T}{T_n}\cdot\frac{\omega_n}{\Delta\omega}\cdot\frac{T_n}{\omega_n} = D_p^{\text{pu}}\cdot\frac{P_n}{\omega_n^2} \tag{10.10}$$

with D_p^{pu} being the normalized frequency droop coefficient defined as

$$D_p^{\text{pu}} = -\frac{\Delta T}{T_n}\cdot\frac{\omega_n}{\Delta\omega}.$$

If 100% increase of real power corresponds to 1% drop of frequency then $D_p^{\text{pu}} = 100$. This is the value used later in this chapter.

The wind turbine is also able to take part in the regulation of system voltage via regulating the reactive power according to the difference between the terminal voltage U_g and the rated voltage U_n. The voltage droop coefficient is defined as

$$D_q = -\frac{\Delta Q}{\Delta U} = -\frac{\Delta Q}{Q_n}\cdot\frac{U_n}{\Delta U}\cdot\frac{Q_n}{U_n} = D_q^{\text{pu}}\cdot\frac{Q_n}{U_n},$$

where

$$D_q^{\text{pu}} = -\frac{\Delta Q}{Q_n}\cdot\frac{U_n}{\Delta U}$$

is the normalized voltage droop coefficient. If 100% increase of reactive power corresponds to 10% of voltage drop then $D_q^{\text{pu}} = 10$. This is the value used later in this chapter.

According to (Zhong and Weiss 2011), the virtual inertia J_s can be chosen as

$$J_s = D_p \cdot \tau_f,$$

where τ_f is the desired time constant of the frequency loop, and the parameter K_s can be chosen as

$$K_s = D_q\omega_n\tau_v,$$

where τ_v is the desired time constant of the voltage loop. Considering the capacity of the back-to-back converters is roughly 30% of the generator capacity, D_{gs} and J_{gs} can be chosen as $D_{gs} = 0.3D_p$ and $J_{gs} = 0.3J_s$.

Table 10.2 Parameters of the DFIG-VSG simulated.

Symbol	Description	Value
R_s, L_s	Stator resistance, leakage inductance	0.023 pu, 0.018 pu
R_r, L_r	Rotor resistance, leakage inductance	0.16 pu, 0.016 pu
L_m	Mutual inductance	2.9 pu
P_n, Q_n	Rated real and reactive power	1.5 MW, 1.2 Mvar
U_n, U_r	Rated stator and rotor voltages	690 V, 2200 V
f_n	Rated frequency	50 Hz
R_f, L_f	Filter resistance and inductance	0.05 pu, 0.1 pu
C, U_{dc}	DC-bus capacitance and voltage	10000 μF, 2000 V
J_s, J_{gs}	RSC and GSC virtual inertia	22.8, 6.84 $kg \cdot m^2$
τ_f	Time constant of frequency loops	0.015 s
J	Turbine inertia	11.7 $kg \cdot m^2$
K_s, K_{gs}	RSC and GSC voltage regulation	$3.3 \cdot 10^3$, $5.4 \cdot 10^4$
τ_v	Time constant of RSC voltage loop	0.006 s
	Time constant of GSC voltage loop	0.33 s
	RSC self-synchronization PI controller	$1 \cdot 10^{-10}$, $1 \cdot 10^{-9}$
	GSC self-synchronization PI controller	$3 \cdot 10^{-10}$, $3 \cdot 10^{-9}$
	DC-bus voltage PI controller	8, 60
R_v, L_v	Impedance for synchronization	0.02Ω, 0.4mH

10.8 Simulation Results

The system parameters are given in Table 10.2. The switching frequency is 5 kHz.

The simulation was carried out according to the following sequence of actions:

1. At 0 s, the system was initialized with the wind speed of 8 m s^{-1} and the turbine rotor initial speed of 0.8 pu. All IGBT switches were OFF. Both circuit breakers S_s and S_r were OFF. All mode switches were at position 2.

2. At 0.1 s, the rotor circuit breaker S_r was turned ON. And at 0.3 s, the PWM signals to the GSC were enabled and the mode switches S_1 and S_2 were turned to position 1. The GS-VSM was started to regulate the DC-bus voltage.

3. At 1 s, the PWM signals to the RSC were enabled and the RS-VSG started to synchronize the DFIG with the grid. At 2 s, the circuit breaker S_s was turned ON and the mode switches S_3, S_4 and S_5 were turned to position 1. The DFIG started to inject power into the system at the maximum power point $P_g^{ref} = P_{max}$.

4. At 4 s, the grid voltage dropped by 5% and, at 6 s, it returned back to normal.

5. At 8 s, the wind speed increased to 14 m s^{-1}.

6. At 10 s, the grid frequency dropped to 49.75 Hz, and at 12 s, it returned to 50 Hz.

In order to clearly show the dynamic response, only phase *a* of the instantaneous AC voltages and currents are shown in the simulation results.

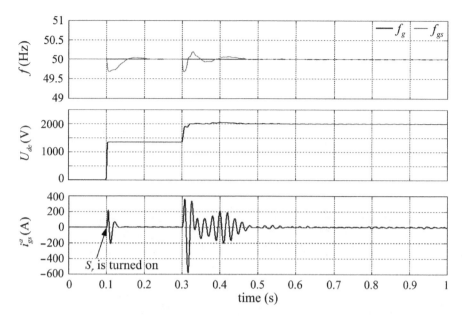

Figure 10.7 Connection of the GS-VSM to the grid.

Figure 10.7 shows the results before and after the connection of the GSC to the grid. When the rotor circuit breaker S_r was turned ON at 0.1 s, the GSC was connected to the grid and functioned as a diode rectifier to charge the DC bus U_{dc} to about 1400 V with an inrush current in i_{gs}^a. Note that no special measure was taken to reduce the inrush current in the simulation, which can be addressed with ease in practice. The GS-VSM frequency f_{gs} started synchronizing with the grid due to the self-synchronization process. At 0.3 s, the GS-VSM started to operate the GSC as a VSM and the DC-bus voltage was established and maintained at 2000 V. The GS-VSM frequency f_{gs} fluctuated around f_g to establish the DC-bus voltage. There are some transient currents as expected.

Figure 10.8 shows the dynamic performance before and after connecting the RS-VSG to the grid. When the RS-VSG was enabled at 1 s, the rotor current i_{rs} increased to establish the excitation flux and reactive power needed. At the same time, the self-synchronization was started and the stator voltage u_s quickly synchronized with the grid voltage u_g. The stator frequency f_s and the DC-bus voltage U_{dc} both had some fluctuations around their nominal values but then quickly recovered. When the voltage across the stator circuit breaker S_s is small enough, e.g. less than 1%, the DFIG-VSG is ready for grid connection. When the circuit breaker S_s was turned ON at 2 s, the DFIG started to send power into the grid, gradually, without inrush, currents occurred on the stator windings. Note that no PLL was adopted. The rotor current increased accordingly. Note that the frequency of the rotor current was about 10 Hz, which corresponds to the slip of 0.2. The stator frequency f_s increased initially in order to increase the real power output but then returned back to the grid frequency f_g. Similarly, the DC-bus voltage increased initially and then returned back to normal.

Figure 10.9 shows the response of the DFIG-VSG during the whole process. After the GSC was connected to the grid at 0.1 s, both the active power and the reactive power had some small spikes but both were below the capacity. At 0.3 s, the DC-bus voltage was established.

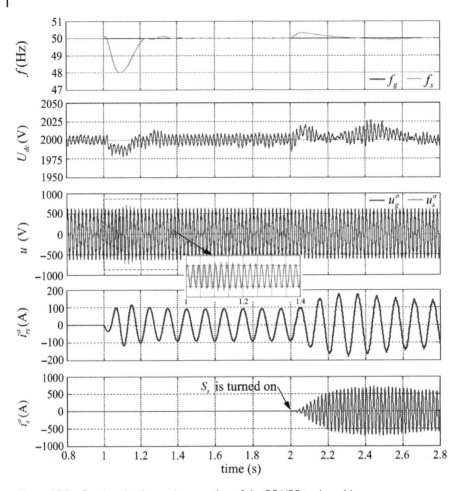

Figure 10.8 Synchronization and connection of the RS-VSG to the grid.

After the RSC was enabled at 1 s, a small amount of real power was drawn from the grid by the GSC to maintain the DC-bus voltage. After the stator circuit breaker S_s was turned ON at 2 s, the real power sent to the grid gradually increased to the maximum power corresponding to the wind speed of 8 m s^{-1}. The rotor speed increased to 0.84 pu and the slip speed reduced to 0.16 pu. The reactive power had some coupling effect but returned to zero. Note that the response of the GSC is faster than the response of the RSC, which is in line with the assumption that the GSC has a faster dynamics in order to better maintain the DC-bus voltage. At 4 s the grid voltage dropped by 5%. The reactive power of the DFIG stator sent to the grid increased to 0.6 MVAr (50% of the rated reactive power according to the voltage droop coefficient); the GSC also sent some reactive power initially but then regulated it to zero. Hence, both channels contributed to the voltage support initially but the stator took the main responsibility. This led to temporary reduction of the real power sent to the grid by the stator windings, which helps reduce the stress on the machine. The real power drawn by the GSC did not change much. When the voltage recovered at 6 s,

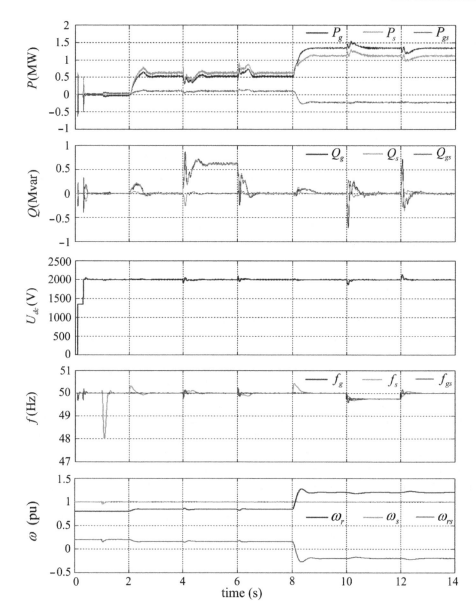

Figure 10.9 Operation of the DFIG-VSG.

the GSC drew some reactive power from the grid and the stator stopped sending reactive power to the grid. Once again, the response of the GSC was much faster than that of the RSC. The GS-VSM frequency did not change much but the stator frequency did respond to the voltage change accordingly. When the wind speed increased at 8 s, the real power sent to the grid was increased. The rotor speed increased from 0.84 pu to 1.2 pu and the slip speed reduced to −0.2 pu. As a result, the GS-VSM also started injecting real power to the grid, i.e. the real power became negative. The reactive power had some small dynamics but then returned to zero. The GS-VSM frequency did not have any noticeable change but the stator frequency increased and then returned to the grid frequency. At 10 s, when the grid frequency dropped abruptly by 0.25 Hz, the GS-VSM quickly followed the grid frequency and maintained the stability of the DC-bus voltage. The stator frequency also followed the grid frequency, which forces the DFIG stator to release some kinetic energy to support the grid frequency as the rotor speed temporarily dropped by 0.03 pu. At 12 s, the grid frequency returned to 50 Hz. The DFIG stored some kinetic energy by increasing the rotor speed and temporarily reducing the real power to support the grid frequency.

Hence, the DFIG-VSG has demonstrated the feasibility of increasing the equivalent inertia and taking advantage of the kinetic energy stored in the turbine rotor to support the grid frequency. During the whole process, the DC-bus voltage was maintained very well because the GS-VSM was designed to have faster responses than the RS-VSG. Indeed, the GS-VSM frequency f_{gs} tracked the grid frequency f_g faster than the stator frequency f_s, even when the grid frequency dropped and recovered. Note that no PLL was needed.

10.9 Experimental Results

The control strategy was tested with an experimental system. The DFIG machine is a 3 kW DFIG driven by a 4 kW squirrel-cage induction machine. The parameters of the DFIG are given in Table 10.3. The DFIG stator is connected to a 110 V/50 Hz three-phase grid with its frequency and voltage changeable manually. The back-to-back converters are two 2 kW three-phase inverters (GSC and RSC) connected through a common DC-bus at 350 V.

Figure 10.10 shows the synchronization and grid connection processes. Initially, the DC-bus was established at 350 V with the rotor shaft speed at 30 Hz. At 0.2 s, the self-synchronization was started. The DC-bus voltage dropped and then recovered. The stator frequency went through some transients. The rotor current was quickly established and the stator voltage synchronized with the grid voltage in four cycles although it took longer for the frequency to synchronize. At 1.0 s, the stator circuit breaker S_s was turned ON. The stator current gradually increased and settled down and the rotor current also increased. No inrush current was injected into the grid. There were no visible transients in the DC-bus voltage and the stator frequency.

Table 10.3 Parameters of the experimental DFIG system.

Symbol	Description	Value
P_{ng}	Rated power	3 kW
f_n	Rated frequency	50 Hz
P_g	Pole pairs	2
L_{sg}	Stator inductance	13 mH
L_{rg}	Rotor inductance	7 mH
R_{sg}	Stator resistance	2.5 Ω
R_{rg}	Rotor resistance	1.35 Ω
Q_{ng}	Rated reactive power	1.2 kVar

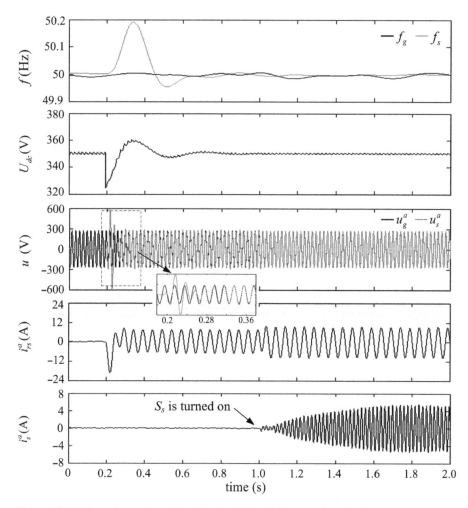

Figure 10.10 Experimental results of the DFIG-VSG during synchronization process.

Figure 10.11 shows the experimental results when the DFIG-VSG under various disturbances in the rotor speed, the grid frequency, and the grid voltage. Initially, the system was operated with 0.2 p.u. slip frequency in the steady state injecting 1.25 kW into the grid.

At 2.0 s, the rotor speed increased from 0.8 pu to 1.1 pu. The stator active power increased but the RSC active power decreased to negative because the rotor speed exceeded the nominal speed. Hence, the power flow in the rotor circuit changed direction. The total active power injected into the grid increased smoothly to 2.0 kW according to the maximum power command. There was some dynamics in the reactive power. The DC-bus voltage changed a bit and then recovered. The stator frequency increased a bit in order to increase the active power but then settled down to the grid frequency.

At 4.0 s, the grid frequency decreased to 49.75 Hz rapidly. The total active power increased to support the grid frequency. Noticeably, there was a drop in the rotor speed and the RSC injected more active power, which means the rotor released additional energy to support

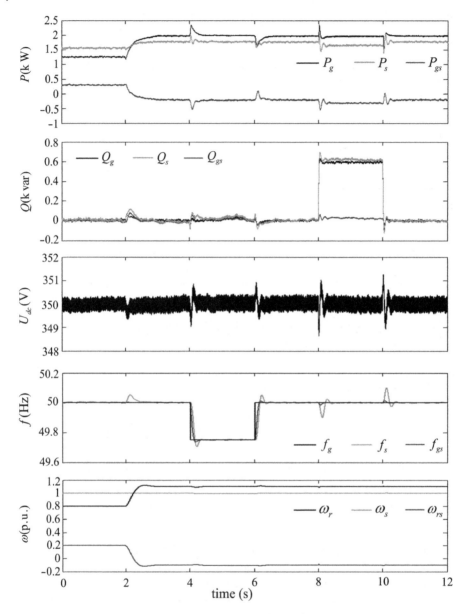

Figure 10.11 Experimental results during the normal operation of the DFIG-VSG.

the grid. Again, there were some transients in the reactive power but they were relatively small. The stator frequency tracked the grid frequency and so did the GSC.

At 6.0 s, the grid frequency returned to 50 Hz. The total active power injected to the grid decreased to support the grid and the additional kinetic energy was absorbed by the rotor shaft, causing the rotor speed to increase slightly.

At 8.0 s, the grid voltage amplitude decreased by 5%. The total reactive power exchanged with the grid was increased to 600 Var to support the grid voltage according to the droop coefficient. The GSC reactive power was still regulated at around zero so the reactive power was supplied by the stator circuit. There were some dynamics in the active power and the frequency but relatively small.

At 10.0 s, the voltage amplitude returned to 110 V. The reactive power exchanged with the grid decreased to 0 to support the grid voltage.

10.10 Summary

Based on (Zhong 2016a, 2017c; Zhong et al. 2018c), a strategy is presented to make a DFIG system behave like a VSG, even when the speed of the rotor shaft changes. A distinctive feature is that the rotor circuit voltages and currents are not used for control. Instead, the voltages and the currents at the PCC are used for control so that the overall system behaves as a VSG, by regulating the real power and reactive power sent to the grid. The GSC is controlled as a VSM to maintain a stable DC-bus voltage without drawing reactive power in the steady state. Both the GSC and the RSC are equipped with a self-synchronization function so there is no need to have a dedicated synchronization unit, such as a PLL. Such a system is referred to as a DFIG-VSG. It can release the kinetic energy stored in the rotor/turbine shaft to support the system frequency and also to provide reactive power to support the system voltage. Simulation results with 1.5 MW connected to a weak grid as well as experimental results from a DFIG-VSG prototype system are presented.

11

Synchronverter Based Transformerless Photovoltaic Systems

In this chapter, the synchronverter technology is applied to a transformerless photovoltaic (PV) system, which consists of an independently controlled neutral leg and an inversion leg. The presence of the neutral leg enables the direct connection between the ground of PV panels and the neutral line of the grid, removing the need of having an isolation transformer. This significantly reduces the leakage current because the stray capacitance between the PV panels and the grid neutral line (ground) is bypassed. Another benefit is that the voltage of the PV is only required to be higher than the peak value of the grid voltage, which is the same as that required by conventional full bridge inverters. In addition, the synchronverter technology is applied to control the inversion leg, making the PV inverter grid-friendly. The control of the two legs are independent of each other. Real-time simulation results are provided to demonstrate the operation of the system.

11.1 Introduction

In recent years, the penetration of PV systems in the grid has been constantly growing (Deo et al. 2015; Femia et al. 2005; Liu et al. 2015; Montoya et al. 2016).

A typical topology for grid-tied PV systems has two processing stages consisting of a DC/DC converter and a DC/AC converter (Zhong and Hornik 2013). The DC/DC converter, usually a boost converter, converts the variable DC voltage from the PV source into a constant DC voltage (Deo et al. 2015; Femia et al. 2008; Montoya et al. 2016; Pilawa-Podgurski and Perreault 2013). Then, the DC/AC converter converts the DC power into AC and interacts with the AC grid. Since sunlight varies all the time, it is essential to use power electronic converters to extract the maximum power available with MPPT strategies; see, e.g., (Ghaffari et al. 2014; Killi and Samanta 2015; Teng et al. 2016). For a two-stage grid-tied PV system, the MPPT function is usually embedded into the control of the DC/DC converter, and the DC-bus voltage is regulated by the DC/AC converter; see, e.g. (Deo et al. 2015; Femia et al. 2005; Liu et al. 2015; Montoya et al. 2016). A PI controller is then adopted for the DC-bus voltage regulation to generate a current reference for the DC/AC converter, according to which the current injected to the grid is directly controlled by the dq-transformation. This often causes problems when connecting the inverter to weak grids. Moreover, because these PV inverters are designed to follow the grid frequency through PLL or their variants, this often causes instability or performance degradation. Such PV inverters can easily trip when

Power Electronics-Enabled Autonomous Power Systems: Next Generation Smart Grids,
First Edition. Qing-Chang Zhong.
© 2020 John Wiley & Sons Ltd. Published 2020 by John Wiley & Sons Ltd.

voltage and frequency deviates away from the given ranges, which are fairly tight. Because of the large-scale utilization of solar PV, it is very important for PV inverters to provide frequency and voltage support.

In addition to the problems on the control side mentioned above, there are also challenges on the hardware side. First of all, the most popular topology adopted in the industry is still the bridge topology. This often causes large leakage currents and common-mode voltages, which requires the addition of common-mode filters and/or isolation transformers. Of course, this increases the size, weight, and cost of the PV system. Moreover, a residential PV inverter often has a single phase, which often requires large amount of electrolytic capacitors on the DC bus to handle the pulsating power. Electrolytic capacitors are known to be heavy and bulky, with limited lifetime and reliability. Another issue is the so-called potential induced degradation (Omron 2013), which reduces the output power of a PV module after just a few years of service by up to 70%. This problem is critical because PV plants are usually financially planned for a 25+ year life time.

In this chapter, a solution that is able to address these issues will be discussed.

11.2 Leakage Currents and Grounding of Grid-tied Converters

11.2.1 Ground, Grounding, and Grounded Systems

In electrical engineering, ground refers to the reference point in an electrical circuit from which voltages are measured, a common return path for electric currents, or a direct physical connection to the earth. For grid-tied converters, separate reference points are often adopted for the DC and AC parts of the circuit, respectively. As a result, there are often three (or more) grounds in a grid-tied converter, i.e. the earth, the AC ground and the DC ground.

Grounding (Wiles 2012) is connecting, whether intentionally or accidentally, an electrical circuit or equipment to some conducting body that serves in place of the earth or to the earth itself, which is then often called earthing. In this chapter, only intentional grounding is considered. In general, there are two kinds of grounding: equipment grounding and system grounding. Equipment grounding refers to connecting any exposed non-current carrying electrical conductors or equipment to the earth. By doing this, the voltage from the conductors or equipment to the earth can be limited, which avoids potential electrical shock. Proper equipment grounding is essential for any electrical equipment as the electrical conductors or equipment may be energized because of failed insulation. System grounding refers to connecting one of the (current-carrying) electrical conductors to the earth. For a grid-tied converter, the (current-carrying) electrical conductors that are used for grounding are normally the AC ground or the DC ground. Since a converter is often connected to the grid neutral line, the AC ground is often connected to the earth. However, DC grounds may or may not be connected to the earth, depending on the topology and the control strategy of the converter. If the DC ground is not connected to the AC ground (the earth), then a leakage current appears, which will be discussed later.

Grounded systems are those with one of the DC conductors (either positive or negative) connected to the earth or to some conducting body that serves in place of the earth (Wiles 2012). Otherwise, systems are ungrounded. As a result, electrical systems can be classified into grounded systems and ungrounded systems, according to the arrangement for system grounding. Note that system grounding is optional but equipment grounding is essential for safety reasons. For example, for grid-tied PV inverters, the pre-2005 edition of the US National Electrical Code (NEC) required that all PV systems have one of the DC circuit conductors grounded whenever the maximum system voltage was higher than 50 V (NEC 690.41). However, ungrounded PV systems are now allowed according to NEC 690.35, which was added to the 2005 NEC (Wiles 2012).

The International Standard IEC 60364 defines three earthing networks TN, TT, and IT (IEC 1993; Sallam and Malik 2011). The first letter indicates the connection between earth and the power-supply equipment (generator or transformer): "T": direct connection of a point with earth (Latin: terra); "I": no point is connected with earth (isolation), except perhaps via a high impedance (IEC 1993; Sallam and Malik 2011). The second letter indicates the connection between earth and the electrical device being supplied: "T": direct connection of a point with earth; "N": direct connection to neutral at the origin of installation, which is connected to the earth (IEC 1993; Sallam and Malik 2011). Different areas may have different requirements on the earthing networks during different periods (Lacroix and Calvas 1995). For example, old houses in the UK often have the TN-S network while most modern houses in the UK have the TN-C-S network. More details about earthing systems can be found in (Lacroix and Calvas 1995; Sallam and Malik 2011).

Three earthing networks widely used in English-speaking countries are shown in Figure 11.1. Note that the equipment grounding is defined as the connection between the equipment and the line PE but not the line N because the line N is designed to be a current return path and it cannot be used for equipment grounding, although the lines PE and N have the same voltage potential when ignoring the line impedance. System grounding is not depicted in the networks, which can be achieved by connecting either the positive or the negative pole of the DC bus to the line N (if allowed).

Network	TN-S	TN-C	TN-C-S
Structure	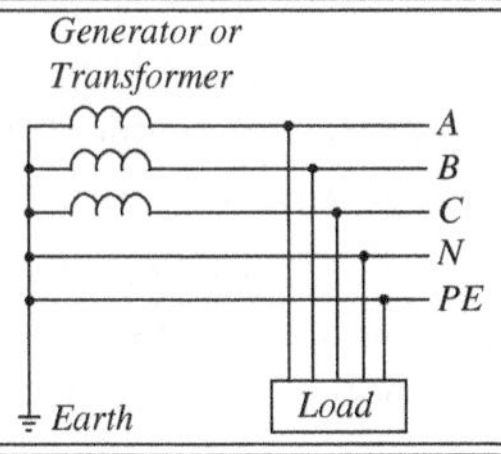	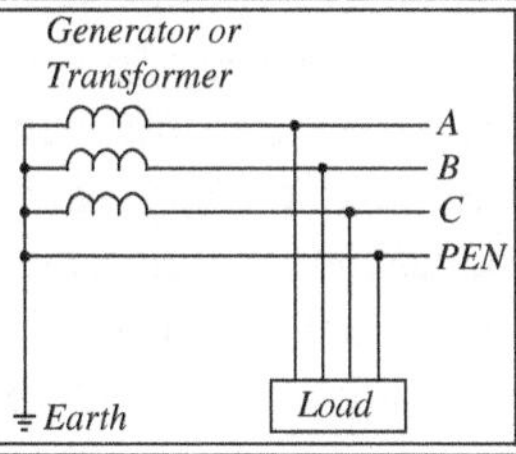	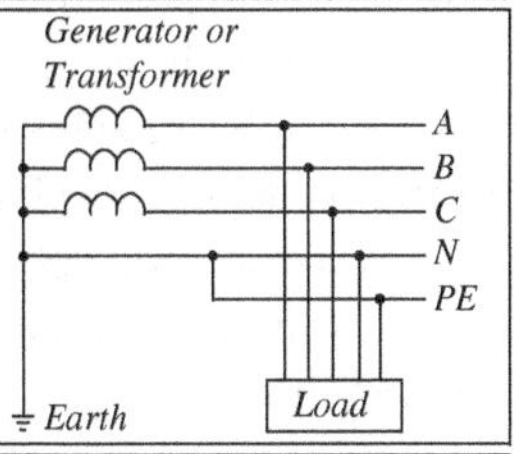
Features	Separate protective earth (PE) and neutral (N) conductors from transformer to the load.	Combined PE and N conductor all the way from the transformer to the the load.	Combined PEN conductor from transformer to building distribution point, but separate PE and N conductors in fixed indoor wiring and flexible power cords.

Figure 11.1 Three typical earthing networks in low-voltage systems.

11.2.2 Leakage Currents in a Grid-tied Converter

In a grid-tied converter, there often exist common-mode (CM) voltages caused by the switching of power switches and other reasons. Because there exists parasitic capacitance between the device and the earth, leakage currents may appear. Figure 11.2 illustrates a generic equivalent circuit for analyzing leakage currents, which consists of a CM voltage source v_{CM}, a filter L and a parasitic capacitor C_p between the DC ground and the earth. As mentioned above, the AC ground of the converter is often connected to the earth via the neutral line when an isolating transformer is not used. This closes the loop for the leakage current i_l so that it can flow through the parasitic capacitor C_p. If the switching frequency is high enough, the leakage current i_l can be very large even if the parasitic capacitor C_p is relatively small.

In general, the leakage current i_l mainly depends on both the behavior of the CM voltage and the impedance in the path of the leakage current. In light of this, there are four main approaches in the literature to reduce leakage currents. One approach is to reduce high-frequency components in the voltage v_p across the parasitic capacitor C_p. This can be achieved by reducing the high-frequency components in v_{CM}, e.g. changing the modulation strategies (Bae and Kim 2014; Freddy Tan et al. 2015) and decoupling the AC and DC sides of converters during freewheeling phases (Buticchi et al. 2014; Xiao et al. 2014). It can also be achieved by putting passive or active components (Barater et al. 2014) in series or in parallel with v_{CM}. Although these methods can reduce leakage currents to some extent, they suffer from reduced voltage utilization ratio (Cavalcanti et al. 2010), reduced ability to process reactive power (Gu et al. 2013), and/or degraded performance caused by parasitic capacitors of switches (Gu et al. 2013).

The second approach is to increase the impedance in the path of the leakage current. For example, this can be achieved by increasing the filter L. Such a filter is often called a CM filter (Heldwein et al. 2010), which has large inductance to mitigate leakage currents. However, due to the large size and heavy weight, the system power density is reduced.

The third approach is to provide another current path between the AC ground and the DC ground so that most of the leakage current flows through this path instead of the parasitic capacitor. For example, this can be achieved by connecting the AC ground to the midpoint of the DC-bus split capacitors (Xiao and Xie 2012). This makes the split capacitors in parallel with the parasitic capacitor. Because the split capacitors are normally much larger than the parasitic capacitor C_p, the leakage current is reduced. For example, converters based on the half-bridge topology, which are equipped with split capacitors, have the potential of bypassing the leakage current (Srinivasan and Oruganti 1998). The three-phase four-wire

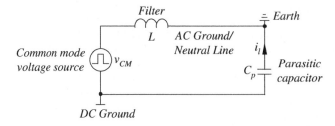

Figure 11.2 Generic equivalent circuit for analyzing leakage currents.

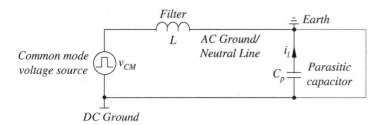

Figure 11.3 Equivalent circuit for analyzing leakage current of a grid-tied converter with a common AC and DC ground.

power converters proposed in (Zhong and Hornik 2013) are equipped with DC-bus split capacitors and hence have a bypassing current path to reduce leakage current. Another way is to connect the neutral point of the AC capacitors of the LC filters to the DC-bus split capacitors (Dong et al. 2012). In this case, the AC capacitors are in series with the DC-bus split capacitors in the current path, which are in parallel with the parasitic capacitor C_p.

The last approach is to directly connect the AC ground and the DC ground together, if possible to do so, as shown in Figure 11.3. This almost eliminates the leakage current i_l completely because the parasitic capacitor is short-circuited without any other efforts (when ignoring the parasitic inductance of the cables). One such converter is the Karschny inverter (Karschny 1998). Unfortunately, this topology suffers from a complex structure, an increased number of power switches and the need of large capacitors. Another one was proposed in (Gu et al. 2013), where the concept of a virtual DC-bus is used to enable the direct connection between AC and DC grounds. The virtual DC bus is achieved by adding an extra active switch and an extra DC capacitor into a conventional full-bridge converter. Benefiting from its topological structure, the converter is capable of exchanging both real and reactive power with the power grid. The operation of the two converters are similar to that of conventional full-bridge converters. Moreover, for split-phase systems, it is also possible to connect the AC ground and the mid-point of split capacitors, treated as the DC ground, together (Breazeale and Ayyanar 2015). Unfortunately, this topology also suffers from its complex structure and an increased number of power switches.

11.2.3   Benefits of Providing a Common AC and DC Ground

Directly connecting the AC ground and the DC ground together or providing a common AC and DC ground brings a lot of benefits. It not only eliminates the leakage current, as discussed above, but also improves electromagnetic compatibility. Moreover, a common AC and DC ground also makes it always possible to have a grounded system, improving safety. Furthermore, providing a common AC and DC grid will enable the negative pole of DC conductors (i.e. the DC ground) to have almost the same voltage potential as the earth. If the inverter is for PV applications, then the potential induced degradation effect of solar panels can be prevented.

It is recommended always to adopt a converter topology that is able to provide a common AC and DC ground, e.g., the Beijing converter (Zhong et al. 2017) and the θ-converter (Ming and Zhong 2017; Zhong and Ming 2016).

11.3 Operation of a Conventional Half-bridge Inverter

The conventional half-bridge inverter shown in Figure 11.4(a) has several advantages such as fewer power switches, reduced leakage currents and relatively low costs. However, it suffers from a major drawback for PV applications, which is the requirement of a high input voltage (Gu et al. 2013). Nevertheless, its operation is instrumental for understanding the topology to be discussed later.

Assume that V_+ and V_- are the voltages across the capacitors C_+ and C_- with respect to the neutral point N and the negative point of the DC bus, respectively. Then the DC input voltage is

$$V_{DC} = V_+ + V_-.$$

Moreover, assume the output current of the inverter is sinusoidal

$$i_o = I_o \sin(\omega t + \varphi) \tag{11.1}$$

without any harmonics and the grid voltage is

$$v_o = V_o \sin \omega t,$$

where V_o and I_o are the peak values of the output voltage and output current, respectively, and ω is the angular line frequency. Because the two switches are operated in a complementary way, the average circuit model of the half-bridge inverter can be obtained as shown in Figure 11.4(b), according to the procedures in (Srinivasan and Oruganti 1998; Tymerski et al. 1989). The switches Q_1 and Q_2 are replaced by a current source $i_o(1 - d_2)$ and a voltage source $V_{DC}(1 - d_2)$, respectively, where d_2 is the duty cycle of Q_2 given by

$$d_2 = \frac{V_+}{V_{DC}} - \frac{V_o}{V_{DC}} \sin \omega t$$

when the switching frequency is much higher than the line frequency.

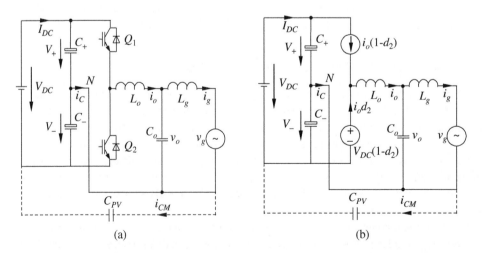

(a) (b)

Figure 11.4 A conventional half-bridge inverter. (a) Topology. (b) Average circuit model.

11.3.1 Reduction of Leakage Currents

The voltage across the parasitic capacitor is clamped by the voltages across the DC capacitors because the grid neutral line is directly connected to the midpoint of the DC bus. In this case, most of the high frequency CM currents would flow through the DC capacitors but not the parasitic capacitor. Although low-frequency components still exist, the leakage current is significantly reduced.

11.3.2 Output Voltage Range

In order to ensure successful operation, each of the voltages across the DC-bus capacitors must be higher than the output voltage, which means

$$V_+ > V_0 \text{ and } V_- > V_0.$$

In other words, the required DC-bus voltage should be at least twice that of the conventional full-bridge inverter for single-phase applications. A possible solution is to put two DC sources in parallel with the split capacitors but the additional power source leads to increased cost. Moreover, balancing the two power sources can be a serious problem. Furthermore, it is not possible to just put one DC power source in parallel with one of the capacitors because capacitors cannot provide the return path for a DC current.

11.4 A Transformerless PV Inverter

11.4.1 Topology

Figure 11.5(a) shows a transformerless PV inverter. It adds an additional leg with Q_3 and Q_4 to the half-bridge converter discussed in the previous section, which actually forms an independently controlled neutral leg proposed in (Zhong and Hornik 2013). Through independent control of Q_3 and Q_4 from Q_1 and Q_2, it is possible to operate the converter with one power source. Moreover, the negative pole of the power source is connected to the AC ground. As a result, it becomes easy to ground the PV panels and to have a common AC and DC ground.

The main function of the neutral leg is to balance the voltages of the capacitors, provide a return path for the DC source current, and maintain a stable reference point for the inversion leg. The function of the inversion leg remains unchanged, which is to operate as a half bridge to convert the DC bus voltage into AC voltage in high power quality and to control the reactive and active power exchange between the DC bus and the grid. Note that the inversion leg is controlled independently from that of the neutral leg. It is also possible to add two more phases for three-phase applications.

11.4.2 Control of the Neutral Leg

Similarly, the average model of the inverter can be obtained as shown in Figure 11.5, where

$$d_3 = \frac{V_-}{V_{\text{DC}}}$$

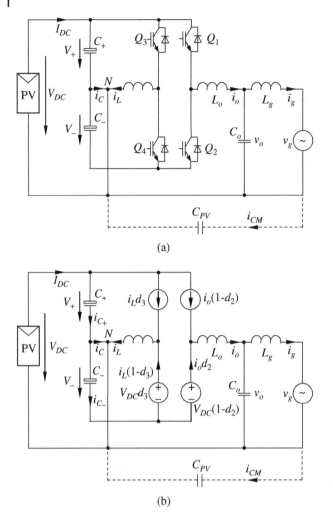

Figure 11.5 A transformerless PV inverter. (a) Topology. (b) Average circuit model.

is the duty ratio of switch Q_3. According to Kirchhoff's law, there is

$$i_C + i_L = I_{DC} - i_o. \tag{11.2}$$

The right-hand side of the equation is the combination of a DC component and an AC component. As is well known, capacitors cannot provide DC currents. Hence, i_C cannot contain any DC component and the DC component has to be contained in i_L, i.e. the current flowing through the additional neutral leg. At the same time, the voltage ripples on the DC-bus capacitors should be maintained small, which means i_C should be controlled to be

$$i_C = 0.$$

As a result,

$$i_L = I_{DC} - i_o,$$

containing the DC current I_{DC} and the output current i_o. In other words, the neutral leg does not only provide the current return path for the DC source current but also provides the current return path for the output current. This should be taken into consideration when determining the ratings of the switches.

In order to regulate i_C to be around zero, i_C should be measured as a feedback signal. Many current controllers, such as a hysteresis controller (Tilli and Tonielli 1998) with a variable switching frequency and a repetitive controller (Hornik and Zhong 2011a) with a fixed switching frequency, can be applied to regulate the current i_C to zero. Because of the excellent harmonic rejection performance of the repetitive controller, it is adopted here, as shown in Figure 11.6. It consists of a proportional controller K_r and an internal model

$$C(s) = \frac{1}{1 - \frac{\omega_i}{s+\omega_i}e^{-\tau_d s}},$$

where τ_d is designed to be slightly smaller than the fundamental period τ as

$$\tau_d = \tau - \frac{1}{\omega_i}$$

and ω_i is the cut-off frequency of the low-pass filter. For a 50 Hz system, $\tau = 0.02$s and ω_i can be chosen as $\omega_i = 2550$ rad s^{-1}. The gain K_r can be chosen as

$$K_r = \omega_i L_r,$$

where L_r is the neutral line inductor, according to (Hornik and Zhong 2011a). More details about repetitive control can be found in (Zhong and Hornik 2013).

Another important task is to make sure that $V_- > V_o$ because the inversion leg is operated as a half bridge. This can be achieved with a PI controller to maintain V_- at its reference value V_-^*, which can be set higher than V_o. Note that although the inversion leg is operated as a half bridge, the source voltage required is only half of the DC-bus voltage because it is only connected in parallel with one capacitor.

The outputs of the PI controller and the repetitive controller are added together to form the control signal for the neutral leg, as shown in Figure 11.6. Note that the PI controller is responsible for a DC component while the repetitive controller is responsible for harmonic components. Hence, the two loops can be separately tuned.

Figure 11.6 Controller for the neutral leg.

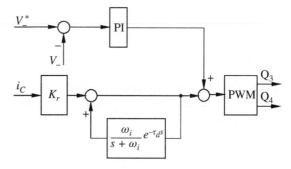

11.4.3 Control of the Inversion Leg as a VSM

In order to make the PV inverter grid-friendly, the inverter leg is controlled according to the self-synchronized synchronverter (Zhong et al. 2014) discussed in Chapter 8. The controller is shown in Figure 11.7. It consists of a real power-frequency channel, a reactive power-voltage channel, and some blocks for self-synchronization. Because the power in a single-phase system is pulsating, there is a need to add a low-pass filter to filter T_e and Q.

The operation modes of the system are summarized in Table 11.1. The set mode is further classified as the P-mode to regulate the real power P to the set point P_{set} and the Q-mode to regulate the reactive power Q to the set point Q_{set}. The droop mode is further classified as the f_D-mode for the frequency droop and the V_D-mode for the voltage droop. These modes belong to the normal operation mode. Moreover, there is also a mode called the synchronization mode for the synchronverter to synchronize with the grid. The change of these modes is implemented by three switches S_P, S_Q, and S_C, shown in Figure 11.7, with the corresponding modes given in Table 11.1.

11.4.3.1 Real Power Control and Frequency Droop
Normally, the power angle of a synchronous generator is controlled by regulating the mechanical torque. Here, a virtual mechanical torque T_m is calculated from the real power

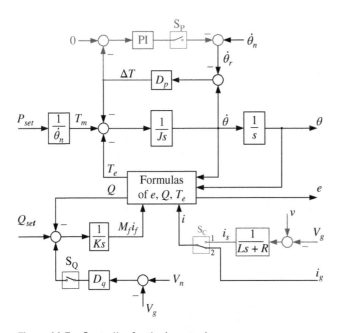

Figure 11.7 Controller for the inverter leg.

Table 11.1 Operation modes of the PV inverter.

S_C	Mode	S_P	Mode	S_Q	Mode
1	Synchronization	ON	P-mode	ON	V_D-mode
2	Normal	OFF	f_D-mode	OFF	Q-mode

set point P_{set}. Taking into account the fact that during normal operation the grid frequency $\dot{\theta}$ only varies within a small range around the rated grid frequency $\dot{\theta}_n$, i.e. $\dot{\theta} \approx \dot{\theta}_n$, there is

$$T_m = \frac{P_{set}}{\dot{\theta}} \approx \frac{P_{set}}{\dot{\theta}_n}, \tag{11.3}$$

which is then fed into a swing equation, as shown in Figure 11.7.

When the switch S_P is ON, the inverter works in the P-mode and the frequency $\dot{\theta}$ is regulated to track $\dot{\theta}_g$. The real power sent to the grid is the same as the real power set point P_{set}. When the switch S_P is OFF, the inverter works in the f_D-mode. The real power P varies around P_{set} according to the frequency droop coefficient D_p, providing frequency support to the grid.

11.4.3.2 Reactive Power Control and Voltage Droop
The reactive power-voltage channel is shown in the lower part of Figure 11.7. When the switch S_Q is OFF, it is operated in the Q-mode and $M_f i_f$ is generated according to the tracking error between Q_{set} and Q only, via the integrator with the gain $1/K$. Therefore, the reactive power Q follows the reactive power set point Q_{set} regardless of the voltage difference between V_n and V_g. When the switch S_Q is ON, the voltage droop is enabled, the voltage error $\Delta V = V_n - V_g$ is added to the reactive power error through the droop coefficient D_q. Hence, the reactive power Q is regulated according to the voltage V_g, providing voltage support to the grid.

11.4.3.3 Synchronization
Before the PV inverter is connected to the grid, it must be synchronized with the grid voltage v_g. The controller in Figure 11.7 is equipped with the self-synchronization strategy. When S_C is in position 1 with with $P_{set} = 0$ and $Q_{set} = 0$, the virtual current i_s generated by passing the voltage difference $v - v_g$ through the virtual impedance $R + sL$ is fed into the controller. The inverter regulates P and Q to zero in the set mode, achieving synchronization with the grid. Here, the voltage v can be e or v_o, with a small difference. When S_C is in position 2, the inverter works in the normal operation and the physical current i_g is fed into the controller.

11.5 Real-time Simulation Results

Real-time simulations were carried out with the solver ode4 (Runge–Kutta) with a fixed step size of 6 μs. The parameters of the system are given in Table 11.2. The parameters of the PI controller for the voltage V_- are $K_P = 0.1$, $K_I = 10$ and the gain of the repetitive controller for the neutral current i_C is $K_r = 20$.

The neutral leg was started before enabling the inversion leg in order to establish the voltage V_-. All the operation modes, including the direct active and reactive power control (P-mode, Q-mode), frequency droop control (f_D-mode), and voltage droop control (V_D-mode) were tested. Initially, the grid frequency was set at $f_g = 50$ Hz and the voltage amplitude was set at $V_g = 1.02V_n$. The sequence of events during the simulation are:
(1) Start the neutral leg at $t = 0$ s
(2) Start the inversion leg with the synchronization command at $t = 0.5$ s

Table 11.2 Parameters of the system.

Parameters	Values
DC input voltage	200 V
Grid voltage (RMS)	110 V
Line frequency	50 Hz
L_s	2.2 mH
DC-bus capacitance C_+	5000 μF
DC-bus capacitance C_-	5000 μF

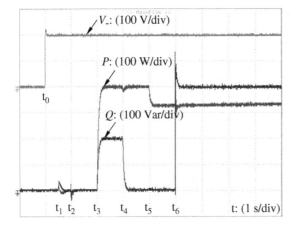

Figure 11.8 Real-time simulation results of the transformerless PV inverter in Figure 11.5.

(3) Connect the inverter to the grid at $t = 1$ s

(4) Apply the set point $P_{set} = 400$ W and $Q_{set} = 200$ Var at $t = 2$ s

(5) Enable the f_D–mode and V_D–mode by switching S_P OFF and S_Q ON at $t = 3$ s

(6) Increase the grid frequency to $f_g = 50.02$ Hz at $t = 4$ s

(7) Drop the grid voltage amplitude to $V_g = 0.98 V_n$ at $t = 5$ s

(8) Stop the simulations at $t = 6$ s.

The simulation results are shown in Figure 11.8. With the independently controlled neutral leg, the voltage across the capacitor C_- was well maintained to be balanced with the DC input voltage regardless of the operation mode of the inverter. For the inversion leg, there was no problem when connecting it to the grid at $t = 1$ s. The real power and reactive power were regulated at zero after connection to the grid because $P_{set} = 0$ and $Q_{set} = 0$ before $t = 2$ s. When $P_{set} = 400$ W and $Q_{set} = 200$ Var were applied at $t = 2$ s, the response of the inverter was fast and smooth.

When the f_D-mode and the V_D-mode were enabled at $t = 3$ s, the grid voltage V_g was 2% higher than V_n. Because D_q was set to drop 100% of reactive power (i.e. 1000 Var) when the grid voltage increases by 10%, the reactive power expected to be dropped was

$$\Delta Q = -1000 \times \frac{2}{10} = -200 \text{ Var} \tag{11.4}$$

from the set point, which matches well with the simulation result shown in Figure 11.8. Since the grid frequency was at its nominal value, the real power did not change. When the grid frequency changed to $f_g = 50.02$ Hz at $t = 4$ s, the real power P dropped by $\Delta P = -70$ W, as expected. When the voltage dropped to $V_g = 0.98V_n$ at $t = 5$ s, the reactive power increased by 400 Var, as expected.

Note that the PV module was simulated with an ideal voltage source. In practice, the MPPT strategy needs to be included and the available real power depends on the irradiation as well.

11.6 Summary

Based on (Ming and Zhong 2014), a transformerless single-phase PV inverter is presented. The problems of grounding and leakage currents are discussed at first. Then, the topology and control strategies of the transformerless PV inverter are discussed. The PV inverter has two prominent features: (1) it has a common AC and DC ground, which eliminates the leakage current and avoids potential-induced degradation of solar panels; (2) it has an independently controlled neutral leg and an inversion leg, which simplifies the control design and brings flexibility. Real-time simulation results are presented to demonstrate its operation. Indeed, it is able to provide support to the grid when the grid frequency and voltage change.

12

Synchronverter Based STATCOM without an Dedicated Synchronization Unit

A synchronous generator that regulates the utility grid voltage without injecting real power is known as a synchronous condenser. In this operating mode, it could continuously generate or consume reactive power by controlling the excitation of the machine. The operating principles of static synchronous compensator (STATCOM) are similar. However, so far, most control methods of STATCOM have not taken into account the internal characteristics of rotational synchronous machines. In this chapter, following the idea of synchronverters, the controller for a STATCOM is designed according to the mathematical model of synchronous generators that are operated in the condenser mode. As a result, no dedicated synchronization unit such as a PLL is needed. Moreover, a third operation mode, i.e. the voltage droop control mode (or the V_D-mode in short), is introduced to the operation of STATCOM, in addition to the conventional voltage regulation mode (or the V-mode in short) and the direct Q control mode (or the Q-mode in short). This allows parallel-operated STATCOMs to share reactive power properly. The control strategy is verified with simulations in MATLAB/Simulink/SimPowerSystems. Different operational scenarios of STATCOM are evaluated.

12.1 Introduction

Synchronous machines have been used for a long time as synchronous compensators or synchronous condensers to control the voltage profile of power systems, especially in transient states, by consuming or generating reactive power. The use of synchronous condensers is considered to be simple and could maintain high flexibility and reliability (Akhmatov and Eriksen 2007; Rush and Smith 1978). They are nowadays still used in many applications, especially when robust operation is required, e.g. in remote and islanded operation of wind farms (Akhmatov and Eriksen 2007) and in high voltage direct current (HVDC) transmission systems (Kirby et al. 2001; Nayak et al. 1994).

Although synchronous condensers have good stable operating characteristics, the loss is usually high and can reach 1–2% of the rated power (Luiz da Silva et al. 2001), which is mainly due to the rotational loss and the heat caused by high reactive currents. As a result, static synchronous condensers (STATCOM), which are effectively voltage-source inverters (VSIs) (Hingorani and Gyugyi 1999; Singh et al. 2009), are becoming an alternative option.

Power Electronics-Enabled Autonomous Power Systems: Next Generation Smart Grids,
First Edition. Qing-Chang Zhong.
© 2020 John Wiley & Sons Ltd. Published 2020 by John Wiley & Sons Ltd.

In a distribution system, STATCOM could also be used to assist in re-balancing the unbalanced voltage sources or riding through voltage sags (Escobar et al. 2004; Hochgraf and Lasseter 1998; Li et al. 2006; Song and Liu 2009). At the load side, STATCOM is very efficient in playing the role of both active filters and displacement power factor correctors for unbalanced, balanced, linear, and nonlinear loads (Bina and Bhat 2008; Singh et al. 2000a, b; Xu et al. 2010). STATCOMs are also found in traction power systems (Horita et al. 2010) and renewable energy applications, especially wind power applications (Chen et al. 2010; El-Moursi et al. 2010; Mohod and Aware 2010; Qiao et al. 2009). The use of STATCOMs in wind farms equipped with fixed-speed induction generators and doubly fed induction generators (DFIGs) offer enhanced stability for the power network, to prevent over-voltages in islanded events under post-fault switching conditions, and to mitigate the voltage fluctuations due to the tower shadow effect (Qiao et al. 2009; Saad-Saoud et al. 1998). For variable-speed wind turbines with full-scale back-to-back inverters, the requirement of STATCOM is not essential (Chen et al. 2009) because the reactive power could be regulated by using appropriate control strategies for the VSI. However, STATCOMs are still being used in several cases, e.g. to support the transient voltage stability in uninterrupted operation of wind turbines during short-term grid faults, to support the regulation of the steady-state grid voltage (Qiao et al. 2009) and to provide reactive power in islanded operation.

The operating principles of STATCOM are similar to those of rotational synchronous condensers externally (Gyugyi 1988; Hingorani and Gyugyi 1999; Mori et al. 1993; Wildi 2005), but most current technologies have not taken advantage of the mathematical model of synchronous machines. In this chapter, STATCOMs are controlled to operate as virtual synchronous condensers, following the idea of the synchronverter. From the grid side, the power system sees STATCOMs as actual synchronous machines operated in the condenser mode. The controller has a built-in self-synchronization mechanism and does not need a PLL to provide the grid frequency and synchronization. A third operation mode, the droop mode, is also introduced to the operation of STATCOM. This allows parallel-operated STATCOMs to share reactive power according to the voltage drop.

The rest of the chapter is organized as follows. The conventional control schemes for STATCOM are outlined in Section 12.2. The synchronverter based controller for STATCOM that does not need a PLL is presented in Section 12.3 and simulation results are shown in Section 12.4, followed by summaries in Section 12.5.

12.2 Conventional Control of STATCOM

Figure 12.1(a) depicts a STATCOM connected to a power system. From the point of common coupling (PCC), it consists of a circuit breaker (CB), a coupling transformer, a three-phase inverter and a DC-bus capacitor. The leakage reactance X_L of the transformer plays two roles: one is to enable the power exchange between the STATCOM and the grid and the other is to filter out the harmonic current components. Its equivalent circuit is shown in Figure 12.1(b), where the STATCOM is modeled as a voltage source $e = E\angle\theta$ in series with the impedance X_L and the power system is modeled as a voltage source v_{eq} in series with an impedance Z_{eq}. The voltage at the PCC, called the grid or terminal voltage, is denoted as $v_g = V_g\angle\theta_g$. Here, V_g and E are the phase voltage (RMS) of v_g and e, respectively.

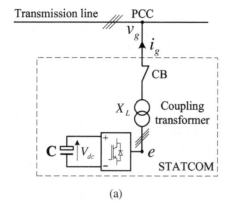

(a)

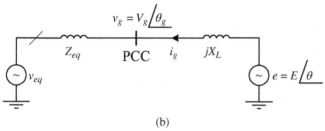

(b)

Figure 12.1 STATCOM connected to a power system. (a) Sketch of the connection. (b) Single-phase equivalent circuit.

12.2.1 Operational Principles

A STATCOM provides or consumes reactive power to obtain a high power factor. It is also able to regulate the PCC voltage v_g. The phase difference between e and v_g is

$$\delta = \theta - \theta_g. \tag{12.1}$$

The real power P and reactive power Q flowing out of the STATCOM are (Singh et al. 2009)

$$P = 3\frac{V_g E}{X_L} \sin \delta, \tag{12.2}$$

$$Q = 3\left(\frac{V_g E}{X_L} \cos \delta - \frac{V_g^2}{X_L}\right). \tag{12.3}$$

Under normal conditions of operation, the STATCOM extracts some real power to cover the losses of the inverter so that the DC-bus voltage V_{dc} is maintained constant. Hence, δ is small (and negative). As a result,

$$\sin \delta \approx \delta \quad \text{and} \quad \cos \delta \approx 1.$$

Equations (12.2) and (12.3) then become as

$$P = 3\frac{V_g E}{X_L} \delta, \tag{12.4}$$

$$Q = 3\frac{V_g}{X_L}(E - V_g). \tag{12.5}$$

Equation (12.5) illustrates that the reactive power Q can be controlled via changing the voltage difference $E - V_g$. When e is controlled with $E > V_g$, the STATCOM provides reactive power and $Q > 0$; when e is controlled with $E < V_g$, the STATCOM absorbs reactive power and $Q < 0$. When $E = V_g$, the STATCOM is floating on the power system and no reactive power is exchanged. It is worth noting that when the reactive power Q is exchanged, the voltage v_g changes slightly as well. This can be used to regulate the voltage at the PCC. Hence, a STATCOM mainly has two different operation modes: one is to provide the desired amount of Q to improve the power factor of the system, which is called the direct Q control mode or the Q-mode in short, and the other is to regulate the PCC voltage for improved voltage regulation, which is called the voltage regulation mode or the V-mode.

Equation (12.4) shows the relationship between the real power P and the phase difference δ. Note that δ and P are negative because the positive sign of the current i_g is flowing out of the STATCOM. It is obvious that

$$\begin{cases} \delta \uparrow \Rightarrow \text{ less real power flows in } \Rightarrow V_{dc} \downarrow, \\ \delta \downarrow \Rightarrow \text{ more real power flows in } \Rightarrow V_{dc} \uparrow, \\ \delta = 0 \Rightarrow P = 0, \; V_{dc} \text{ unchanged.} \end{cases} \tag{12.6}$$

Note that $\delta \uparrow$ means the absolute value of δ is reduced because δ is negative. As a result, the DC-bus voltage V_{dc} can be maintained via controlling the phase difference δ or the phase angle θ of the voltage e.

12.2.2 Typical Control Strategy

Figure 12.2 shows a typical two-axis control scheme in the dq-frame for a PWM-based STATCOM. The instantaneous voltage and currents are converted into their d- and q-components. There are two control loops in this strategy. The outer loop includes two PI controllers: one to regulate the DC voltage V_{dc} and the other to regulate the STATCOM AC output voltage. The inner loop is a current control loop, which also consists of two PI controllers to regulate the d- and q-components of the current i_g to follow I_d^* and I_q^* generated by the outer-loop voltage controllers. The outputs of the current PI controllers are combined via a dq/abc transformation to generate e before feeding into the PWM conversion.

A PLL is adopted to provide the phase θ_g of the PCC voltage so that the STATCOM can be operated properly. The slow response of the PLL used often affects the system stability. When a high power load is switched ON/OFF, it also takes time for the PLL to synchronize with the new voltage phase angle. Hence, it is desirable if a PLL can be removed while keeping the same grade of controllability. A nice attempt is made in (Zhang et al. 2010; Zhang et al. 2011a), but a back-up PLL is still needed.

It has been shown in (Norouzi and Sharaf 2005) that the equivalent impedance Z_{eq} of the power system, referred to as system strength (Hingorani and Gyugyi 1999), plays the role of a positive feed-forward proportional gain inside the current control loop. This gain directly affects system stability. When the power system impedance Z_{eq} is large, i.e. when the system is weak, the system is easier to be destabilized. When the power system impedance Z_{eq} is small, i.e. when the system is strong, the system is more stable but has a slower response. In a normal power system, the impedance Z_{eq} varies with the change of loads etc., which directly affects the operation and stability of the STATCOM.

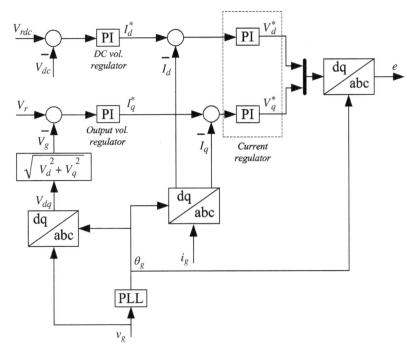

Figure 12.2 A typical two-axis control strategy for a PWM based STATCOM using a PLL.

12.3 Synchronverter Based Control

The synchronverter technology is able to control real power and reactive power indepen-dently with a compact control structure, in which the model of a synchronous machine is embedded. Hence, it can be applied to implement STATCOM if it is operated in the condenser mode, i.e. with $P = 0$. This leads to the controller for a STATCOM shown in Figure 12.3, after dealing with some special aspects and making some necessary changes.

The controller has an upper channel and a lower channel. The upper channel regulates the real power to control the internal frequency $\dot{\theta}$ and the phase θ so that the DC-bus voltage V_{dc} is maintained constant and that the STATCOM tracks the phase of the grid voltage. The lower channel regulates the reactive power and/or the voltage. A particular property is that the STATCOM can be operated in a third mode, called the droop mode or the V_{D}-mode, in addition to the conventional Q-mode and V-mode, so that parallel-operated STATCOMs can share reactive power properly. The V_{D}-mode is the combination of the conventional Q-mode and V-mode, by operating the two switches S_Q and S_V shown in the lower part of Figure 12.3. The positions of the switches S_Q and S_V, together with the corresponding operational modes, are shown in Table 12.1.

12.3.1 Regulation of the DC-bus Voltage and Synchronization with the Grid

As discussed in Section 12.2, the DC-bus voltage can be regulated by controlling the phase θ of the voltage e. A control loop can be added to regulate the DC-bus voltage to generate the

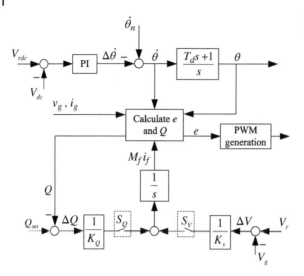

Figure 12.3 A synchronverter based STATCOM controller.

Table 12.1 Operation modes of a STATCOM.

S_Q	S_v	Operation mode
ON	OFF	Q-mode: direct Q control
OFF	ON	V-mode: voltage regulation
ON	ON	V_D-mode: droop control

corresponding T_m needed by the synchronverter, as done in Chapter 9. However, this adds another control loop into the system, which complicates the controller tuning. Another option is to directly control the frequency instead of controlling it via the mechanical torque or the real power, as described below.

Since

$$\delta = \theta - \theta_g,$$

any change in the STATCOM frequency $\dot\theta$ causes V_{dc} to change in the way shown in (12.6) because $\dot\theta \uparrow \Rightarrow \delta \uparrow$ and $\dot\theta \downarrow \Rightarrow \delta \downarrow$. Similarly, any change in the grid frequency $\dot\theta_g$ causes V_{dc} to change in the opposite way, as shown in (12.6), because $\dot\theta_g \uparrow \Rightarrow \delta \downarrow$ and $\dot\theta_g \downarrow \Rightarrow \delta \uparrow$. This means a control loop, as shown in the upper part of Figure 12.3, can be formed to directly regulate the frequency $\dot\theta$ according to the change of the DC-bus voltage. The STATCOM frequency $\dot\theta$ is then

$$\dot\theta = \dot\theta_n - \Delta\dot\theta,$$

where $\Delta\dot\theta$ is the output of the PI controller used to regulate the DC-bus voltage and $\dot\theta_n$ is the rated system frequency. The regulating mechanism can be described as

$$V_{dc} \downarrow \Rightarrow \Delta\dot\theta \uparrow \Rightarrow \dot\theta \downarrow \Rightarrow \theta \downarrow \Rightarrow \delta \downarrow \Rightarrow V_{dc} \uparrow \text{ to } V_{rdc}$$

and

$$V_{dc} \uparrow \Rightarrow \Delta\dot\theta \downarrow \Rightarrow \dot\theta \uparrow \Rightarrow \theta \uparrow \Rightarrow \delta \uparrow \Rightarrow V_{dc} \downarrow \text{ to } V_{rdc}.$$

As a result, there is no need to add an extra PLL to obtain $\dot\theta_g$ as the reference frequency for the STATCOM.

In order to speed up the effect of any change in the frequency $\dot\theta$ on the phase angle θ so that it can quickly synchronize with the grid phase θ_g, the phase angle is obtained with $\frac{T_d s + 1}{s}$ instead of a pure integrator $\frac{1}{s}$ from the frequency $\dot\theta$ as normally done, e.g., in the case of the synchronverter in (Zhong and Weiss 2011). That is, a proportional derivative (PD) unit with the time constant T_d is cascaded with the integrator to speed up the dynamic response of the integrator. The PD unit only plays a role in the transient period and does not affect the steady-state response of the integrator, which is still

$$\theta = \int_0^t \dot\theta \, dt$$

in the time domain. The time constant T_d can be chosen as 5 times the system period, as a rule of thumb.

12.3.2 Operation in the Q-mode to Regulate the Reactive Power

When S_Q is ON and S_V is OFF, i.e. when the voltage control channel in the lower part of Figure 12.3 is disabled, the STATCOM is operated in the Q-mode and the reactive power Q is regulated at the set-point Q_{set}. Therefore, $M_f i_f$ is controlled according to the reactive power error

$$\Delta Q = Q_{set} - Q. \tag{12.7}$$

Since the RMS value $E = \frac{1}{\sqrt{2}} M_f i_f \dot\theta$ of e is proportional to $M_f i_f$, and E is proportional to the generated reactive power Q, according to (12.5), $M_f i_f$ is directly related to the generated Q and can be adopted to control Q. The control effect can be explained as

$$Q_{set} > Q \Rightarrow \Delta Q > 0 \Rightarrow M_f i_f \uparrow \Rightarrow Q \uparrow \text{ to } Q_{set}, \tag{12.8}$$

and

$$Q_{set} < Q \Rightarrow \Delta Q < 0 \Rightarrow M_f i_f \downarrow \Rightarrow Q \downarrow \text{ to } Q_{set}. \tag{12.9}$$

Because of the integrator in the loop, the error ΔQ is eliminated and the reactive power Q tracks the set-point Q_{set} accurately in the steady state.

In practical implementation, the reactive power Q for the feedback can be obtained with e and i_g if the effect of the filter and the transformer is negligible. Otherwise, the reactive power measured at the PCC can be used so that the reactive power at the PCC is regulated to be the set point Q_{set} accurately in the steady state. One option is to calculate the reactive power at the PCC from the measured voltage v_g and current i_g with

$$Q = \langle i_g, v_q \rangle, \tag{12.10}$$

where $\langle, \rangle$ denotes the inner product and v_q has the same amplitude as $v_g = \begin{bmatrix} v_{ga} & v_{gb} & v_{gc} \end{bmatrix}^T$ but with a phase delayed by $\frac{\pi}{2}$ rad from that of v_g. For balanced three-phase systems,

$v_q = \begin{bmatrix} v_{qa} & v_{qb} & v_{qc} \end{bmatrix}^T$ can be obtained as

$$\begin{cases} v_{qa} = \frac{1}{\sqrt{3}}(v_{gc} - v_{gb}), \\ v_{qb} = \frac{1}{\sqrt{3}}(v_{ga} - v_{gc}), \\ v_{qc} = \frac{1}{\sqrt{3}}(v_{gb} - v_{ga}). \end{cases}$$

As a result, the control strategy only takes the measurements of the voltages and currents at the PCC and the DC-bus voltage. Some other signals may need to be measured for the purpose of protection but not for the purpose of control.

12.3.3 Operation in the *V*-mode to Regulate the PCC Voltage

The system strength Z_{eq} of a power system, as shown in Figure 12.1(a), changes with load change on the grid, which causes V_g to drop or increase accordingly. The fluctuation of V_g reduces system stability and directly affects the normal operation of other equipment on the grid. It may also increase the grid loss. Therefore, regulating the voltage V_g at the PCC is an essential function of a STATCOM, which makes the *V*-mode the most common operation mode of STATCOMs. In this mode, the voltage V_g at the PCC is regulated at the set point V_r, which is usually the nominal value of the system.

In order to operate the STATCOM in the *V*-mode, S_Q is OFF and S_V is ON, i.e. the reactive power control channel in the lower part of Figure 12.3 is disabled. Therefore, $M_f i_f$ is controlled according to the voltage error

$$\Delta V = V_r - V_g. \tag{12.11}$$

The control effect can be explained as

$$V_g < V_r \Rightarrow \Delta V > 0 \Rightarrow M_f i_f \uparrow \Rightarrow E \uparrow \Rightarrow V_g \uparrow \text{ to } V_r, \tag{12.12}$$

and

$$V_g > V_r \Rightarrow \Delta V < 0 \Rightarrow M_f i_f \downarrow \Rightarrow E \downarrow \Rightarrow V_g \downarrow \text{ to } V_r. \tag{12.13}$$

In the steady state, $V_g = V_r$ because of the integrator in the loop.

It is worth noting that the voltage error ΔV depends not only on the generated voltage E but also on the system strength Z_{eq}, which is uncertain to the control loop. If the system is very weak then $M_f i_f$ is very insensitive to ΔV and could become very large, which may lead to a large reactive current. Therefore, proper protection mechanisms must be incorporated.

12.3.4 Operation in the V_D-mode to Droop the Voltage

When both S_Q and S_V are turned ON, the STATCOM is operated in the voltage droop mode, denoted the V_D-mode. The generated $M_f i_f$ depends on both the reactive power error ΔQ and the voltage error ΔV. When the STATCOM reaches the steady state in this mode, the input to the integrator to generate the $M_f i_f$ is 0, i.e.,

$$\frac{\Delta Q}{K_Q} + \frac{\Delta V}{K_v} = 0. \tag{12.14}$$

Hence, the voltage droop coefficient $D_q = -\frac{\Delta Q}{\Delta V}$ is

$$D_q = -\frac{\Delta Q}{\Delta V} = \frac{K_Q}{K_v}. \tag{12.15}$$

In practice, the parameters K_Q and K_v can be tuned as follows. The droop coefficient D_q can be determined at first with the specification of the STATCOM. For example, if it is required to generate the rated reactive power when the voltage drops by 10%, then

$$D_q = \frac{Q_n}{10\% V_n}, \tag{12.16}$$

where V_n is the amplitude of the nominal phase voltage of the grid and Q_n is the rated capacity. According to Chapter 4, the time constant of the voltage loop is

$$\tau_v = \frac{K_v}{\dot\theta} \approx \frac{K_v}{\dot\theta_n}, \tag{12.17}$$

from which the gain K_v can be determined as $K_v = \tau_v \dot\theta_n$ after choosing the time constant τ_v. Then, the parameter K_Q can be obtained from (12.15) as

$$K_Q = K_v D_q. \tag{12.18}$$

Since $\frac{1}{K_v} = \frac{D_q}{K_Q}$, which is often much larger than $\frac{1}{K_Q}$, the voltage loop is much faster than the reactive power loop, as expected.

It is worth noting that a different droop coefficient D_q, i.e. the ratio $\frac{K_Q}{K_v}$, results in a different steady-state error ΔV of the voltage in the V_D-mode. The larger the D_q, the smaller the $|\Delta V|$, according to (12.15). However, increasing D_q too much could lead to instability. In other words, the selection of K_Q and K_v, or τ_v and D_q, should take into account the trade-off between voltage regulation and system stability.

12.4 Simulation Results

12.4.1 System Description

The power system shown in Figure 12.4 was simulated in MATLAB/Simulink/SimPower Systems. The detailed model of the STATCOM is shown in Figure 12.5. The STATCOM was connected to the grid via a 1.5/25 kV transformer. The rated system frequency was $f_n = 50$ Hz. The parameters of the system are given below.

- Power source
  - Apparent power: 100 MVA (power system base power)
  - Nominal voltage: 25 kV (power system base voltage)
  - Reactance $X_s = 0.3$ pu, Resistance $R_s = 0.1$ pu
- Transmission line
  - Model: π-section line
  - Length: 50 km
  - Resistance: $R_0 = 0.4\ \Omega\ \text{km}^{-1}$, $R_+ = 0.1\ \Omega\ \text{km}^{-1}$
  - Inductance: $L_0 = 2\ \text{mH km}^{-1}$, $L_+ = 1\ \text{mH km}^{-1}$

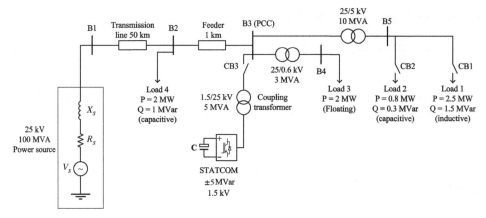

Figure 12.4 Single-line diagram of the power system used in the simulations.

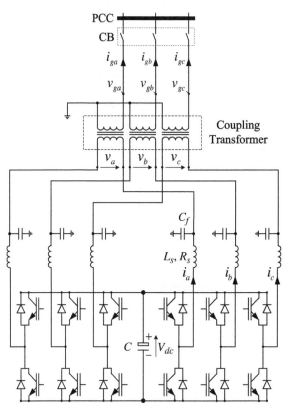

Figure 12.5 Detailed model of the STATCOM used in the simulations.

- Capacitance: $C_0 = 5 \, \text{nF km}^{-1}$, $C_+ = 11 \, \text{nF km}^{-1}$
- Feeder
  - Model: mutual inductance model
  - Length: 1 km
  - Self-impedance: $Z_0 = 0.39 + j0.27\pi \, (\Omega/\text{km}^{-1})$

- Mutual impedance: $Z_+ = 0.11 + j0.10\pi$ (Ω /km^{-1})
- Load 1
    - Rated voltage: 5 kV
    - Active power: 2.5 MW
    - Reactive power: 1.5 MVar (inductive)
- Load 2
    - Rated voltage: 5 kV
    - Active power: 0.8 MW
    - Reactive power: 0.3 MVar (capacitive)
- STATCOM
    - Type: PWM based, double-bridge VSI STATCOM
    - Rated power: ±5 MVar
    - Nominal AC voltage: 1500 V
    - Nominal DC voltage: 2500 V
    - DC-bus capacitor: $C = 10$ mF
    - VSI output inductance: $L_s = 300$ μH
    - Inductor's resistance: $R_s = 1$ mΩ
    - Filter capacitor: $C_f = 220$ μF
    - R_f (parallel to C_f): 10 kΩ.

For the real power control loop, the PI control gains were chosen as $K_P = 0.005$ and $K_I = 0.1$ and the time constant of the PD unit was chosen as $T_d = 0.1$s. For the reactive power control loop, D_q was chosen so that 100% nominal reactive power increment corresponds to 10% of voltage drop, i.e. $D_q = \frac{Q_n}{10\%V_n} = 2449.5$ where $Q_n = 5$ MVar is the rated power of the STATCOM and $V_n = 25\sqrt{\frac{2}{3}}$ kV. The time constant of the voltage loop was chosen as $\tau_v = 0.058$ s and, as a result, $K_v = 18.37$ and $K_Q = 45000$. The simulation step size was chosen as $T_s = 50\mu s$, with the IGBT models running in the average mode.

12.4.2 Connection to the Grid

Before connecting the STATCOM to the grid, the voltage on the DC bus was zero. It is expensive to design a separate pre-charging circuit to charge the DC bus capacitor and a switching mechanism to switch the charged capacitor back to the DC bus. However, when the STATCOM is controlled to operate as a synchronous condenser, this task becomes very easy. The procedure to connect the STATCOM to the grid is described below:

(1) Set the STATCOM to work in the Q-mode with $Q_{set} = 0$.

(2) Reset the STATCOM phase θ to zero when the grid voltage v_g crosses 0 from negative to positive so that e is in phase with v_g.

(3) Turn the circuit breaker CB on to connect the STATCOM to the grid and, at the same time, stop resetting the STATCOM phase θ.

The results when connecting the STATCOM to the grid at $t = 0.5$ s are shown in Figure 12.6. The DC bus was quickly charged to the nominal value 2.5 kV with very small overshoot. A large amount of real power (still below the power rating) was drawn from the grid to charge the DC-bus capacitor. There was some coupling effect in the reactive power, as usual, but the reactive power returned to 0 quickly. The phase θ tracked the grid

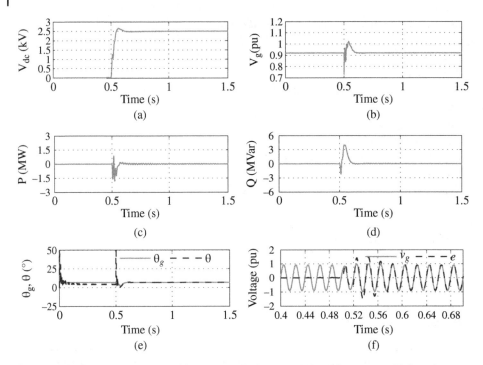

Figure 12.6 Connecting the STATCOM to the grid. (a) V_{dc}. (b) V_g. (c) Real power. (d) Reactive power. (e) θ and θ_g. (f) e and v_g (phase a).

phase θ_g very well after the DC-bus voltage was established. The controller regulated the STATCOM phase voltage very well and the generated AC voltage e quickly tracked the grid voltage v_g. The connection of the STATCOM caused the voltage V_g to drop to about 0.8 pu but it quickly recovered to 0.92 pu, which is the level before connecting the STATCOM, because $Q_{set} = 0$. Note that no PLL was used in the process of connection to the grid.

12.4.3 Normal Operation in Different Modes

The normal operation of the STATCOM was tested in different operation modes, i.e. the Q-mode, the V-mode and the V_D-mode. After connecting the STATCOM to the grid in the Q-mode, the simulation was conducted in the following sequence of events:

(1) At $t = 2$ s, change Q_{set} from 0 to -3 MVar (in the Q-mode)
(2) At $t = 3$ s, switch to the V-mode to regulate V_g to 1 pu
(3) At $t = 4$ s, switch to the V_D-mode
(4) At $t = 4.5$ s, stop the simulation.

The results are shown in Figure 12.7. The V_{dc} was successfully maintained at the set point $V_{rdc} = 2.5$ kV with some dynamics, while the real power P remained at nearly zero and the phase difference δ was kept very small all the time. Although the system frequency was kept constant, the phase of the grid voltage (the constant part) did change when the events happened, as shown in Figure 12.7(e), and the STATCOM successfully tracked the grid phase without a PLL, as shown in the Figure 12.7(f). The transitions between the three operation modes of the STATCOM were very smooth with fast responses and small overshoots. V_g was

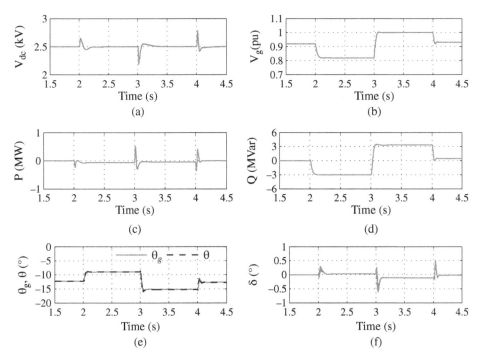

Figure 12.7 Simulation results of the STATCOM operated in different modes. (a) V_{dc}. (b) V_g. (c) Real power. (d) Reactive power. (e) θ and θ_g. (f) $\delta = \theta - \theta_g$.

about 0.92 pu before the STATCOM started regulating the reactive power. The STATCOM quickly responded to the change of $Q_{set} = -3$ MVar at $t = 2$ s, which caused V_g to reduce to about 0.8 pu. When the mode was changed to the voltage regulation mode at $t = 3$ s, V_g was quickly regulated to 1 pu with about $Q = +3.5$ MVar provided. The transition from consuming to generating reactive power is equivalent to switching the STATCOM from working as an inductive load, with i lagging v by $\frac{\pi}{2}$ rad, to working as a capacitive load, with i leading v by $\frac{\pi}{2}$ rad. The v and i during this transition are shown in Figure 12.8(f). It took only one cycle for the current i to change the phase, which was very fast. It took only about three cycles for the whole system to settle down, as shown in Figure 12.8.

When the STATCOM was switched to the V_D-mode at $t = 4$ s, the reactive power generated was reduced to about 0.42 MVar, which caused the V_g to settle down at about 0.932 pu, i.e. with the voltage error of $\Delta V = 6.8\% V_n$. The change of the reactive power is $\Delta Q = 3.42$ MVar from the set point $Q_{set} = -3$ MVar. This is in line with (12.15) because

$$\frac{\Delta Q}{\Delta V} = \frac{3.42 \times 10^6}{6.8\% V_n} \approx \frac{5 \times 10^6}{10\% V_n} = D_q.$$

12.4.4 Operation under Extreme Conditions

12.4.4.1 With a Changing Grid Frequency

The purpose of this simulation is to demonstrate the voltage phase tracking performance of the STATCOM under extreme grid frequency variations. The grid frequency was set at the

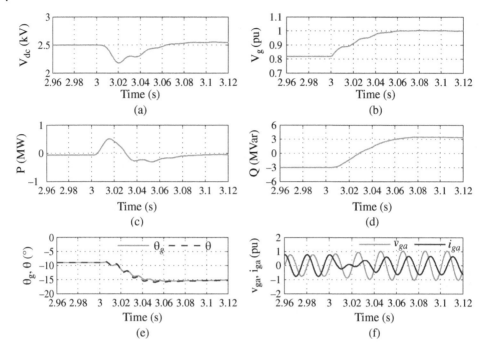

Figure 12.8 Transition from inductive to capacitive reactive power when the mode was changed at $t = 3.0$ s from the Q-mode to the V-mode. (a) V_{dc}. (b) V_g. (c) Real power. (d) Reactive power. (e) θ and θ_g. (f) v_{ga} and i_{ga}.

nominal frequency $f_n = 50$ Hz at the beginning. After connecting the STATCOM to the grid in the Q-mode, the simulation was conducted in the following sequence of events:

(1) Switch to the V-mode at $t = 2$ s and, at the same time, change the grid frequency to $f = 52$ Hz

(2) Change the grid frequency back to the nominal value $f_n = 50$ Hz at $t = 3$ s

(3) Drop the grid frequency to $f = 48$ Hz at $t = 4$ s

(4) Change the grid frequency back to nominal value $f_n = 50$ Hz at $t = 5$ s

(5) Stop the simulation at $t = 6$ s.

The results are shown in Figure 12.9. The DC-bus voltage shown in Figure 12.9(b) reflects the extreme frequency change but was successfully driven back to the set point without any problem. At $t = 2$ s, although the operation mode and the frequency were changed at the same time, the system still responded very well. The phase tracking performance was excellent under the large grid frequency variations, as can be seen from Figure 12.9(e), again, without using a PLL.

The voltage V_g with the STATCOM connected is shown in Figure 12.9(b). Since the STATCOM was operated in the V-mode from $t = 2$ s, the voltage V_g was maintained very well at 1 pu for the whole process. The STATCOM has rejected the effect of the extreme frequency change and has demonstrated excellent self-synchronization capability without a dedicated synchronization unit.

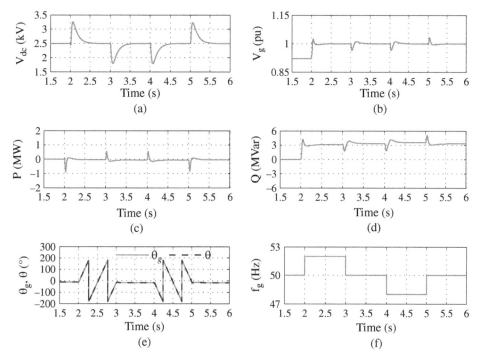

Figure 12.9 Simulation results of the STATCOM operated with a changing grid frequency. (a) V_{dc}. (b) V_g. (c) Real power. (d) Reactive power. (e) θ and θ_g. (f) Grid frequency.

12.4.4.2 With a Changing Grid Voltage

The purpose of this simulation is to demonstrate the voltage regulation capability of the STATCOM under a variable grid voltage. After connecting the STATCOM to the grid, the simulation was conducted in the following sequence of events:

(1) Switch to the V-mode at $t = 2$ s

(2) Change the source voltage so that V_g drops to about 0.9 pu (if the STATCOM is not connected) at $t = 2.5$ s

(3) Switch to the V_D-mode at $t = 3$ s

(4) Change the source voltage so that V_g increases to about 1.05 pu (if the STATCOM is not connected) at $t = 3.5$ s

(5) Switch back to the V-mode at $t = 4$ s

(6) Stop the simulation at $t = 4.5$ s.

The results are shown in Figure 12.10. The DC-bus voltage was maintained well and the phase difference δ was small. The voltage V_g with the STATCOM connected is shown in Figure 12.10(b) and the voltage V_g without the STATCOM connected is shown in Figure 12.10(f). The transitions between the V-mode and the Q-mode were very smooth with small overshoots. In the V_D-mode, there was a steady-state voltage error ΔV on the bus for the reason explained before, which allows the parallel operation of multiple STATCOMs. In the V-mode, the voltage V_g is regulated at the set-point of 1 pu without any steady-state error.

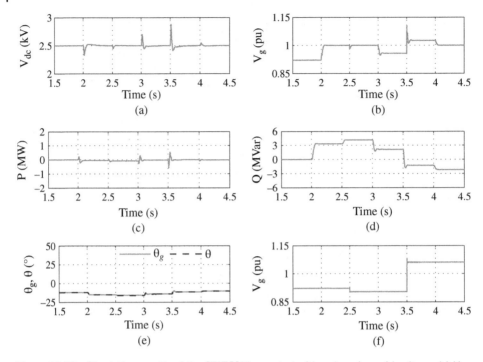

Figure 12.10 Simulation results of the STATCOM operated with a changing grid voltage. (a) V_{dc}. (b) V_g. (c) Real power. (d) Reactive power. (e) θ and θ_g. (f) V_g without the STATCOM.

12.4.4.3 With a Changing System Strength

The purpose of this simulation is to demonstrate the performance of the STATCOM when the system strength changes, e.g., when high power loads are connected or disconnected from the grid. Such operations may cause immediate phase changes at the PCC. If a PLL was used, there would be oscillations at the phase tracked, which causes the STATCOM response to be slow or even unstable (Norouzi and Sharaf 2005). This was avoided because no PLL was needed for the control strategy.

In this simulation, load 1 was connected at the beginning. It was switched OFF and ON to change the system strength. After connecting the STATCOM to the grid, the simulation was conducted in the following sequence of events:

(1) Switch to the V-mode at $t = 2$ s
(2) Switch load 1 OFF at $t = 2.5$ s
(3) Switch load 1 ON at $t = 3$ s
(4) Stop the simulation at $t = 3.5$ s.

The simulation results are shown in Figure 12.11. The voltage V_g is shown in Figure 12.11(b) when the STATCOM was connected and in Figure 12.11(f) when the STATCOM was not connected. The STATCOM was able to eliminate the effect of load changes on the voltage. V_g was quickly regulated to 1 pu without any problem although there were reasonable spikes at V_g when load 1 was connected/disconnected because of the high power changed on the bus. When the load was disconnected, V_g settled down

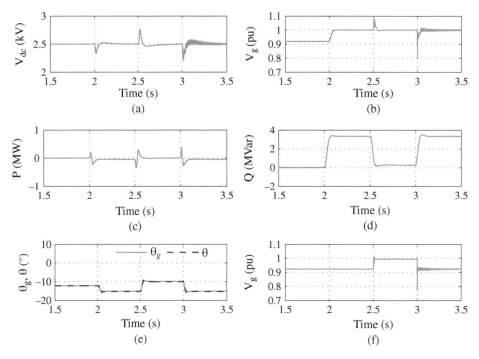

Figure 12.11 Simulation results with a variable system strength. (a) V_{dc}. (b) V_g. (c) Real power. (d) Reactive power. (e) θ and θ_g. (f) V_g without the STATCOM.

within about two cycles, which was very fast. When load 1 was connected, it caused small oscillations on V_g, which led to a longer settling time.

There was no problem for e to track the phase of v_g, as shown in Figure 12.11(e), when the phase of v_g changed. The STATCOM was able to quickly follow the change.

12.5 Summary

Base on (Nguyen et al. 2012), the synchronverter technology is applied to control STATCOM. As a result, a STATCOM is operated as a VSM. The controller is able to track the phase of the PCC voltage quickly and, hence, there is no need to use a PLL. The STATCOM controller also introduces a third operational mode, i.e. the voltage droop control mode or the V_D-mode in short, to the operation of STATCOM, in addition to the conventional direct Q control mode (the Q-mode) and the voltage regulation mode (the V-mode). Simulation results are presented to demonstrate the excellent performance of the control strategy without the need of a PLL under normal operation modes, under extreme conditions with wide variations in the grid frequency, in the grid voltage and in the system strength. The STATCOM is able to track the phase of the grid very well in all cases.

13

Synchronverters with Bounded Frequency and Voltage

Power systems are required to operate within tight ranges of frequencies and voltages. In this chapter, an improved synchronverter is presented to make sure that its frequency and voltage always stay within given ranges, while maintaining the original functions. Furthermore, its stability region is analytically characterized, which guarantees that the improved synchronverter is always stable and converges to a unique equilibrium as long as the power exchanged at the terminal is kept within this region. Extensive real-time simulation results are presented for the improved and the original self-synchronized synchronverters connected to a stiff grid to verify the theoretical development.

13.1 Introduction

It has been shown that a power electronic converter can be operated as a VSM. However, the stability of synchronverters and particularly maintaining both the voltage and the frequency within given ranges have not been established yet. This is not an easy task due to the non-linearity of the controller, e.g. the calculation of the real power and the reactive power, and the coupling between the frequency and the field-excitation current loops. Although local stability results of grid connected inverters can be provided using the small-signal analysis and linearization (Guo et al. 2014; Liu et al. 2016a; Paquette and Divan 2015; Pogaku et al. 2007; Wu et al. 2016), the non-linear dynamics of the system make non-linear analysis essential to achieve global stability results. A non-linear control strategy with a power-damping property was presented to guarantee non-linear system stability (Ashabani and Mohamed 2014), while requiring knowledge of the filter parameters. Several approaches for maintaining the stability of synchronous generators have been presented in the literature, which have the same dynamics with the synchronverter, but the field-excitation current is usually considered as constant (Fiaz et al. 2013; Natarajan and Weiss 2014). Adding a saturation unit is often adopted to maintain given bounds for the voltage and the frequency, but it often leads to instability due to the problem of integrator windup (Paquette and Divan 2015). To overcome this issue, anti-windup methods could be included in the controller to change the original operation but this can no longer guarantee system stability in the original form or require additional knowledge of the system structure and parameters (Bohn and Atherton 1995; Zaccarian and Teel 2004).

In this chapter, an improved version of the synchronverter connected to the grid with an LCL filter is presented. The non-linear model of the system is firstly derived using the

Power Electronics-Enabled Autonomous Power Systems: Next Generation Smart Grids,
First Edition. Qing-Chang Zhong.
© 2020 John Wiley & Sons Ltd. Published 2020 by John Wiley & Sons Ltd.

Kron-reduced network approach (Kundur 1994). Then, both the frequency loop and the field-excitation current loop are implemented by using the bounded integral controller in (Konstantopoulos et al. 2016b). The improved synchronverter approximates the behavior of the original synchronverter under normal operation (near the rated value) and guarantees given bounds for both the frequency and the voltage independently from each other, without the need of additional saturation units that complicates the proof of stability. Moreover, the region where a unique equilibrium exists is obtained and the convergence to the equilibrium is proven for the given voltage and frequency bounds. According to the analysis, the stability of the self-synchronized synchronverter discussed in Chapter 8 and (Zhong et al. 2014), where the synchronization unit is no longer required, is proven as well.

Extensive real-time simulation results are presented to compare the original and the improved self-synchronized synchronverters to verify the strategy under both normal and abnormal conditions (e.g., with errors in the measurement and sudden disturbances) as well as the case of two synchronverters connected to a common bus with one operating in the droop mode as a weak grid. The second scenario investigates both the stand-alone operation of the presented method as well as the operation with a weak grid.

13.2 Model of the Original Synchronverter

As shown in Chapter 4, the complete dynamic model of the synchronverter consists of a power part and a control part. The power part of the synchronverter consists of a three-phase inverter connected to the grid through an LCL filter. Using the Kron-reduced network approach (Kundur 1994), the node of the capacitor bank can be eliminated, which results in the per-phase system of the synchronverter connected to the grid, as shown in Figure 13.1. In this representation, the synchronverter and the grid are connected via a complex admittance $Y = G + jB$ with conductance G and susceptance B, while G_s, G_g and B_s, B_g are the shunt conductance and susceptance of the synchronverter and the grid, respectively. These values can be found using the star-delta transformation of the LCL filter. Thus, the real power and reactive power at the output of the synchronverter can be found as

$$P_s = 3(G_s + G)E^2 - 3EV_g(G\cos\delta + B\sin\delta) \tag{13.1}$$

$$Q_s = -3(B_s + B)E^2 - 3EV_g(G\sin\delta - B\cos\delta), \tag{13.2}$$

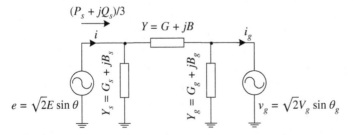

Figure 13.1 Per-phase diagram with the Kron-reduced network approach.

where the power angle $\delta = \theta - \theta_g$ is the phase difference between e and v_g, and it is often small, according to power systems theory (Kundur 1994).

The control part of the synchronverter consists of a frequency $\omega = \dot{\theta}$ loop and a field-excitation current $M_f i_f$ loop. The dynamics of the frequency ω are given by

$$\dot{\omega} = \frac{1}{J}(T_m - T_e) - \frac{D_p}{J}(\omega - \omega_r), \tag{13.3}$$

where T_e is the electromagnetic torque (hence $P_s = T_e \omega$), T_m is the mechanical torque corresponding to the desired real output power $P_{set} = T_m \omega_n$, the reference frequency ω_r is either equal to the grid frequency $\omega_g > 0$ or the rated frequency $\omega_n > 0$ when the frequency droop is disabled or enabled, respectively. Both J and D_p are positive constants.

The time constant of the frequency loop is $\tau_f = \frac{J}{D_p}$ so the inertia J is

$$J = D_p \tau_f, \tag{13.4}$$

where τ_f can be chosen similar or much smaller compared to the case of a physical synchronous generator. The dynamics of the field-excitation current i_f are given by

$$\dot{i}_f = \frac{1}{KM_f}(Q_{set} - Q_s) + \frac{D_q}{KM_f}(V_n - V_g), \tag{13.5}$$

where K and D_q are positive. Note that the capacitor voltage V_c, instead of V_g, can be used as well (Zhong and Weiss 2011). The time constant of the field excitation current loop is $\tau_v \approx \frac{K}{\omega_n D_q}$, which is often chosen much larger than τ_f, so the gain K can be obtained as

$$K = \omega_n D_q \tau_v. \tag{13.6}$$

The RMS phase output voltage of the synchronverter is

$$E = \frac{\omega M_f i_f}{\sqrt{2}}. \tag{13.7}$$

The complete dynamic model of the synchronverter is given by (13.3) and (13.5), together with (13.1), (13.2), and (13.7), taking into account also that $\dot{\delta} = \omega - \omega_g$.

13.3 Achieving Bounded Frequency and Voltage

In this section, an improved synchronverter is presented to maintain given bounds around the rated values for the voltage and the frequency at all times (transients, disturbances, etc.) and guarantee the stability of the closed-loop system. Particularly, a bounded dynamic controller is designed for the frequency and field-excitation dynamic loops to achieve the desired bounded performance without introducing any additional saturation units or suffering from integrator wind-up. These continuous-time bounded dynamics allow the investigation of the area of existence of a unique equilibrium point and facilitate the stability proof for convergence to the equilibrium.

13.3.1 Control Design

According to utility regulations, the frequency ω should be maintained within a range around the rated frequency ω_n, i.e. $\omega \in [\omega_n - \Delta\omega_{max}, \omega_n + \Delta\omega_{max}]$, where there is normally $\Delta\omega_{max} \ll \omega_n$. A common approach is to use a saturation unit at the output of the integrator (13.3) but this can cause integrator wind-up and instability (Paquette and Divan 2015). Here, the bounded integral controller (Konstantopoulos et al. 2016b) is modified to suit the needs of the frequency dynamics for the synchronverter.

13.3.1.1 Maintaining the Frequency within a Given Range

The frequency loop (13.3) is modified and implemented as

$$\dot{\omega} = -k\left(\frac{(\omega - \omega_n)^2}{\Delta\omega_{max}^2} + \omega_q^2 - 1\right)(\omega - \omega_n) + \omega_q^2\left(\frac{1}{J}(T_m - T_e) - \frac{D_p}{J}(\omega - \omega_r)\right) \quad (13.8)$$

$$\dot{\omega}_q = -k\left(\frac{(\omega - \omega_n)^2}{\Delta\omega_{max}^2} + \omega_q^2 - 1\right)\omega_q - \frac{\omega_q(\omega - \omega_n)}{\Delta\omega_{max}^2}\left(\frac{1}{J}(T_m - T_e) - \frac{D_p}{J}(\omega - \omega_r)\right)$$

$$\quad (13.9)$$

with the initial control states $\omega_0 = \omega_n$, $\omega_{q0} = 1$, and k being a positive constant gain.

In order to understand the controller (13.8)–(13.9), consider the Lyapunov function

$$W = \frac{(\omega - \omega_n)^2}{\Delta\omega_{max}^2} + \omega_q^2. \quad (13.10)$$

Taking the time derivative of W while considering (13.8)–(13.9), after some calculations, results in

$$\dot{W} = -2k\left(\frac{(\omega - \omega_n)^2}{\Delta\omega_{max}^2} + \omega_q^2 - 1\right)W.$$

Given the initial conditions $\omega_0 = \omega_n$ and $\omega_{q0} = 1$, there is

$$\dot{W} = 0 \Rightarrow W(t) = W(0) = 1, \; \forall t \geq 0.$$

Hence, ω and ω_q will start and stay thereafter on the ellipse

$$W_\omega = \left\{\omega, \omega_q \in R : \frac{(\omega - \omega_n)^2}{\Delta\omega_{max}^2} + \omega_q^2 = 1\right\}.$$

Note that the ellipse W_ω is centered at $(\omega_n, 0)$ on the ω–ω_q plane, as shown in Figure 13.2(a), which means the frequency is bounded within a range around the rated value,

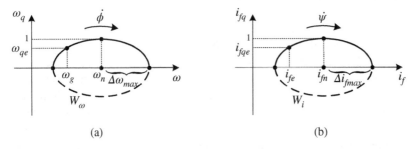

(a) (b)

Figure 13.2 Phase portraits of the controller. (a) The frequency dynamics. (b) The field-excitation current dynamics.

i.e. $\omega \in [\omega_n - \Delta\omega_{max}, \omega_n + \Delta\omega_{max}]$, independently of the field-excitation current i_f and the function $\frac{1}{J}(T_m - T_e) - \frac{D_p}{J}(\omega - \omega_r)$ that needs to be regulated to 0.

Using the mathematical transformation $\omega = \omega_n + \Delta\omega_{max} \sin\phi$ and $\omega_q = \cos\phi$ and taking into account that ω and ω_q operate on W_ω, then from (13.8) it is yielded that

$$\dot{\phi} = \frac{\omega_q}{\Delta\omega_{max}} \left(\frac{1}{J}(T_m - T_e) - \frac{D_p}{J}(\omega - \omega_r) \right), \tag{13.11}$$

which means that ω and ω_q travel on W_ω with angular velocity $\dot{\phi}$. Hence, when $\frac{1}{J}(T_m - T_e) - \frac{D_p}{J}(\omega - \omega_r) = 0$, as required at the steady state, there is $\dot{\phi} = 0$ and both controller states ω and ω_q converge at the equilibrium point (ω_g, ω_{qe}), as illustrated in Figure 13.2(a), assuming $\omega_{qe} > 0$. It should be highlighted that starting from point $(\omega_n, 1)$, both ω and ω_q are constrained on the upper semi-ellipse of W_ω, since if at any case (e.g. during transient), the trajectory tries to reach the horizontal axis, then $\omega_q \to 0$ and from (13.11) there is $\dot{\phi} \to 0$ independent of the non-linear expression $\frac{1}{J}(T_m - T_e) - \frac{D_p}{J}(\omega - \omega_r)$. This means that the controller states would slow down until the system reacts, changes the sign of the angular velocity from the term $\frac{1}{J}(T_m - T_e) - \frac{D_p}{J}(\omega - \omega_r)$, and forces them to converge to the desired equilibrium. Hence, no oscillation around the whole ellipse can occur, which is an important property.

Since ω and ω_q operate on the upper semi-ellipse of W_ω and there are $\omega \approx \omega_n$ and $\omega_q \approx 1$ around the nominal operational point, then equation (13.8) becomes

$$J\dot{\omega} = \omega_q^2((T_m - T_e) - D_p(\omega - \omega_r)) \approx T_m - T_e - D_p(\omega - \omega_r). \tag{13.12}$$

A direct comparison of this equation with (13.3) implies that the improved synchronverter approximates the dynamics of the original synchronverter around the nominal operational point, while additionally guaranteeing a given bound for the frequency at all times. Note also that when $\omega \to \omega_n \pm \Delta\omega_{max}$, i.e. the frequency tries to reach the upper or lower limits, then from (13.12) it holds true that $\dot{\omega} \to 0$, which means the integration slows down. Therefore, the controller (13.8)–(13.9) inherits an anti-wind-up structure in a continuous-time manner that facilitates the investigation of stability, while at the same time maintains the original performance of the synchronverter around the rated values.

13.3.1.2 Maintaining the Voltage within a Given Range

Utility regulations also require the RMS output voltage E be maintained within a range around the rated voltage V_n, i.e. $E \in [E_{min}, E_{max}] = [(1 - p_c)V_n, (1 + p_c)V_n]$, where p_c is often around 10%. In other words, according to (13.7), the condition

$$(1 - p_c)V_n \leq \frac{\omega M_f i_f}{\sqrt{2}} \leq (1 + p_c)V_n \tag{13.13}$$

should hold. Since the frequency ω is proven to satisfy $\omega \in [\omega_n - \Delta\omega_{max}, \omega_n + \Delta\omega_{max}]$ by (13.8), there is,

$$\frac{(1 - p_c)V_n\sqrt{2}}{(\omega_n + \Delta\omega_{max})M_f} \leq i_f \leq \frac{(1 + p_c)V_n\sqrt{2}}{(\omega_n - \Delta\omega_{max})M_f}. \tag{13.14}$$

This can be rewritten as

$$|i_f - i_{fn}| \leq \Delta i_{fmax}, \tag{13.15}$$

with

$$i_{fn} = \frac{V_n\sqrt{2}(\omega_n + p_c\Delta\omega_{max})}{M_f(\omega_n + \Delta\omega_{max})(\omega_n - \Delta\omega_{max})} \tag{13.16}$$

and

$$\Delta i_{fmax} = \frac{V_n\sqrt{2}(p_c\omega_n + \Delta\omega_{max})}{M_f(\omega_n + \Delta\omega_{max})(\omega_n - \Delta\omega_{max})}. \tag{13.17}$$

Since $\Delta\omega_{max} \ll \omega_n$ normally, there are

$$i_{fn} \approx \frac{V_n\sqrt{2}}{\omega_n M_f} \quad \text{and} \quad \Delta i_{fmax} \approx \frac{pV_n\sqrt{2}}{\omega_n M_f}.$$

In order to achieve (13.15), similarly to the frequency dynamics, the field-excitation loop (13.5) can be modified and implemented as

$$\dot{i}_f = -k\left(\frac{(i_f - i_{fn})^2}{\Delta i_{fmax}^2} + i_{fq}^2 - 1\right)(i_f - i_{fn}) + i_{fq}^2\left(\frac{Q_{set} - Q_s}{KM_f} + \frac{D_q(V_n - V_g)}{KM_f}\right) \tag{13.18}$$

$$\dot{i}_{fq} = -k\left(\frac{(i_f - i_{fn})^2}{\Delta i_{fmax}^2} + i_{fq}^2 - 1\right)i_{fq} - \frac{i_{fq}(i_f - i_{fn})}{\Delta i_{fmax}^2}\left(\frac{Q_{set} - Q_s}{KM_f} + \frac{D_q(V_n - V_g)}{KM_f}\right) \tag{13.19}$$

with initial control states $i_{f0} = i_{fn}$ and $i_{fq0} = 1$.

A similar Lyapunov analysis can show that the states i_f and i_{fq} would start and stay thereafter on the ellipse

$$W_i = \left\{i_f, i_{fq} \in R : \frac{(i_f - i_{fn})^2}{\Delta i_{fmax}^2} + i_{fq}^2 = 1\right\}$$

centered at $(i_{fn}, 0)$ on the i_f–i_{fq} plane, as shown in Figure 13.2(b), which means the field-excitation current is bounded within the range given by (13.15), independent of the frequency ω and the function $\frac{1}{KM_f}(Q_{set} - Q_s) + \frac{D_q}{KM_f}(V_n - V_g)$ that needs to be regulated to 0. In the same framework, i_f and i_{fq} travel only on the upper semi-ellipse of W_i with angular velocity

$$\dot{\psi} = \frac{i_{fq}}{\Delta i_{fmax}}\left(\frac{1}{KM_f}(Q_{set} - Q_s) + \frac{D_q}{KM_f}(V_n - V_g)\right),$$

as shown in Figure 13.2(b).

The above design has actually resulted in an improved synchronverter with its frequency ω and voltage E satisfying $\omega \in [\omega_n - \Delta\omega_{max}, \omega_n + \Delta\omega_{max}]$ and $E \in [(1 - p_c)V_n, (1 + p_c)V_n]$, respectively. These are crucial properties for guaranteeing the system stability.

Since the given bounds are established independently from the non-linear functions of the field-excitation current loop and the frequency loop, this fact also applies to the boundedness of the self-synchronized synchronverter discussed in Chapter 8, which no longer requires a PLL. This can be achieved, as shown in Figure 13.3, by replacing the dynamics of the frequency loop and the field-excitation current loop of the self-synchronized synchronverter with the control laws (13.8)–(13.9) and (13.18)–(13.19), respectively. Since the dynamics (13.8)–(13.9) and (13.18)–(13.19) introduce bounded outputs independently from

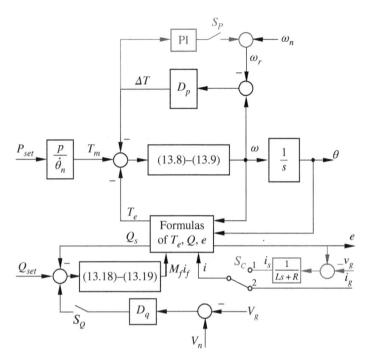

Figure 13.3 The controller to achieve bounded frequency and voltage.

the inputs (zero gain property) (Konstantopoulos et al. 2016b), the voltage and frequency bounds are guaranteed independently of the PI control block at the frequency loop or the first-order system used to create the virtual current i_s shown in Figure 13.3.

13.3.2 Existence of a Unique Equilibrium

The improved synchronverter has a bounded closed-loop solution for the frequency and the voltage (resulting from the field excitation current). Here, the existence of a unique equilibrium (with frequency ω_e and voltage E_e) and the convergence to this point will be shown analytically.

13.3.2.1 Theoretical Analysis

Assume that the grid is stiff with a constant grid voltage V_g and a constant frequency ω_g to facilitate the analysis. Then, at the steady state, there should be $\omega_e = \omega_g$ because the frequency of the complete system should be the same. It has been shown that the control laws do not change the synchronverter operation at the steady state under normal operation. Then, from the equations of the improved synchronverter control part at the steady state, i.e. (13.18)–(13.19) and (13.8)–(13.9) or (13.5) and (13.3) respectively, the real power P_s and reactive power Q_s delivered by the synchronverter are

$$P_s = \frac{\omega_g}{\omega_n} P_{set} - D_p \omega_g (\omega_g - \omega_r), \tag{13.20}$$

$$Q_s = Q_{set} + D_q (V_n - V_g), \tag{13.21}$$

which are constant for given constant references P_{set}, Q_{set}.

From the controller operation, it is guaranteed that $\omega \in [\omega_n - \Delta\omega_{max}, \omega_n + \Delta\omega_{max}]$. Since $\omega_e = \omega_g$ at the steady state, the maximum frequency deviation $\Delta\omega_{max} = 2\pi\Delta f_{max}$ should be selected so that the grid frequency, which usually slightly deviates from ω_n, falls into the range $[\omega_n - \Delta\omega_{max}, \omega_n + \Delta\omega_{max}]$, although a smaller Δf_{max} can guarantee a tighter frequency bound. Usually, it is enough to choose $\Delta f_{max} = 0.5$Hz.

Moreover, the steady-state value of the voltage E_e, resulting from the field excitation current value i_{fe} and the frequency ω_e from (13.7), should be unique and remain inside the given range $E \in [E_{min}, E_{max}] = [(1 - p_c)V_n, (1 + p_c)V_n]$. To this end, rewrite (13.1) and (13.2) as

$$P_s - 3(G_s + G)E^2 = -3EV_g(G\cos\delta + B\sin\delta), \tag{13.22}$$

$$Q_s + 3(B_s + B)E^2 = -3EV_g(G\sin\delta - B\cos\delta). \tag{13.23}$$

Taking the sum of the squares of (13.22) and (13.23), then it yields

$$(P_s - 3(G_s + G)E^2)^2 + (Q_s + 3(B_s + B)E^2)^2 = 9E^2V_g^2(B^2 + G^2), \tag{13.24}$$

which results in the following second order equation of E^2 with respect to P_s and Q_s:

$$P_s^2 + Q_s^2 - (6P_s(G_s + G) - 6Q_s(B_s + B) + 9V_g^2(B^2 + G^2))E^2$$
$$+ 9((G_s + G)^2 + (B_s + B)^2)E^4 = 0. \tag{13.25}$$

As a result,

$$E^2 = \frac{2\gamma P_s - 2\eta Q_s + 3\alpha V_g^2}{6\beta} \pm \frac{\sqrt{\Delta}}{6\beta}, \tag{13.26}$$

where

$$\Delta = -4(\gamma P_s + \eta Q_s)^2 + \alpha V_g^2(12\gamma P_s - 12\eta Q_s + 9\alpha V_g^2) \geq 0, \tag{13.27}$$

in order to obtain a real solution E_e. Here, $\alpha = B^2 + G^2$, $\gamma = G_s + G$ and $\eta = B_s + B$ with $\beta = \gamma^2 + \eta^2$. Note that for a typical LCL filter there is $\gamma > 0$ and $\eta < 0$. Since $\Delta \geq 0$, then if

$$\frac{2\gamma P_s - 2\eta Q_s + 3\alpha V_g^2}{6\beta} > 0, \tag{13.28}$$

the solution with the $+$ sign, denoted as E_+^2, is positive and hence E_+ exists. The negative one $(-E_+)$ is not of interest and can be ignored. In order for E_+ to fall into the given range, there should be

$$(1 - p_c)^2V_n^2 \leq \frac{2\gamma P_s - 2\eta Q_s + 3\alpha V_g^2}{6\beta} + \frac{\sqrt{\Delta}}{6\beta} \leq (1 + p_c)^2V_n^2. \tag{13.29}$$

Since a unique solution is required in the given range, then if

$$0 < \frac{2\gamma P_s - 2\eta Q_s + 3\alpha V_g^2}{6\beta} \leq (1 - p_c)^2V_n^2, \tag{13.30}$$

which includes inequality (13.28), then the solution with the $-$ sign, denoted as E_-^2, satisfies

$$E_-^2 \leq (1 - p_c)^2V_n^2.$$

Hence, if E_- exists then it would be outside of the range. As a result, under conditions (13.27), (13.29), and (13.30), there exists a unique equilibrium E_e inside the given range $[(1 - p_c)V_n, (1 + p_c)V_n]$ with $0 \leq p_c < 1$ for the synchronverter voltage E and it is

$$E_e = E_+ = \sqrt{\frac{2\gamma P_s - 2\eta Q_s + 3\alpha V_g^2}{6\beta} + \frac{\sqrt{\Delta}}{6\beta}}.$$

Note that, when p_c is large, from (13.30) this may result in an area on the $P_s - Q_s$ plane that does not contain the origin $P_s = Q_s = 0$ at which $E_+ = V_g\sqrt{\frac{\alpha}{\beta}}$ and $E_- = 0$. Practically, the origin should be included to represent the operation before connecting to the grid. For $P_s = Q_s = 0$, the inequality (13.30) can be simplified as

$$0 \leq p_c \leq 1 - \frac{V_g}{V_n}\sqrt{\frac{\alpha}{2\beta}} < 1, \tag{13.31}$$

which provides a maximum practical value for p_c.

Since $P_s = Q_s = 0$ results in $E_+ = V_g\sqrt{\frac{\alpha}{\beta}}$, this also gives important information for the LCL filter design, i.e. there should be $\frac{\alpha}{\beta} \approx 1$ in order to have a smooth connection with the grid ($E_+ \approx V_g$). For the parameters in Table 13.1, $\frac{\alpha}{\beta} = 1.0006$. Indeed this is the case.

13.3.2.2 A Numerical Example

According to conditions (13.27), (13.29), and (13.30), for a given voltage range, the unique solution E_e of the synchronverter voltage inside the voltage range can be calculated from the values of P_s and Q_s. Then the area where there exists a unique equilibrium can be plotted on the P_s-Q_s plane. In order to demonstrate this further, the system with parameters given in Table 13.1 is taken as an example. Both solutions E_+ and E_- are plotted for different values of P_s and Q_s and shown in Figure 13.4. The white curve between the surfaces of E_+ and E_- defines the values of P_s and Q_s for which $E_+ = E_-$, i.e. $\Delta = 0$.

The contour curves of the surface are shown in Figure 13.5 on the P_s-Q_s plane for $p_c = 10\%$ around the rated voltage V_n. Note that $p_c = 10\%$ satisfies (13.31), which gives $p_c \leq 0.2927$ when $V_g = V_n$. The areas characterized by E_+ and E_- within $(0.9-1.1)V_n$ can be clearly seen. Conditions (13.27), (13.29), and (13.30) characterize the area with a unique equilibrium point $E_+ = E_e$, which is inside the area characterized by E_+ lines and below the dashed line in Figure 13.5. The area where E_- exists is excluded by the gray dashed line because it would lead to excessive power that exceeds the capacity of the synchronverter.

Table 13.1 Parameters of a synchronverter.

Parameter	Value	Parameter	Value
L_s	0.15 mH	L_g	0.15 mH
R_s	0.045 Ω	R_g	0.045 Ω
C	22 μF	Nominal frequency	50 Hz
R (parallel to C)	1000 Ω	V_n	12 Vrms
Rated power	100 VA	DC-link voltage	42 V

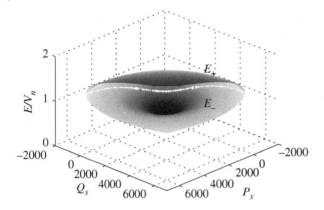

Figure 13.4 F_+ surface (upper) and E_- surface (lower) with respect to P_s and Q_s.

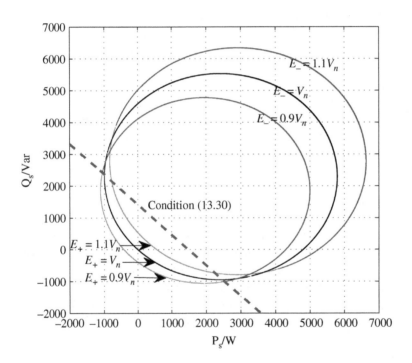

Figure 13.5 Illustration of the areas characterized by E_+ lines and E_- lines.

The area where there exists a unique equilibrium E_e inside the voltage range is zoomed in and shown in Figure 13.6(a) for two different voltage ranges with $p_c = 0.05$ and $p_c = 0.1$, i.e. 5% or 10% around the rated voltage V_n, respectively, when the grid voltage is equal to the rated voltage ($V_g = V_n$). When the grid voltage is 5% lower than the rated voltage, the area where there exists a unique equilibrium E_e inside the voltage range is shown in Figure 13.6(b). The area is shifted towards the first quadrant, which means for the same voltage range more power could be sent out but less power could be drawn. This is expected and reasonable. When the grid voltage is 5% higher than the rated voltage, the area shifts towards the third quadrant, as shown in Figure 13.6(c). Hence, in practice, when determining the capacity of the synchronverter the maximum variation of the grid voltage should be considered as well.

13.3.3 Convergence to the Equilibrium

Based on the analysis of Sections 13.3.1 and 13.3.2 with P_s and Q_s satisfying conditions (13.27), (13.29), and (13.30), it has been shown that by using the improved synchronverter (13.8)–(13.9) and (13.18)–(13.19), it is guaranteed that the synchronverter connected to the grid results in a bounded voltage $E(t)$ and frequency $\omega(t)$ in a given range where a unique equilibrium (E_e, ω_e) exists. Additionally, by choosing $\tau_f \ll \tau_v$, then according to (Khalil 2001, chapter 11), the frequency and field excitation current dynamics can be viewed as a two-time-scale dynamic system where the frequency dynamics are fast and the field-excitation dynamics are slow with respect to each other. The stability of these dynamics can be shown below.

13.3.3.1 Stability of the Fast Dynamics
According to (13.3), (13.1), and (13.7), the fast frequency dynamics around the equilibrium point (δ_e, ω_g) are given as

$$\begin{bmatrix} \Delta\dot{\delta} \\ \Delta\dot{\omega} \end{bmatrix} = \begin{bmatrix} 0 & 1 \\ \frac{3M_f i_f V_g (B\cos\delta_e - G\sin\delta_e)}{\sqrt{2}J} & -\left(\frac{D_p}{J} + \frac{3(G_s+G)M_f^2 i_f^2}{2J} \right) \end{bmatrix} \begin{bmatrix} \Delta\delta \\ \Delta\omega \end{bmatrix}, \tag{13.32}$$

because according to the two-time-scale analysis and the boundedness of i_f from (13.18), $i_f > 0$ and can be regarded as a constant with respect to ω. Hence, for $\delta_e \in \left(\tan^{-1}\left(\frac{B}{G}\right), \frac{\pi}{2} \right]$, the equilibrium point of the fast frequency dynamic system (13.32) can be easily proven to be asymptotically stable uniformly in i_f because i_f is restricted in a positive set. Note that $B < 0$ for typical *LCL* filters. Based on the fact that the bounded system (13.8) maintains the asymptotic behaviour of the original system (Konstantopoulos et al. 2016b) with a given tight bound for the frequency ω, the equilibrium point of the fast frequency dynamics (13.8) is asymptotically stable since ω, $\omega_g \in [\omega_n - \Delta\omega_{max}, \omega_n + \Delta\omega_{max}]$, i.e. the equilibrium point exists inside the given bound.

13.3.3.2 Stability of the Slow Dynamics
Once the frequency settles down quickly, according to (13.32), there is $\dot{\delta} = 0$ and $\dot{\omega} = 0$. As a result, $\omega = \omega_g$. For the field excitation dynamics (13.18)–(13.19), the slow dynamics of i_f in (13.18)–(13.19) results in an autonomous system $[i_f \ i_{qf}]^T = f(i_f, i_{qf})$ having the same equilibrium point as the original non-linear system, because it is unique inside the bounded range. The structure of the field excitation current (13.18)–(13.19) prohibits the existence of limit cycles across the whole closed curve W_i, as explained in Section 13.3.1, and the fact that i_f and i_{fq} operate on the ellipse W_i and stay exclusively above the horizontal axis, i.e. $i_{fq} \geq 0$. If i_f and i_{fq} pass the equilibrium point and try to reach the horizontal axis, as shown in Figure 13.2(b), their angular velocity approaches zero, i.e. they slow down until the system acts and changes the sign of the angular velocity, forcing the states to oscillate around the equilibrium point and not continuously oscillate around the whole ellipse W_i. As a

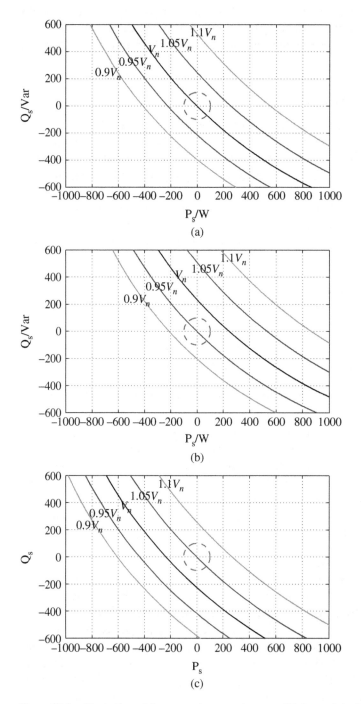

Figure 13.6 Illustration of the area where a unique equilibrium exists. (a) When the grid voltage is at the rated value. (b) When the grid voltage is 5% lower than the rated value. (c) When the grid voltage is 5% higher than the rated value.

result, the field-excitation current dynamics (13.18)–(13.19) is described by a second-order autonomous system that cannot have a periodic solution, corresponding to a closed orbit on the i_f–i_{fq} plane. Additionally, no chaotic solution exists according to the Poincaré–Bendixon theorem (Wiggins 2003) and the solution of the system asymptotically converges to the unique equilibrium point (i_{fe}, i_{fqe}) corresponding to the desired equilibrium of the system with (E_e, ω_e) (Khalil 2001).

As a result, for $\tau_f \ll \tau_v$ and for P_s and Q_s given from (13.20)–(13.21) and satisfying conditions (13.27), (13.29), and (13.30), which can be achieved from the synchronverter design, the improved grid-connected synchronverter is stable with given bounds for the voltage $E \in [(1 - p_c)V_n, (1 + p_c)V_n]$ and the frequency $\omega \in [\omega_n - \Delta\omega_{max}, \omega_n + \Delta\omega_{max}]$, and asymptotically converges to a unique equilibrium point, when $\omega_g \in [\omega_n - \Delta\omega_{max}, \omega_n + \Delta\omega_{max}]$ and p_c satisfies (13.31).

In practice, the capacity of a synchronverter, which can be represented as a circle centered at the origin of the P_s–Q_s plane, is limited and pre-defined at the design stage. If this circle, e.g. the dashed ones shown in Figure 13.6, falls into the area where a unique equilibrium exists, then the stability of the synchronverter is always guaranteed within the voltage range. If the real power and/or the reactive power of the synchronverter exceed the circle but still remain inside the bounded voltage area then the synchronverter is still stable but it could damage itself. In order to avoid damage due to overloading, the power handled by the synchronverter should be limited, e.g. to 125% for 10 min, 150% surge for 10 s. This is an excellent property that could be adopted to enhance the fault-ride through capability of grid-connected inverters, e.g. when there is a need to send (controlled) reactive power to the grid in case of a fault on the grid and to maintain the safe operation of power systems. This offers the potential for all synchronverters to work at the maximum capacity without causing instability.

13.4 Real-time Simulation Results

Real-time simulations are carried out to compare the original grid-tied self-synchronized synchronverter and the improved self-synchronized synchronverter shown in Figure 13.3. The system parameters are the same as those shown in Table 13.1. The controller parameters are chosen as $k = 1000$, $\Delta i_{fmax} = 0.15 i_{fn}$ (15% difference of the rated value) and $\Delta\omega_{max} = \pi$ rad s^{-1}, i.e. $\Delta f_{max} = 0.5$ Hz. This leads to $E_{max} \approx 1.15V_n$ and $E_{min} \approx 0.85V_n$ as the upper and lower bounds for the synchronverter voltage.

The system starts operating in the self-synchronization mode with $P_{set} = 0$ and $Q_{set} = 0$ at $t = 5$ s, with the switch S_C (Figure 13.3) set at position 1, S_P turned ON, S_Q turned OFF and the circuit breaker turned OFF. The initial transient observed is due to the initial conditions and the calculation of the amplitude but it does not affect the system since both synchronverters have not been connected to the grid yet. Note that the grid voltage is set to be 2% higher than the rated value in order to test the case where the grid voltage differs from the rated voltage. As can be seen from Figure 13.7(b), during the self-synchronization mode, the synchronverter frequency is quickly synchronized with the grid frequency. At $t = 6$ s, the circuit breaker is turned ON, thus connecting the synchronverter to the grid, and S_C is changed to position 2. Very little transient is observed. At $t = 10$ s, the real power

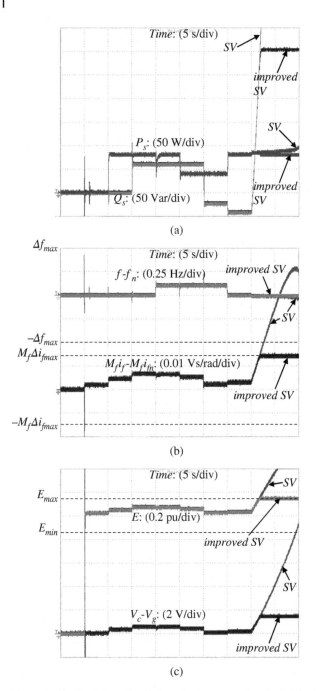

Figure 13.7 Real-time simulation results comparing the original (SV) with the improved self-synchronized synchronverter (improved SV). (a) Real power P_s and reactive power Q_s. (b) Frequency f and field-excitation current ($M_f i_f$). (c) Voltage E (normalized) and the amplitude $V_c - V_g$.

reference is changed to P_{set} = 80 W and at t = 15 s, the reactive power reference is changed to Q_{set} = 60 Var. Both types of synchronverters respond to the reference values well, as shown in Figure 13.7(a). At t = 20 s, a 0.2% step increase at the grid frequency f_g, i.e. from 50 Hz to 50.1 Hz, is assumed. Both types of synchronverters respond to the frequency change well. The transient change in the real power is small. At t = 25 s, switch S_P is turned OFF to enable the frequency droop mode, which leads to a drop of the real power P_s. Switch S_Q is turned ON at t = 30 s to enable the voltage droop mode, which results in a drop of the reactive power Q_s. This forces the synchronverter voltage to regulate closer to the rated value (1 pu) as is clearly shown in the normalized voltage amplitude E in Figure 13.7(c). At t = 35 s, the grid frequency is changed back to 50 Hz. Both synchronverters behave similarly to increase the real power. Actually, during the first 40 s, both the original and the improved synchronverter have behaved almost the same, which verifies the fact that the improved synchronverter maintains the original performance during normal operation. At t = 40 s, an increasing error is assumed at the grid voltage sensor with the rate of 10%/s^{-1}, i.e. the measured voltage becomes 10%/s^{-1} less than its actual value, which constitutes an abnormal operation and forces the power to increase, leading the original synchronverter to instability. As shown in Figure 13.7(b), the frequency of the original synchronverter starts diverging away from the grid frequency, and the field-excitation current increases dramatically. On the other hand, the frequency and the field-excitation current of the improved synchronverter remain inside the given ranges as expected. In particular, the field excitation current and the voltage smoothly converge to their upper bounds, as shown in Figure 13.7(b) and 13.7(c), respectively. The frequency of the improved synchronverter is maintained equal to the grid frequency, opposed to the original synchronverter, and the reactive power converges to the upper limit as described in Figure 13.6 due to the convergence of the field-excitation current to the upper bound. Hence, both the frequency and the field-excitation current remain inside their given bounds at all times.

In order to further verify the theory, the trajectories of the control states $M_f i_f$, $M_f i_{fq}$ and $f = \frac{\omega}{2\pi}, f_q = \omega_q$ are shown in Figure 13.8 using the data from the real-time simulations. It is shown that they indeed stay on the desired ellipses as explained in Section 13.3.1.

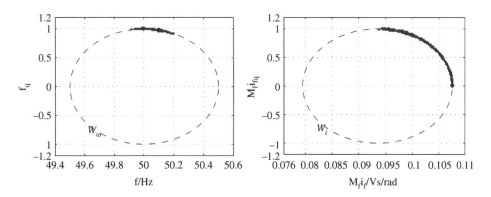

Figure 13.8 Phase portraits of the controller states in real-time simulations. (a) The frequency dynamics. (b) The field-excitation current dynamics.

13.5 Summary

Based on (Konstantopoulos et al. 2015; Zhong et al. 2018a), an improved synchronverter is presented to bound its frequency and voltage within given ranges. It is proven that the original behavior of the synchronverter is maintained near the rated value and additional given bounds are guaranteed for the frequency and the field-excitation current (voltage). Moreover, the stability and convergence to the unique equilibrium are established for a given voltage range using the non-linear model description, under the assumption of constant grid voltage and frequency. The complete region where the unique equilibrium exists is characterized. The improved synchronverter is always stable as long as the real power and the reactive power do not exceed this region.

Although the assumption of constant grid voltage and frequency is made for the purpose of deriving the existence of an equilibrium point, the impact of grid voltage and frequency variations seems insignificant from numerical examples.

14

Virtual Inertia, Virtual Damping, and Fault Ride-through

Virtual synchronous machine (VSM) technology enables friendly grid integration of distributed energy resources through the provision of virtual inertia. This requires a certain amount of energy available at a certain speed of response. In this chapter, it is shown at first that the concept of inertia has two aspects of meaning: the inertia time constant that characterizes the speed of the frequency response and the inertia constant that characterizes the amount of energy stored. It is then shown that, while the energy storage aspect of the virtual inertia of a VSM can be met by storage units, the inertia time constant that can be provided by a VSM may be limited because a large inertia time constant may lead to oscillatory frequency responses. A VSM with a reconfigurable inertia time constant is then introduced by adding a low-pass filter to the real power channel. Moreover, a virtual damper is introduced to provide the desired damping ratio, e.g., 0.707, together with the desired inertia time constant. Two approaches are presented to implement the virtual damper: one through impedance scaling with a voltage feedback controller and the other through impedance insertion with a current feedback controller. A by-product from this is that the fault ride-through capability of the VSM can be designed as well. Simulation and experimental results are presented to demonstrate these features. The technology is patent pending at the time when this book goes for publication and is expected to be patented.

14.1 Introduction

Power systems are going through a paradigm change from centralized generation to distributed generation (Zhong, 2017e,f). More and more distributed energy resources (DERs), including renewables, electric vehicles, and energy storage systems, are being integrated into the grid (Chen et al. 2017; Pogaku et al. 2007; Xie et al. 2017). The integration of DERs presents unprecedented challenges and has become an important research topic in recent years (Ahrens 2017; Lu et al. 2015; Mcelroy and Chen 2017; Pinson et al. 2017; Zhong 2017e). One challenge is that a DER unit often means low inertia or even no inertia (Guerrero et al. 2004). The large-scale utilization of DERs would cause a significant decrease of inertia (Lalor et al. 2005; Morren et al. 2006b; Olivares et al. 2014), which brings considerable concerns about grid stability (Dag and Mirafzal 2016; Tielens and Hertem 2017; Winter et al. 2015), because inertia has been regarded as a critical factor affecting grid stability.

Power Electronics-Enabled Autonomous Power Systems: Next Generation Smart Grids,
First Edition. Qing-Chang Zhong.
© 2020 John Wiley & Sons Ltd. Published 2020 by John Wiley & Sons Ltd.

Since synchronous machines (SMs) can provide large inertia because of the large kinetic energy stored in the rotors (Kundur 1994), a lot of effort has been expended in recent years to provide additional energy when needed to mimic the inertia. For example, in (Delille et al. 2012; Zhao et al. 2019), a fast-response battery energy storage system is adopted to inject additional power when needed. In (Im et al. 2016), the inertia of a PV system is increased by adjusting the DC-link voltage and the PV array output. In (Ekanayake and Jenkins 2004; Morren et al. 2006a; Zhao et al. 2016), the kinetic energy stored in the rotor of a wind turbine is utilized for wind plants to participate in system frequency regulation.

Another important trend, as the focus of this book, is to operate power electronic converters in DER units as a VSM, which emulate the major features, including the inertia, of traditional SMs. The inertia of a VSM is realized by the virtual moment of inertia in the swing equation. This is directly linked to the well known energy storage aspect of the inertia. Another less well known aspect of inertia is that it indicates or affects the speed of frequency responses.

Unlike the moment of inertia in a traditional SM, which is inherent from the rotating mass and proportional to the machine nominal power (Elizondo and Kirtley 2014; Kundur 1994), the VSM virtual moment of inertia in the swing equation is a control parameter, which affects the speed of the frequency response, and it should be designed to guarantee stability and performance (Zhong and Weiss 2011). The trade-off in control parameter design may limit the range of the VSM inertia (Wu et al. 2016; Zhong and Nguyen 2012; Zhong et al. 2014). Another important parameter related to SMs is called damping, which is the capability of reducing, restricting or preventing oscillations in system responses (Ogata and Yang 1970). For a physical SM, the damping is often provided by damper windings and viscous friction (Kundur 1994). While it is possible to copy the exact model of an SM, including the damping terms when implementing a VSM, this increases the complexity of the controller. In an synchronverter, the frequency droop coefficient plays the role of the damping (Zhong and Weiss 2011), which provides excellent frequency regulation performance but may cause oscillatory frequency responses when the inertia is large.

In this chapter, it is shown at first that the inertia has two different aspects, i.e. the speed of the frequency response and the amount of energy stored. The amount of the energy stored can be provided by an energy storage system on the DC bus while the speed of the frequency response can be adjusted by designing the controller. Based on this, a VSM with reconfigurable inertia time constant and damping is then described. As a result, the inertia time constant of a VSM can be reconfigured in a large range while maintaining a desired damping ratio, as long as there is enough energy stored in the DC bus. This makes it possible for VSMs to offer inertia that is comparable to or even larger than conventional SMs.

14.2 Inertia, the Inertia Time Constant, and the Inertia Constant

As is well known from Newton's first law of motion (Newton 2014), an object has a natural tendency to resist changes in its state of motion. Inertia is the natural resistance an object has to a change in its state of motion. In linear kinetics, the inertia of an object is its mass m and is characterized by the Newton's second law $F = ma$ after measuring the acceleration *a* resulting from a net external force *F*. In rotational kinetics, the term of rotational inertia, widely known as the moment of inertia, is used. The moment of inertia of a rigid body depends on the mass distribution of the body and the rotational axis chosen. For example, the moment of inertia *J* of a point mass *m* is the mass times the square of the perpendicular distance *r* to the rotation axis, i.e. $J = mr^2$. The longer the perpendicular distance *r*, the larger the moment of inertia *J*. For example, spinning skaters often take advantage of this to reduce their moment of inertia by pulling in their arms, allowing them to spin faster. Tightrope walkers often use a long rod to increase the moment of inertia of the body for enhanced balance when walking along the rope. The moment of inertia of a rigid composite system is the sum of the moments of inertia of its component subsystems with respect to the same axis. It can be defined as the second moment of mass with respect to the distance *r* from the axis, i.e. $J = \int r^2 dm$ integrating over the entire mass.

The moment of inertia *J*, usually called the inertia for simplicity, characterizes the ratio of the amount of applied torque *T* to the resulting angular acceleration (the rate of change in the angular velocity ω), i.e.

$$J\frac{d\omega}{dt} = T.$$

The inertia *J* of an SM consists of the combined moments of inertia of the generator and the prime mover that rotate together. It satisfies (Kundur 1994)

$$J\frac{d\omega}{dt} = T_m - T_e - D_p\omega, \tag{14.1}$$

where D_p is the damping factor derived from the damping torque of the rotor and affected by the mechanical friction, the stator loss and the damping winding; T_m and T_e are the mechanical torque and the electromagnetic torque, respectively, and ω is the rotational speed or the angular velocity of the machine. Without loss of generality, it can be assumed that ω is also the frequency of the electricity generated under the assumption that the pair of poles per phase for the magnetic field is one. The equation (14.1) is often called the swing equation (Kundur 1994). When a synchronous machine works within a tight range around the rated frequency ω_n, the swing equation can also be written in a form involving the active power (Du et al. 2013b; Guan et al. 2015; Soni et al. 2013) as

$$J\frac{d\omega}{dt} = \frac{P_m - P_e}{\omega_n} - D_p\omega, \tag{14.2}$$

where P_m and P_e are the mechanical power and electrical power, respectively. Denote $\Delta P = P_e - P_m$ and the corresponding frequency change as $\Delta\omega$. Then, there is

$$\frac{\Delta\omega}{\Delta P} = -\frac{1}{\omega_n D_p} \cdot \frac{1}{\tau_\omega s + 1}, \tag{14.3}$$

where $\tau_\omega = \frac{J}{D_p}$ is the time constant of the transfer function. It is called the *inertia time constant* for easy references and it determines the response of the motion, which is the frequency ω in this case, with respect to the change of the torque (or power). The larger the moment of inertia *J*, the larger the inertia time constant τ_ω, the slower the response, and the more the power (torque) needed to speed up or slow down the system.

Note that the inertia time constant τ_ω is the time constant of the transfer function from the change of power (torque) to the change of frequency and it characterizes the speed of the frequency response. It is different from the widely-used *inertia constant H* (Kundur 1994) defined as

$$H = \frac{\text{kinetic energy } \frac{1}{2}J\omega_n^2 \text{ of the rotating masses}}{\text{VA rating } S_n},$$

which reflects the time needed to fully discharge the stored kinetic energy at the rated power. It characterizes the amount of energy stored. For a conventional SM, H is in the order of several seconds.

Both the inertia time constant τ_ω and the inertia constant H are related to the inertia J but they reflect two different aspects of the inertia J: the time constant or speed of the frequency response and the amount of the kinetic energy $\frac{1}{2}J\omega_n^2$ stored in the machine, respectively. On one hand, the smaller the inertia J, the smaller the inertia time constant τ_ω and the faster the frequency response. On the other hand, the larger the inertia J, the more the energy stored and the larger the frequency disturbances allowed. For a physical SM, these two aspects are normally coupled together because both the inertia time constant $\tau_\omega = \frac{J}{D_p}$ and the inertia constant H (or the kinetic energy $\frac{1}{2}J\omega_n^2$) are directly in proportion to the same inertia J. However, for a VSM, these two aspects are decoupled (Zhong and Weiss 2011), which is an advantage of VSMs over conventional SMs. The amount of the energy stored can be provided by an energy storage system on the DC bus. At the same time, the speed, i.e. the inertia time constant, of the frequency response can be changed by designing the control algorithm. Since there is no technical limit on the capacity of energy storage systems that can be added, there is no limit on the potential inertia constant technically. In practice, the capacity of the energy storage system is limited for economic reasons but this does not cause a problem for the purpose of providing inertia comparable to conventional SM. For example, for a 600 kW VSM to provide the inertia at the full power for 9 s, the required energy to be stored is only $600 \times 9/3600 = 1.5$ kWh. This can be easily met by using 16 of 12 V 16Ah batteries, taking into account the depth of discharge of 50%. The amount of energy stored can be increased without considering the inertia time constant as long as it is economically viable. As a result, the inertia time constant can be designed separately without considering the energy stored as long as there is enough energy storage available. This chapter focuses on the time constant aspect of the inertia, i.e. the speed of the frequency response, with the assumption that the inverter has sufficient energy storage. For this reason, the inverters in the simulation and experimental sections have a constant DC-bus voltage.

14.3 Limitation of the Inertia of a Synchronverter

In order to facilitate the presentation, the controller of the synchronverter discussed in Chapter 4 is re-drawn in Figure 14.1 with some notation changes.

The controller has two channels: the active power (torque)-frequency channel and the reactive power-voltage channel. The damping factor D_p in the swing equation actually plays the role of the frequency droop control and hence it is called the frequency droop coefficient. The term D_q serves the purpose of voltage droop control. The integrator $\frac{1}{Ks}$ generates the field excitation $M_f i_f$. In the steady state, there are

$$\omega = \omega_{\text{ref}} + \frac{1}{D_p}(T_m - T_e),$$

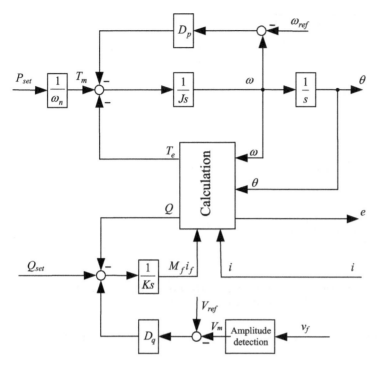

Figure 14.1 The controller of the original synchronverter.

$$V_{\mathrm{m}} = V_{\mathrm{ref}} + \frac{1}{D_{\mathrm{q}}}(Q_{\mathrm{set}} - Q).$$

As a result, a synchronverter is able to take part in the regulation of system frequency and voltage. It was shown in Chapter 8 and in (Zhong et al. 2014) that the commonly needed phase-locked loop can be removed so it is not shown in the controller.

The frequency droop coefficient D_{p} is defined in the same way as in Chapter 4, i .e. for the frequency drop of α (%) to cause the torque (hence, the active power) increase of 100%. Then,

$$D_{\mathrm{p}} = \frac{S_{\mathrm{n}}/\omega_{\mathrm{n}}}{\alpha\omega_{\mathrm{n}}} = \frac{S_{\mathrm{n}}}{\alpha\omega_{\mathrm{n}}^2}, \tag{14.4}$$

where ω_{n} is the rated angular frequency and S_{n} is the rated power. Similarly, the voltage droop coefficient D_{q} is defined for the voltage drop of β (%) to cause the reactive power increase of 100% as

$$D_{\mathrm{q}} = \frac{S_{\mathrm{n}}}{\beta\sqrt{2}V_{\mathrm{n}}}, \tag{14.5}$$

where V_{n} is the rated RMS phase voltage. The $\sqrt{2}$ is due to the fact that the voltage amplitude V_{m} instead of the RMS voltage is fed back.

Following the small-signal analysis carried out in (Wu et al. 2016), the two channels can be decoupled by design under some conditions. The resulting active power-frequency channel can be illustrated as shown in Figure 14.2(a). The gain K_{pd} reflects the amplification

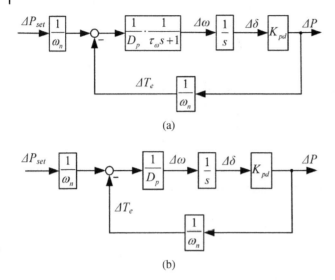

Figure 14.2 Active power regulation in a conventional synchronverter after decoupling. (a) The whole active power channel. (b) The simplified active power channel.

from the change of the power angle, $\Delta\delta$, to the change of the active power and can be represented as

$$K_{\mathrm{pd}} = N\frac{V_{\mathrm{n}}^2}{X},$$

where $X = \omega_{\mathrm{n}}L_{\mathrm{s}}$ is the impedance of the inductor L_{s}, and N represents the number of phases, i.e. $N = 3$ for a three-phase VSM or $N = 1$ for a single-phase VSM. The transfer function from ΔP to $\Delta\omega$ is then

$$\frac{\Delta\omega}{\Delta P} = -\frac{1}{\omega_{\mathrm{n}}D_{\mathrm{p}}} \cdot \frac{1}{\tau_\omega s + 1}, \tag{14.6}$$

where $\tau_\omega = J/D_{\mathrm{p}}$ is called the inertia time constant as mentioned before. It can be seen from (14.3) and (14.6) that the synchronverter has the same active power-frequency relationship as conventional SM. Naturally, $J = \tau_\omega D_{\mathrm{p}}$ is often regarded as the virtual inertia. However, as pointed out in Chapter 4 and in the original synchronverter paper (Zhong and Weiss 2011), it also determines the time constant of the frequency loop, which may impose a limit on the inertia J, as explained below.

The transfer function from ΔP_{set} to ΔP can be found as

$$\frac{\Delta P}{\Delta P_{\mathrm{set}}} = \frac{1}{\tau_{\mathrm{p}}(\tau_\omega s + 1)s + 1}, \tag{14.7}$$

where

$$\tau_{\mathrm{p}} = \frac{D_{\mathrm{p}}\omega_{\mathrm{n}}}{K_{\mathrm{pd}}} = \frac{D_{\mathrm{p}}\omega_{\mathrm{n}}X}{NV_{\mathrm{n}}^2}. \tag{14.8}$$

Substituting (14.4) into it, then

$$\tau_{\mathrm{p}} = \frac{S_{\mathrm{n}}X}{\alpha NV_{\mathrm{n}}^2\omega_{\mathrm{n}}}. \tag{14.9}$$

Assume the base voltage is chosen as the RMS phase voltage V_n and the base power is chosen as the rated power S_n. Then the base impedance is

$$Z_{base} = NV_n^2/S_n$$

and the per-unit output impedance of the synchronverter is

$$X^{pu} = \frac{X}{Z_{base}} = \frac{S_n X}{NV_n^2}.$$

Hence, there is

$$\tau_p = \frac{X^{pu}}{\alpha\omega_n}. \tag{14.10}$$

Since ω_n and α are both specified by the grid code, it is obvious that, for a given power system, τ_p is only determined by and proportional to X^{pu}. Once the synchronverter hardware is designed, the corresponding τ_p is fixed.

The system in Figure 14.2(a) or the transfer function (14.7) has two eigenvalues

$$\lambda_{1,2} = \frac{-\tau_p \pm \sqrt{\tau_p^2 - 4\tau_p\tau_\omega}}{2\tau_p\tau_\omega} = \frac{-1 \pm \sqrt{1 - 4\tau_\omega/\tau_p}}{2\tau_\omega}.$$

Figure 14.3(a) illustrates the path of the eigenvalues when τ_ω increases. When $0 < \tau_\omega \ll \tau_p$, the two eigenvalues are on the real axis, at $\lambda_1 \approx -\frac{1}{\tau_\omega}$ and $\lambda_2 \approx -\frac{1}{\tau_p}$. The system is dominated by λ_2 because λ_1 is too far away from the imaginary axis. When τ_ω increases, the eigenvalues move towards each other on the real axis. Then the two eigenvalues become conjugate when $\tau_\omega > 0.25\tau_p$, with the imaginary part initially increasing and then decreasing. The imaginary part reaches the maximum when $\tau_\omega = 0.5\tau_p$. The two eigenvalues move toward the imaginary axis when τ_ω increases further. The damping ratio of the active-power loop is

$$\zeta = \frac{1}{2\sqrt{\tau_\omega/\tau_p}} \tag{14.11}$$

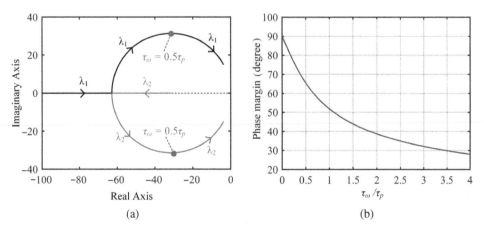

(a)

(b)

Figure 14.3 Properties of the active power loop of a conventional synchronverter with $X^{pu} = 0.05$, $\omega_n = 100\pi$ rad s^{-1}, and $\alpha = 0.5\%$. (a) Closed-loop eigenvalues. (b) Phase margin.

when $\tau_\omega > 0.25\tau_p$. Note that $\zeta = \frac{1}{\sqrt{2}}$ when $\tau_\omega = 0.5\tau_p$. In other words, increasing τ_ω makes the system response oscillatory and reduces the stability margin. Indeed, this can be seen from the phase margin of a sample system shown in Figure 14.3(b). When τ_ω is very small, the phase margin of the system is close to 90°. The phase margin decreases to 52° when $\tau_\omega = \tau_p$ and further to 28° when $\tau_\omega = 4\tau_p$. This means there is an upper limit on the inertia time constant τ_ω for a given system, which has a fixed τ_p as given in (14.10). In other words, the virtual inertia provided by J is limited.

Indeed, since τ_ω is the time constant of the frequency loop, it is recommended in (Zhong and Weiss 2011) to chose it to be much smaller than the fundamental period, e.g., as $0.1/f_n$ s. As a result, the condition $0 < \tau_\omega \ll \tau_p$ often holds and the active-power loop shown in Figure 14.2(a) can be simplified as shown in Figure 14.2(b). Correspondingly, the transfer function from ΔP_{set} to ΔP given in (14.7) can be simplified as

$$\frac{\Delta P}{\Delta P_{set}} = \frac{1}{\tau_p s + 1}, \tag{14.12}$$

which means the active-power response is dominated by τ_p.

14.4 Reconfiguration of the Inertia Time Constant

14.4.1 Design and Outcome

In order to be able to reconfigure the inertia time constant of a VSM, a virtual inertia block $H_v(s)$ can be added to the torque channel of the VSM, as shown in Figure 14.4. It can be

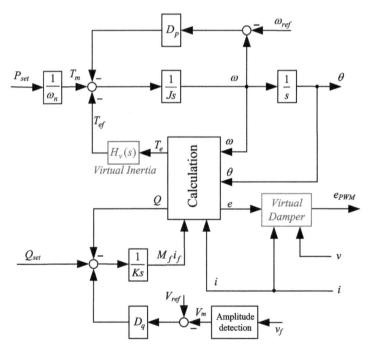

Figure 14.4 VSM with virtual inertia and virtual damping.

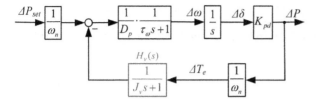

Figure 14.5 The small-signal model of the active-power loop with a virtual inertia block $H_v(s)$.

implemented via a low-pass filter, e.g., the one given by

$$H_v(s) = \frac{1}{J_v s + 1}, \tag{14.13}$$

where J_v is the inertia time constant required. There is normally a low-pass filter to remove the ripples in the torque but its time constant is often much smaller than J_v. The virtual block H_v can be put in series with it or replace it.

The corresponding small-signal model of the active-power loop is shown in Figure 14.5. The transfer function from ΔP to $\Delta \omega$ is

$$\frac{\Delta \omega}{\Delta P} = -\frac{1}{\omega_n D_p} \cdot \frac{1}{\tau_\omega s + 1} \cdot \frac{1}{J_v s + 1}.$$

This is a second-order system with a small τ_ω, as explained in the previous section. It can be simplified as

$$\frac{\Delta \omega}{\Delta P} \approx -\frac{1}{\omega_n D_p} \cdot \frac{1}{(J_v + \tau_\omega)s + 1}. \tag{14.14}$$

If $J_v \gg \tau_\omega$, then the equivalent inertia time constant is J_v. This makes it possible to reconfigure the inertia time constant in a wide range, even up to a level comparable to a conventional SM, without changing τ_ω.

Note that it is not possible to directly increase $\tau_\omega = J/D_p$ to this level because τ_ω is limited as explained in the previous section. While it is possible to add an additional damping channel to change the control structure, this increases the complexity of the controller so it is not considered here.

14.4.2 What is the Catch?

The characteristic equation of the active-power loop shown in Figure 14.5 can be found as

$$1 + \tau_p (J_v s + 1)(\tau_\omega s + 1)s = 0. \tag{14.15}$$

If a relatively large inertia time constant $J_v \gg \tau_\omega$ is desired, then the term $\frac{1}{\tau_\omega s + 1}$ can be ignored and the characteristic equation (14.15) can be simplified as

$$\tau_p (J_v s + 1)s + 1 = 0.$$

This is the same as the characteristic equation of the system in (14.7) but with τ_ω replaced by J_v. According to (14.11), the damping ratio of the system when $J_v > 0.25\tau_p$ is

$$\zeta = \frac{1}{2\sqrt{J_v/\tau_p}}. \tag{14.16}$$

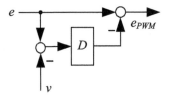

Figure 14.6 Implementations of a virtual damper.
(a) Through impedance scaling with a voltage controller.
(b) Through impedance insertion with a current controller.

(a)

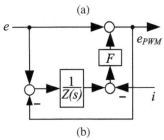

(b)

If the inertia time constant J_v is configured to be large with comparison to τ_p, then the damping of the system is small, resulting in oscillatory responses. This may lead to large transient currents, which might overload or even damage the converter (Coelho et al. 1999; Guerrero et al. 2004). It is critical for the damping of the VSM to be large enough.

Assume the desired damping ratio is ζ_0. Then, according to (14.16), the corresponding τ_p is

$$\tau_p = 4\zeta_0^2 J_v$$

and, according to (14.10), the equivalent pu impedance X_v^{pu} is

$$X_v^{pu} = \alpha\omega_n\tau_p = 4\zeta_0^2 J_v\alpha\omega_n. \tag{14.17}$$

This requires the corresponding inductance L_v to be

$$L_v = \frac{X_v^{pu}}{\omega_n}Z_{base} = 4\zeta_0^2 J_v\alpha Z_{base} = \frac{4\zeta_0^2 J_v\alpha NV_n^2}{S_n}. \tag{14.18}$$

Apparently, this is much different from the inductance L_s existing in the hardware.

14.5 Reconfiguration of the Virtual Damping

As discussed in the previous section, there is a need to reconfigure the damping of the VSM to avoid oscillatory frequency responses, which boils down to replacing inductance L_s with L_v. This can be achieved via putting the generated voltage e through a virtual damper before sending it for PWM conversion, as shown in Figure 14.4. The virtual damper also takes the voltage v and/or the current i as inputs. In this section, two implementations as shown in Figure 14.6, one through impedance scaling with a voltage feedback controller and the other through impedance insertion with a current feedback controller, are discussed.

14.5.1 Through Impedance Scaling with an Inner-loop Voltage Controller

14.5.1.1 Controller Design and Analysis

Following the idea presented in (Zhong et al. 2012a) and (Zhong and Hornik 2013, Ch. 8) to change the impedance of a converter, the inner-loop voltage controller as shown in Figure 14.6(a) is an implementation of the virtual damper. It consists of a voltage feedback controller to scale the voltage difference $e - v$ with the factor D and subtracts it from the original signal e to form the new control signal e_{PWM}. Hence,

$$e_{\text{PWM}} = e - D(e - v), \tag{14.19}$$

which, in lieu of e, is converted to PWM signals to drive the switches in the power part. Since the switching frequency of the converter is normally much higher than the system frequency, there is

$$e_{\text{PWM}} \approx v + v_{\text{s}}, \tag{14.20}$$

where v_{s} is the voltage across the inductor L_{s} when considering the average values over a switching period for the PWM signals. Combining (14.19) and (14.20), then there is

$$e \approx v + \frac{1}{1 - D} v_{\text{s}}. \tag{14.21}$$

In other words, the function of the virtual damper is to replace the inductor L_{s} with $\frac{1}{1-D} L_{\text{s}}$ or to scale the original impedance by $\frac{1}{1-D}$. Hence, this technique is called impedance scaling. In order to scale L_{s} to L_{v}, D should be chosen as

$$D = 1 - \frac{L_{\text{s}}}{L_{\text{v}}}$$

or, substituting (14.18) into it, as

$$D = 1 - \frac{S_{\text{n}} L_{\text{s}}}{4 \zeta_0^2 J_{\text{v}} \alpha N V_{\text{n}}^2}.$$

14.5.1.2 Enhancement of Voltage Quality

The gain $D = 1 - \frac{L_{\text{s}}}{L_{\text{v}}}$ is static but D can also be chosen dynamic as well because the function of the virtual damper is to scale the original impedance by $\frac{1}{1-D}$, which can be designed to include specific frequency characteristics. For example, it can be chosen as

$$D(s) = 2 - \frac{L_{\text{s}}}{L_{\text{v}}} - \Pi_h \frac{s^2 + 2\zeta_h h\omega_n K_h s + (h\omega_n)^2}{s^2 + 2\zeta_h h\omega_n s + (h\omega_n)^2}$$

where ζ_h can be chosen as $\zeta_h = 0.01$ to accommodate frequency variations and h can be chosen to cover the major harmonic components in the current, e.g. the third, fifth and seventh harmonics. The scaling factor is $\frac{1}{1-D} = \frac{1}{1-(2-\frac{L_{\text{s}}}{L_{\text{v}}}-1)} = \frac{L_{\text{v}}}{L_{\text{s}}}$ at low and high frequencies and $\frac{1}{1-D} = \frac{1}{1-(2-\frac{L_{\text{s}}}{L_{\text{v}}}-K_h)} = \frac{1}{K_h - 1 + \frac{L_{\text{s}}}{L_{\text{v}}}}$ at the hth harmonic frequency. While it meets the requirement of the virtual damping, it also scales the impedance at the harmonic frequencies by a factor of $\frac{1}{K_h - 1 + \frac{L_{\text{s}}}{L_{\text{v}}}}$. If $K_h > 2 - \frac{L_{\text{s}}}{L_{\text{v}}}$, then the impedance at the hth harmonic frequency is reduced, which enhances the quality of the VSM voltage v, according to (Zhong and Hornik 2013; Zhong et al. 2012a).

14.5.2 Through Impedance Insertion with an Inner-loop Current Controller

Instead of using the voltage feedback controller shown in Figure 14.6(a), it is also possible to adopt a current feedback controller to implement the virtual damper and generates the new control signal e_{PWM}, as shown in Figure 14.6(b). The voltage difference $e - e_{\mathrm{PWM}}$ is passed through an impedance $Z(s)$ to generate a current reference, of which the difference with the feedback current i is scaled by a factor F and added to the original signal e to form the new control signal e_{PWM}. Hence,

$$e_{\mathrm{PWM}} = e + F\left(\frac{1}{Z(s)}(e - e_{\mathrm{PWM}}) - i\right), \tag{14.22}$$

which, in lieu of e, is converted to PWM signals to drive the switches in the power part. This is equivalent to having

$$e - e_{\mathrm{PWM}} + \frac{F}{Z(s) + F}Z(s)i.$$

Choose F as a positive large number and

$$Z(s) = sL_{\mathrm{v}}.$$

Then $\frac{F}{sL_{\mathrm{v}}+F}$ is a low-pass filter with a small time constant and $\frac{F}{sL_{\mathrm{v}}+F} \approx 1$ over a wide range of frequencies. As a result,

$$e = e_{\mathrm{PWM}} + \frac{F}{sL_{\mathrm{v}} + F}sL_{\mathrm{v}}i \approx e_{\mathrm{PWM}} + sL_{\mathrm{v}}i.$$

In other words, the function of the virtual damper shown in Figure 14.6(b) is to insert an inductor L_{v} between e and e_{PWM} with the current i flowing through it, meeting the requirement on the equivalent inductance. Hence, this technique is called impedance insertion. Here, it is assumed that e_{PWM} is adopted as the feedback voltage v_{f}. If $v_{\mathrm{f}} = v$, then the inductance L_{s} between e_{PWM} and v should be considered, via choosing $Z(s) = s(L_{\mathrm{v}} - L_{\mathrm{s}})$.

14.6 Fault Ride-through

14.6.1 Analysis

The fault ride-through capability of a VSM is very important. The worst case is that there is a ground fault across the capacitor, i.e. $v = 0$. In this case, the whole voltage e is dropped on the corresponding equivalent inductance L_{v}. Since the voltage v is dropped to nearly zero, the corresponding reactive power in the steady state (assuming $Q_{\mathrm{set}} = 0$) is

$$Q = D_{\mathrm{q}}V_{\mathrm{ref}} = \frac{S_{\mathrm{n}}}{\beta\sqrt{2V_{\mathrm{n}}}}\sqrt{2}V_{\mathrm{n}} = \frac{S_{\mathrm{n}}}{\beta}$$

when the parasitic resistance R_{s} is negligible. This is the reactive power of the equivalent inductance L_{v}. Hence, the corresponding pu RMS voltage $E_{\mathrm{fault}}^{\mathrm{pu}}$ is

$$E_{\mathrm{fault}}^{\mathrm{pu}} = \sqrt{\frac{Q\omega L_{\mathrm{v}}}{N}}/V_{\mathrm{n}} = \sqrt{\frac{S_{\mathrm{n}}\omega L_{\mathrm{v}}}{\beta N V_{\mathrm{n}}^2}}.$$

Note that this voltage is a number and it is not a physical voltage so even if it is large it is not a problem. Substituting (14.18) into it and considering that $\omega \approx \omega_n$ and (14.17), then

$$E_{\text{fault}}^{\text{pu}} \approx \sqrt{\frac{S_n \omega}{\beta N V_n^2} \frac{4\zeta_0^2 J_v \alpha N V_n^2}{S_n}} = \sqrt{\frac{4\zeta_0^2 J_v \alpha \omega_n}{\beta}} = \sqrt{\frac{X_v^{\text{pu}}}{\beta}}.$$

As a result, the corresponding pu RMS fault current is

$$I_{\text{fault}}^{\text{pu}} = \frac{E_{\text{fault}}^{\text{pu}}}{X_v^{\text{pu}}} = \frac{1}{\sqrt{\beta X_v^{\text{pu}}}}. \tag{14.23}$$

In practice, the actual $E_{\text{fault}}^{\text{pu}}$ would be larger and the actual $I_{\text{fault}}^{\text{pu}}$ would be smaller because of the parasitic resistance R_s. The larger the voltage droop coefficient β, the smaller the fault current; the larger the equivalent pu impedance X_v^{pu} (or the inertia) the smaller the fault current.

14.6.2 Recommended Design

For a desired fault current $I_{\text{fault}}^{\text{pu}}$, according to (14.23), the required X_v^{pu} is

$$X_v^{\text{pu}} = \frac{1}{\beta (I_{\text{fault}}^{\text{pu}})^2}$$

and the corresponding L_v is

$$L_v = \frac{X_v^{\text{pu}} Z_{\text{base}}}{\omega_n} = \frac{1}{\beta (I_{\text{fault}}^{\text{pu}})^2} \frac{N V_n^2}{\omega_n S_n}.$$

According to (14.17), the corresponding inertia J_v is

$$J_v = \frac{X_v^{\text{pu}}}{4\zeta_0^2 \alpha \omega_n} = \frac{1}{4\zeta_0^2 \alpha \omega_n \beta (I_{\text{fault}}^{\text{pu}})^2}.$$

This is a very fundamental formula. It links together most of the key parameters of a VSM, including the damping ratio ζ_0, the virtual inertia J_v, the frequency droop coefficient α, the voltage droop coefficient β, the fault current level $I_{\text{fault}}^{\text{pu}}$, and the rated system frequency ω_n. They are all linked with each other and the selection of each parameter needs to take into account the others.

In implementation, it is possible to limit the integrator output $M_f i_f$ to limit $E_{\text{fault}}^{\text{pu}}$ and, hence, the fault-current level $I_{\text{fault}}^{\text{pu}}$. If e is adopted as the voltage feedback v_f, it is also possible to reduce the fault-current level $I_{\text{fault}}^{\text{pu}}$.

14.7 Simulation Results

The parameters of the single-phase converter used in the simulations are given in Table 14.1. Simulations are carried out with MATLAB/Simulink R2014b using average models. The solver used in the simulations is ode23tb with a relative tolerance of 10^{-3} and a maximum

Table 14.1 Parameters of the system under simulation.

Parameter	Value	Parameter	Value
τ_ω	0.002s	L_s	0.23 mH
J_v	0.02s	P_{set}	50 W
D_p	0.2026	Q_{set}	0 Var
D_q	117.88	Nominal power	100 VA
V_n	12 V	f_n	50 Hz

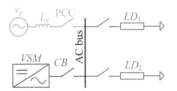

Figure 14.7 A VSM in a microgrid connected to a stiff grid.

step size of 0.2 ms. The frequency droop coefficient is chosen as $\alpha = 0.5\%$, which leads to $D_p = 0.2026$, and the voltage droop coefficient is chosen as $\beta = 5\%$, which leads to $D_q = 117.88$. The desired damping ratio is chosen as $\zeta_0 = 0.707$. As a result,

$$L_v = \frac{2J_v \alpha N V_n^2}{S_n} = \frac{2J_v \times 0.5\% \times 12^2}{100} \times 1000 = 14.4 J_v \text{ mH}.$$

14.7.1 A Single VSM

The system under simulation is a microgrid as shown in Figure 14.7. The capacity of the grid v_g is large enough so it can be treated as a stiff grid. The VSM is connected to the AC bus via a circuit breaker CB. Two local loads LD_1 and LD_2 are connected to the AC bus. The pu active power consumed by LD_1 and LD_2 are both equal to 0.4.

14.7.1.1 Reconfigurability of Virtual Inertia and Virtual Damping
In this case, the PCC circuit breaker is turned OFF and the CB circuit breaker is turned ON so the VSM operates in the stand-alone mode to supply both LD_1 and LD_2 via the AC bus. Before $t = 0$, both LD_1 and LD_2 are connected to the AC bus so the load power is 0.8 pu. At $t = 0$, LD_2 is disconnected from the AC bus. The damping is configured as $\zeta_0 = 0.707$.

The frequency responses of the VSM with different virtual inertia are shown in Figure 14.8. When the load LD_2 is disconnected, the frequency increases. Apparently, the frequency response behaves as expected: increasing the virtual inertia J_v indeed slows the frequency response down. Moreover, because of the damping ratio is designed to be $\zeta_0 = 0.707$, the frequency response is very smooth, without visible overshoot.

14.7.1.2 Effect of the Virtual Damping
In this case, both the PCC circuit breaker and the CB circuit breaker are turned ON with both loads connected to the AC bus. Before $t = 0$, the active power set point of the VSM is 0. At $t = 0$, it is changed to $\Delta P_{set} = 50$ W. The simulation results with virtual inertia $J_v = 0.2$

Figure 14.8 Normalized frequency response of a VSM with reconfigurable inertia and damping.

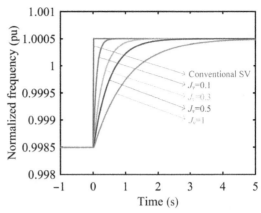

Figure 14.9 Effect of the virtual damping ($J_v = 0.2$ s).

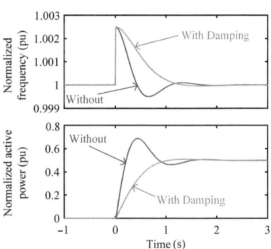

are shown in Figure 14.9 for the cases with and without the virtual damping. When the virtual damping is not enabled, the responses are oscillatory (denoted "Without" in the figure). When it is enabled, the responses are very smooth (denoted "With Damping" in the figure).

14.7.2 Two VSMs in Parallel Operation

The stiff grid in Figure 14.7 is replaced with another VSM having the same parameters, as illustrated in Figure 14.10. The per-unit active power consumed by LD_1 and LD_2 are both equal to 0.4. Both VSMs are configured to have the same damping $\zeta_0 = 0.707$.

Figure 14.10 A microgrid with two VSMs.

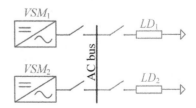

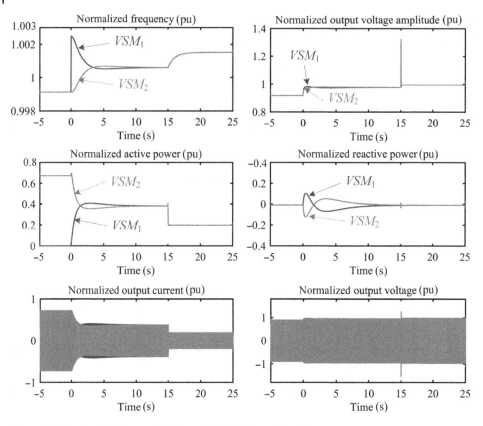

Figure 14.11 Two VSMs operated in parallel with $J_{v1} = J_{v2} = 1$ s.

14.7.2.1 Case 1: $J_{v1} = J_{v2} = 1$ s

Before $t = 0$, VSM_2 supplies LD_1 and LD_2 alone. VSM_1 is connected to the AC bus at $t = 0$. At $t = 15$ s, the load LD_2 is disconnected. The results are shown in Figure 14.11. The system responses are slow, as expected, because the virtual inertia of both VSMs is configured as 1s. At $t = 0$ s, VSM_1 gradually takes some active power over from VSM_2. There are some transients in the reactive power but both settle down within expected time range. The voltage increases a bit because each VSM now supplies less power. At $t = 15$ s, the frequency of both VSMs increases, without noticeable difference, because of the cut-off of LD_2. It is worth noting that the active power changes immediately when the load is cut off but the frequencies change gradually according to the virtual inertia, rather than abruptly. The voltages and currents are very smooth, except some normal voltage spikes when cutting off LD_2.

14.7.2.2 Case 2: $J_{v1} \neq J_{v2}$

The inertia of VSM_1 is changed to $J_{v1} = 0.5$ s so it is half of J_{v2}. The same sequence of events as in case 1 is followed. Although J_{v1} and J_{v2} are not the same, the system works very well, as can be seen from the properly damped responses shown in Figure 14.12. The VSM frequencies in the steady state are equal, but they are different during the transient process because of the different virtual inertia. Again, the voltages and currents are very smooth, except some normal voltage spikes when the load is cut off.

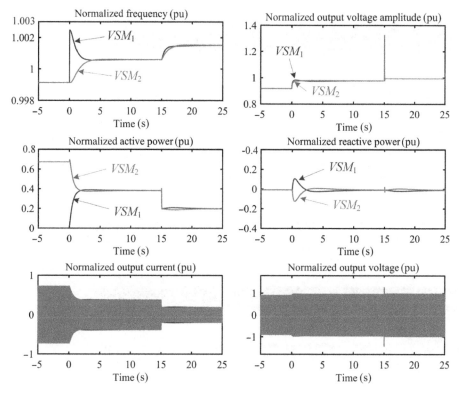

Figure 14.12 Two VSMs operated in parallel with $J_{v1} = 0.5$ s and $J_{v2} = 1$ s.

14.7.2.3 Fault Ride-through

In this case, at $t = 0$, a ground fault occurs at the AC bus of the microgrid shown in Figure 14.10 and lasts for 0.5 s. The frequency responses with different inertia J_v under the ground fault are shown in Figure 14.13. Before the ground fault occurs, both VSMs have

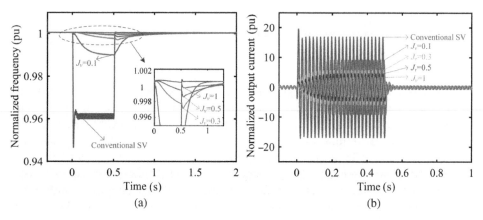

Figure 14.13 Simulation results under a ground fault with $J_v = 0.1$, 0.3, 0.5, 1 s. (a) Normalized frequency. (b) Normalized fault current.

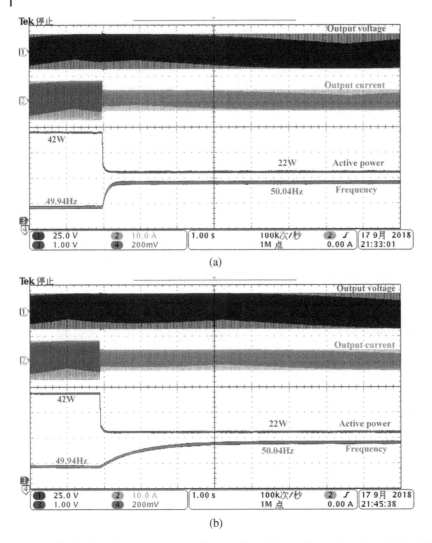

Figure 14.14 Experimental results with reconfigurable inertia and damping. (a) $J_v = 0.1$ s. (b) $J_v = 1$ s.

the same frequency. Because of the ground fault, the frequencies drop and over-currents appear. The larger the inertia, the smaller the frequency drop and the smaller the over-current. With $J_v = 1$ s, the current is about 230%, which is slightly less than the value of 252% calculated from (14.23). After the fault is cleared at $t = 0.5$ s, the frequency and the current return to normal. The frequency excursion and the over-currents with the conventional synchronverter are the largest among all the cases because its inertia is the smallest. The provision of reconfigurable inertia and damping can significantly reduce the frequency drop and the over-currents, and enhance the fault ride-through capability.

14.8   Experimental Results

The results from a prototype VSM with reconfigurable inertia and damping are shown in this section. The controller is implemented with a TI DSC TMS320F28335. The experiment parameters are the same as the parameters in Table 14.1 and the microgrid for experiment is the same as the one illustrated in Figure 14.7. The stiff grid is provided by a grid simulator.

14.8.1   A Single VSM

A single VSM is connected via a circuit breaker CB to the AC bus, which is connected to the stiff grid v_g through a circuit breaker PCC. Two local loads LD_1 and LD_2 with 4 Ω each can be connected to the AC bus.

14.8.1.1   Reconfigurability of the Inertia Time Constant and Damping
At the beginning, the VSM supplies both LD_1 and LD_2 in the islanded mode, with the PCC circuit breaker turned OFF. Figure 14.14 shows the transient waveforms for two different J_v when disconnecting LD_2 from the AC bus. The damping is configured as $\zeta_0 = 0.707$. Indeed, the frequency responds properly under different inertia time constants. When the inertia time constant increases, the frequency response slows down. In all cases, the responses are very smooth without noticeable overshoot.

The results from the original synchronverter are shown in Figure 14.15 for comparison. It can be seen that its inertia time constant is very small. The strategy can indeed reconfigure the speed of the frequency response by tuning J_v.

14.8.1.2   Effect of the Virtual Damping
This experiment demonstrates the function of the virtual damping block. The inertia time constant is $J_v = 0.2$ s. The circuit breaker PCC is turned ON after self-synchronization. A step change of the active-power set point by 0.5 pu is applied at about $t = 0.8$ s.

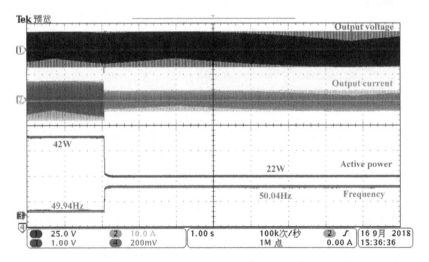

Figure 14.15   Experimental results from the original synchronverter for comparison.

Figure 14.16(a) shows the performance of the VSM without the virtual damping block, i.e. the signal e is directly sent to the PWM generation block. A large overshoot occurs in the frequency and active power responses because of the low damping ratio. If the inertia time constant J_v is increased further, the overshoot would increase further. The performance of the VSM with the virtual damping set at 0.707 is shown in Figure 14.16(b). Both the frequency and the active power become very smooth.

14.8.2 Two VSMs in Parallel Operation

The results of two VSMs (VSM$_1$ and VSM$_2$) having the same parameters operated in parallel to supply the loads LD_1 and LD_2 connected to the AC bus, with the circuit breaker PCC turned OFF, are shown in Figure 14.17 for the case with $J_{v1} = J_{v2} = 0.5$ s and $\zeta_0 = 1$, and in Figure 14.18 for the case with $J_{v1} = J_{v2} = 1$ s and $\zeta_0 = 0.707$, respectively. Initially, the load

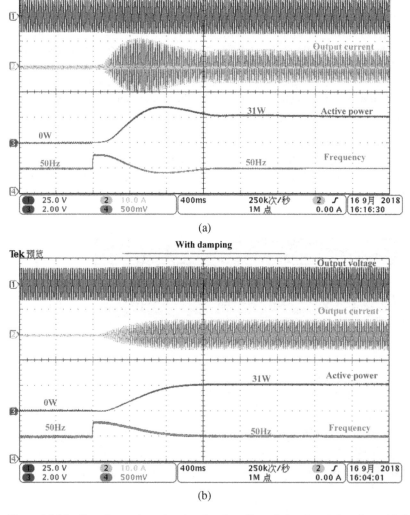

(a)

(b)

Figure 14.16 Experimental results showing the effect of the virtual damping with $J_v = 0.2$ s. (a) Without virtual damping. (b) With virtual damping.

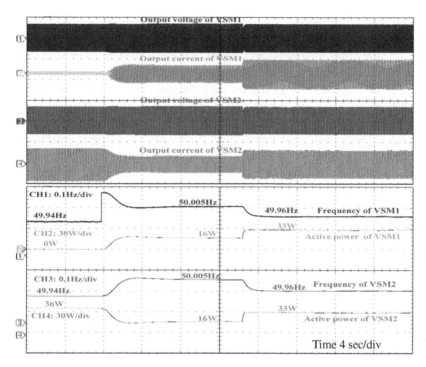

Figure 14.17 Experimental results when two VSMs with the same inertia time constant are in parallel operation: $J_{v1} = J_{v2} = 0.5$ s and $\zeta_0 = 1$.

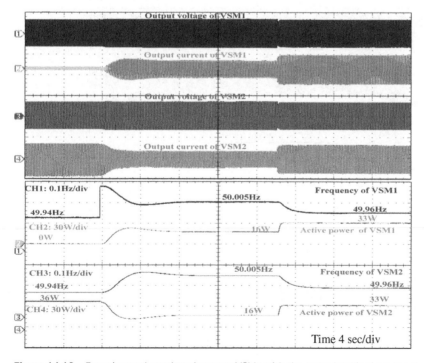

Figure 14.18 Experimental results when two VSMs with the same inertia time constant are in parallel operation: $J_{v1} = J_{v2} = 1$ s and $\zeta_0 = 0.707$.

LD_1 is supplied by VSM_2; then VSM_1 is put into parallel operation with VSM_2 and, finally, load LD_2 is connected to the AC bus.

As shown in the figures, the operation is very smooth, even when VSM_1 is put into parallel operation and the load LD_2 is connected to the AC bus. When VSM_1 is put into parallel operation, the current of VSM_2 gradually reduces while the current of VSM_1 gradually increases. For the case with $J_{v1} = J_{v2} = 0.5$ s, the current of VSM_1 has no visible overshoot because the damping is $\zeta_0 = 1$. There is visible (but acceptable) overshoot in the current of VSM_1 for the case with $J_{v1} = J_{v2} = 1$, because the damping is $\zeta_0 = 0.707$ s. After the transient, the load is shared by VSM_1 and VSM_2 equally as designed. When the load LD_2 is connected to the AC bus, both the currents of VSM_1 and VSM_2 all increases while maintaining the sharing ratio.

Figure 14.19 shows the results when VSM_1 and VSM_2 have different inertia time constants and damping ratios, i.e. $J_{v1} = 0.5$ s, $\zeta_{01} = 1$, $J_{v2} = 1$ s, and $\zeta_{02} = 0.707$. The responses of VSM_1 are very close to the case shown in Figure 14.17 while the responses of VSM_2 are very close to the case shown in Figure 14.18.

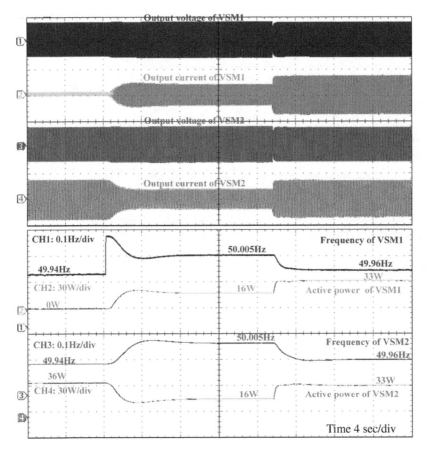

Figure 14.19 Experimental results when two VSMs with different inertia time constants operated in parallel.

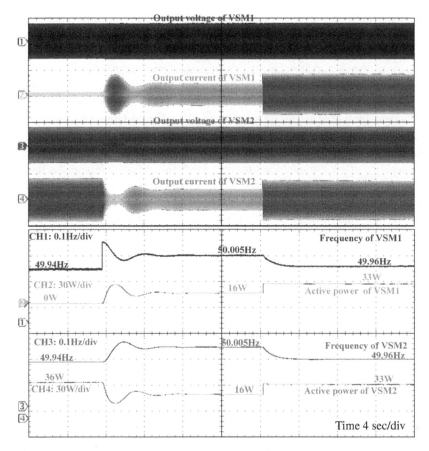

Figure 14.20 Experimental results when the two VSMs operated as the original SV in parallel operation with $\tau_{\omega 1} = \tau_{\omega 2} = 1$ s for comparison.

When VSM$_1$ is put into parallel operation, the active power changes gradually according to the inertia time constant but it changes suddenly when the load LD_2 is connected, as expected. Although the active power changes suddenly when the load is connected, the frequencies do not. For the case with $J_{v1} = J_{v2} = 0.5$ s, it takes about 2.0 s for the frequencies to settle down. This is consistent with the inertia of $J_{v1} = J_{v2} = 0.5$ s because $0.5 \times 4 = 2.0$ s. For the case with $J_{v1} = J_{v2} = 1$ s, it takes about 4.0 s for the frequencies to settle down. This is consistent with the inertia of $J_{v1} = J_{v2} = 1$ s.

Figure 14.20 shows the results from the strategy with the original synchronverter when VSM$_1$ and VSM$_2$ are operated as the original synchronverter with $\tau_{\omega 1}$ and $\tau_{\omega 2}$ increased to 1 s. Clearly, the frequency responses are oscillatory when VSM$_1$ is put into parallel operation.

14.9 Summary

Based on (Zhong 2018; Zhong et al. 2018b), the critical concept of inertia has been clarified for SMs and VSMs. It has two aspects, i.e. the inertia time constant that characterizes the speed of the frequency response and the inertia constant that characterizes the amount of energy stored. For a physical SM, these two aspects are coupled together and directly determined by the inertia. For a VSM, these two are decoupled and can be designed separately. The energy storage aspect of the inertia can be implemented by adding an energy storage unit without considering the inertia time constant of the frequency response. The inertia time constant can be reconfigured to achieve the desired speed of frequency response and a virtual inertia has been proposed for this purpose. Moreover, a virtual damper is introduced to reconfigure the damping of the VSM and avoid oscillatory responses due to the reconfigured inertia. Two different approaches, one through impedance scaling and the other through impedance insertion, are described to implement the virtual damping. Furthermore, the fault ride-through capability can be designed to meet the requirement on fault currents. Simulation and experimental results have demonstrated the feasibility of virtual inertia, virtual damping, and the associated fault ride-through capability.

Part III

2G VSM: Robust Droop Controller

15

Synchronization Mechanism of Droop Control

It is well known that droop control is fundamental to the operation of power systems, and now the parallel operation of inverters, while phase-locked loops (PLL) are widely adopted in modern electrical engineering. In this chapter, it is shown at first that droop control and PLLs structurally resemble each other. In other words, the synchronization mechanism inherently exists in droop control, making droop control a natural technical route to implement SYNDEM smart grids. This is then applied to operate a conventional droop controller for inverters with inductive impedance to achieve the function of a PLL, without having a dedicated synchronization unit. Extensive experimental results are provided to validate the theoretical analysis.

15.1   Brief Review of Phase-Locked Loops (PLLs)

15.1.1   Basic PLL

A basic PLL adopts a control loop to track the phase of an input signal. It can often provide the frequency information of the signal as well, but normally without the information of the voltage amplitude.

The operational principle of a PLL is shown in Figure 15.1(a). It consists of a phase detection (PD) unit, a loop filter (LF) and a voltage-controlled oscillator (VCO). The PD unit generates a non-zero DC component, often polluted with ripples, when the phase difference between the input signal and the re-produced output signal is not the same. The DC component is extracted and amplified by the LF before being passed to the VCO, which is often a PI controller, to generate the frequency for the output signal. In the steady state, the input to the PI controller is forced to be zero so the phase difference between the input signal and the output signal is zero. As a result, the phase of the output signal is locked with that of the input signal.

Figure 15.1(b) shows the implementation of a basic PLL, where the PD unit is a multiplier, the LF is a low-pass filter (LPF) and the VCO consists of a PI controller, an integrator, and a sinusoidal function generator. For an input signal $v = V_m \cos \theta_g$ with phase $\theta_g = \omega_g t + \phi_g$

Power Electronics-Enabled Autonomous Power Systems: Next Generation Smart Grids,
First Edition. Qing-Chang Zhong.
© 2020 John Wiley & Sons Ltd. Published 2020 by John Wiley & Sons Ltd.

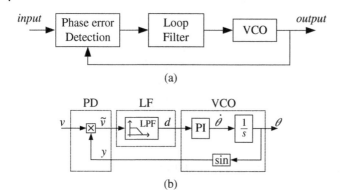

Figure 15.1 Block diagrams of a conventional PLL. (a) Operational concept. (b) A simple PLL.

and an output signal $y = \sin\theta$ with phase $\theta = \omega t + \phi$, the output of the PD unit is

$$\tilde{v} = vy = V_m \sin\theta \cos\theta_g$$
$$= \frac{V_m}{2}\sin(\theta - \theta_g) + \frac{V_m}{2}\sin(\theta + \theta_g). \tag{15.1}$$

The first term is a low-frequency component that contains the phase difference between v and y and the second term is a high-frequency component, which can be filtered out by the low-pass LF. The output $d = \frac{V_m}{2}\sin[(\omega - \omega_g)t + (\phi - \phi_g)]$ of the LF is then fed into a PI controller to generate the estimated frequency $\omega = \dot{\theta}$ until $d = 0$. In the steady state, d is driven to zero and $\theta = \theta_g$, i.e. $\omega = \omega_g$ and $\phi = \phi_g$. The phase of the output signal y is locked with that of the input signal v. In other words, the signal y is synchronized with v.

15.1.2 Enhanced PLL (EPLL)

Although the basic PLL is able to lock the phase but no amplitude information about the input signal is available. In order also to obtain the amplitude information of the input signal, an EPLL (Karimi-Ghartemani and Iravani, 2002, 2001) can be adopted. This method was introduced with several different names, e.g. the sinusoidal tracking algorithm (STA) (Ziarani and Konrad 2004), the amplitude phase model (APM) and amplitude phase frequency model (APFM) (Karimi-Ghartemani and Ziarani 2003). It is able to extract the fundamental component of a periodic signal and, at the same time, to estimate its amplitude, phase, and frequency.

The EPLL can be designed by using the gradient descent method (Giordano and Hsu 1985). Assume that a typical periodic voltage $v(t)$ has the general form of

$$v(t) = \sum_{i=0}^{\infty} \sqrt{2}V_i \sin\theta_{gi} + n(t)$$

where V_i and $\theta_{gi} = \omega_{gi}t + \delta$ are the RMS value and phase of the ith harmonic component of the voltage, and $n(t)$ represents the noise on the signal. The objective of a PLL can be regarded as extracting the component $e(t)$ of interest, which is usually the fundamental

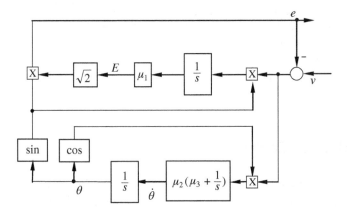

Figure 15.2 Enhanced phase-locked loop (EPLL) or sinusoidal tracking algorithm (STA).

component, from the input signal $v(t)$. Denote the estimated or recovered signal $e(t)$ as

$$e(t) = \sqrt{2}E(t)\sin\left(\int_0^t \omega(\tau)d\tau + \delta(t)\right),$$

where $E(t)$ is the estimated RMS voltage, $\omega(t)$ is the estimated frequency and $\theta(t) = \int_0^t \omega(\tau)d\tau + \delta(t)$ is the estimated phase of $e(t)$. Then the problem of designing a PLL can be formulated as finding the optimal vector $\psi(t) = \begin{bmatrix} E(t) & \omega(t) & \delta(t) \end{bmatrix}^T$ that minimizes the cost function

$$J(\psi(t), t) = \frac{1}{2}d^2(t) = \frac{1}{2}[v(t) - e(t)]^2,$$

where $d(t) = v(t) - e(t)$ is the tracking error. According to the gradient descent method (Giordano and Hsu 1985), this optimization problem can be solved via formulating

$$\frac{d\psi(t)}{dt} = -\mu \frac{\partial J(\psi(t), t)}{\partial \psi(t)}$$

where μ is the diagonal matrix diag$\{\frac{1}{2}\mu_1, \frac{1}{2}\mu_2, \mu_3\}$ chosen to minimize J along the direction of $-\frac{\partial J(\psi(t), t)}{\partial \psi(t)}$. The resulting set of differential equations can be found as (Karimi-Ghartemani and Ziarani 2003; Ziarani and Konrad 2004)

$$\begin{cases} \frac{dE(t)}{dt} = \mu_1 d \sin\theta, \\ \frac{d\omega(t)}{dt} = \mu_2 Ed \cos\theta, \\ \frac{d\theta(t)}{dt} = \omega + \mu_3 \frac{d\omega}{dt}. \end{cases} \qquad (15.2)$$

Since the variation of E is relatively small with comparison to the variation of d, the major dynamics of $\omega(t)$ is from d and the effect of E can then be combined with the proper selection of μ_2. As a result, the enhanced PLL can be constructed as shown in Figure 15.2.

Comparing the enhanced PLL shown in Figure 15.2 to the basic PLL shown in Figure 15.1(b), it can be seen that the enhanced PLL contains a voltage channel to estimate the amplitude of the input signal, in addition to the frequency channel that is very similar to the basic PLL.

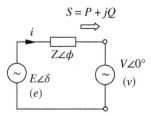

Figure 15.3 Power delivery to a voltage source through an impedance.

15.2 Brief Review of Droop Control

Figure 15.3 illustrates a voltage source $e = \sqrt{2}E \sin \theta$ with $\theta = \omega t + \delta$ delivering power to another voltage source (terminal) $v = \sqrt{2}V \sin \omega t$ through an impedance $Z \angle \phi$. The voltage source could be a conventional synchronous generator or a voltage-controlled inverter. Since the current flowing through the impedance is

$$\bar{I} = \frac{E \angle \delta - V \angle 0^\circ}{Z \angle \phi}$$
$$= \frac{E \cos \delta - V + jE \sin \delta}{Z \angle \phi},$$

the real power and reactive power delivered by the source to the terminal via the impedance can be obtained as

$$P = (\frac{EV}{Z} \cos \delta - \frac{V^2}{Z}) \cos \phi + \frac{EV}{Z} \sin \delta \sin \phi,$$
$$Q = (\frac{EV}{Z} \cos \delta - \frac{V^2}{Z}) \sin \phi - \frac{EV}{Z} \sin \delta \cos \phi,$$

where δ is the phase difference between the supply and the terminal, often called the power angle. This is the basis of the droop control (Brabandere et al. 2007; Guerrero et al. 2008; Guerrero et al. 2006; Guerrero et al. 2005; Yao et al. 2011; Zhong and Hornik 2013), which is widely adopted in power systems and recently in parallel-operated inverters.

When the impedance is inductive, $\phi = 90^\circ$. Then

$$P = \frac{EV}{Z} \sin \delta \quad \text{and} \quad Q = \frac{EV}{Z} \cos \delta - \frac{V^2}{Z}.$$

When δ is small, there are

$$P \approx \frac{EV}{Z} \delta \quad \text{and} \quad Q \approx \frac{V}{Z} E - \frac{V^2}{Z},$$

and, roughly,

$$P \sim \delta \quad \text{and} \quad Q \sim E.$$

As a result, the conventional droop control strategy for an inductive Z takes the form

$$E = E^* - nQ, \tag{15.3}$$

$$\omega = \omega^* - mP, \tag{15.4}$$

where E^* is the rated RMS system voltage. This strategy, as shown in Figure 15.4(a), consists of the $Q - E$ and $P - \omega$ droop, i.e. the voltage E is regulated by controlling the reactive power Q and the frequency f is regulated by controlling the real power P.

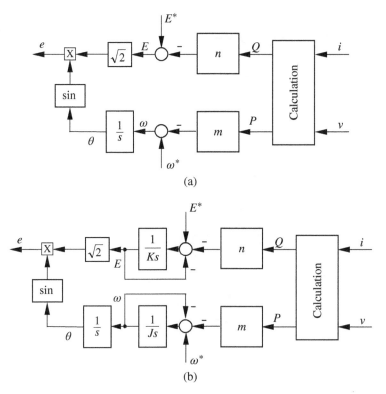

Figure 15.4 Conventional droop control scheme for an inductive impedance. (a) Without considering the integral effect. (b) With the hidden integral effect explicitly considered.

The droop control strategy takes different forms when the impedance is of different types (Zhong and Hornik 2013; Zhong and Zeng 2016), as shown in Figure 15.5. When the impedance is capacitive, the droop control still takes the form of $Q - E$ and $P - \omega$ droop but with positive signs. When the impedance is resistive, the droop control takes the form of $Q - \omega$ and $P - E$ droop. Note that the (output) impedance of an inverter is normally inductive but can be changed to resistive or capacitive and the corresponding inverters are called L-inverters, R-inverters, and C-inverters. See (Zhong and Hornik 2013; Zhong and Zeng 2014) for more detail. The conventional droop control strategy has some fundamental limitations and is not able to maintain accurate sharing of both real power and reactive power when there are component mismatches, parameter shifts, numerical error, disturbances and noise, etc. The robust droop controller presented in (Zhong 2013c; Zhong and Hornik 2013) will be discussed in detail in the next chapter to overcome these issues. It will be shown in the chapter after next that the robust droop controller for converters with resistive impedance is actually universal for inverters with different types of output impedance (Zhong and Zeng 2016). These do not affect the discussions in this chapter so the analysis will be based on the conventional droop control strategy for the sake of simplicity.

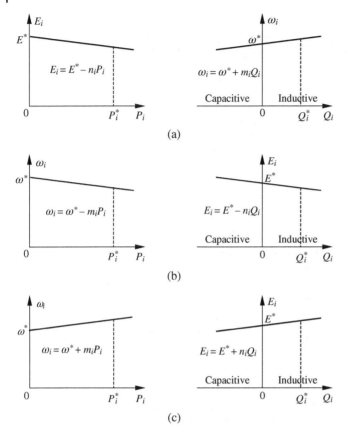

Figure 15.5 Conventional droop control strategies. (a) For resistive impedance. (b) For inductive impedance. (c) For capacitive impedance..

15.3 Structural Resemblance between Droop Control and PLL

15.3.1 When the Impedance is Inductive

One insightful observation about droop control pointed out in (Zhong 2013c) is that the voltage droop control actually includes an integrator because E can be obtained via dynamically integrating

$$\Delta E \triangleq E^* - E - nQ$$

until $\Delta E = 0$ instead of setting $E = E^* - nQ$ statically. This is also true for the frequency droop control, where the frequency ω can be obtained via integrating

$$\Delta \omega \triangleq \omega^* - \omega - mP$$

until $\Delta \omega = 0$. The droop control strategy shown in Figure 15.4(a) is shown in Figure 15.4(b) with this hidden integral effect explicitly considered, where the integral time constants are chosen as J and K for the frequency and voltage channels, respectively. This is equivalent to adding a low-pass filter $\frac{1}{Js+1}$ to the frequency (real power) channel and a low-pass filter

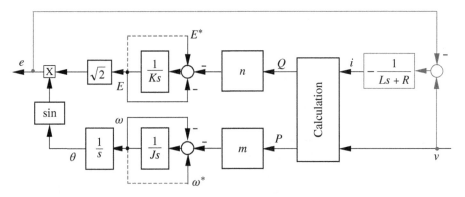

Figure 15.6 Linking the droop controller in Figure 15.4(b) and the (inductive) impedance.

$\frac{1}{Ks+1}$ to the voltage (reactive power) channel shown in Figure 15.4(a), respectively. In the steady state, the inputs to the integrators are zero, which recovers the droop control strategy (15.3) and (15.4). Apparently, Figure 15.4(b) becomes Figure 15.4(a) when the integral time constants are chosen as $K = 0$ and $J = 0$.

The current i flowing through the impedance $Z = Ls + R$ in Figure 15.3 is

$$i = -\frac{v - e}{Ls + R}.$$

This can be adopted to close the loop between v and e in Figure 15.4(b), as shown in Figure 15.6. Note that $i = 0$ when $e = v$ and, in this case, the voltage e accurately recovers or estimates the voltage v. In other words, the voltage e is synchronized with the input v.

Normally, the real power P and reactive power Q are calculated via measuring the terminal voltage v and the current i. Actually, it is better to use the voltage e than the terminal voltage v for this purpose because e is available internally. This leaves out the power losses of the filter inductor but it does not matter. The physical meaning of this is to droop the voltage and frequency according to the real power and the reactive power generated by the voltage source e. To some extent, this is more reasonable than using the terminal voltage v because it reflects the genuine real power and reactive power delivered by the voltage source e. In this case, the real power is

$$P = \frac{1}{T} \int_{t-T}^{t} e \times i \, dt, \tag{15.5}$$

where T is the period of the system. Applying the Laplace transform, this is equivalent to passing the instantaneous real power $p = e \times i$ through the hold filter

$$H(s) = \frac{1 - e^{-Ts}}{Ts}$$

to obtain the (averaged) real power P. The reactive power can be obtained similarly. The ghost e_g of the voltage e is

$$e_g = \sqrt{2}E \sin(\theta + \frac{\pi}{2}) = \sqrt{2}E \cos \theta,$$

and the reactive power can be calculated as

$$Q = -\frac{1}{T} \int_{t-T}^{t} e_g \times i \, dt. \tag{15.6}$$

For example, for the current $i = \sqrt{2}I \sin \theta_i$, there is

$$Q = -\frac{1}{T} \int_{t-T}^{t} 2EI \sin(\theta + \frac{\pi}{2}) \sin \theta_i dt = EI \sin(\theta - \theta_i),$$

which is indeed the reactive power generated by $e = \sqrt{2}E \sin(\theta)$ and i. Note that it is not compulsory to use the hold filter. A low-pass filter with the appropriate bandwidth could be used as well. This does not affect the main reasoning here.

When the droop controller is operated in the droop mode, the voltage set point E^* and the frequency set-point ω^* can be set as the rated system values when it is operated in the grid-connected mode or in the standalone mode. They can also be set as the grid voltage E and the grid frequency ω for grid-connected applications to send the desired real power P_{set} and reactive power Q_{set} to the grid (this is not shown in Figure 15.6 but can be easily implemented by changing $-P$ to $P_{set} - P$ and $-Q$ to $Q_{set} - Q$). If E^* is set as E and ω^* is set as ω^1, as shown in Figure 15.6 by the dashed lines, then the voltage e is the same as v in the steady state. This effectively cancels the loop around the integrators $\frac{1}{Js}$ and $\frac{1}{Ks}$. Hence, the block diagram shown in Figure 15.6 can be redrawn as shown in Figure 15.7(a), after connecting the dashed lines and calculating the power by using e, as described in (15.5) and (15.6). The gains are lumped as $K_e = \frac{n}{K}$ and $K_f = \frac{m}{J}$. This is similar to the widely used enhanced PLL (Karimi-Ghartemani and Iravani 2001, 2002) or the sinusoid tracking algorithm (Karimi-Ghartemani and Ziarani 2004; Ziarani and Konrad 2004) (which are essentially the same) shown in Figure 15.2, apart from three major differences: (1) the sin and cos functions are swapped; (2) there is a low-pass filter $\frac{1}{Ls+R}$, or an integrator when $R = 0$; (3) there is a negative sign in the amplitude channel of Figure 15.7(a). The hold filter $H(s)$ is to filter out the ripples and could/should be inserted into the EPLL/STA to improve the performance so it does not cause any major difference.

15.3.2 When the Impedance is Resistive

When the impedance Z is resistive, $\phi = 0°$. Then

$$P = \frac{EV}{Z} \cos \delta - \frac{V^2}{Z} \quad \text{and} \quad Q = -\frac{EV}{Z} \sin \delta.$$

When δ is small, there are

$$P \approx \frac{EV}{Z} \delta - \frac{V^2}{Z} \quad \text{and} \quad Q \approx -\frac{EV}{Z} \delta,$$

and, roughly,

$$P \sim E \quad \text{and} \quad Q \sim -\delta.$$

As a result, the conventional droop control strategy for resistive impedance takes the form

$$E = E^* - nP,$$

$$\omega = \omega^* + mQ.$$

---

1 Note that this just changes the operational point of the controller, without changing its structure.

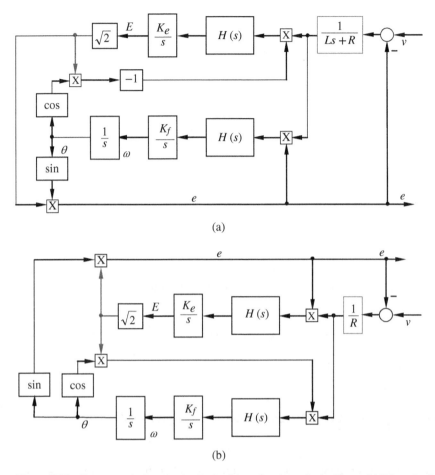

(a)

(b)

Figure 15.7 Droop control strategies in the form of a phase-locked loop. (a) When the impedance is inductive. (b) When the impedance is resistive.

The difference from the inductive case is that the positions of P and Q are swapped and the sign before Q is changed to positive.

Following the same reasoning in the previous section, this droop controller can be described in the form of a phase-locked loop as shown in Figure 15.7(b). Comparing it with the enhanced PLL or the STA shown in Figure 15.2, they are structurally the same, without any major differences. As explained before, the hold filter $H(s)$ is to filter out the ripples and could/should be included in the STA or EPLL to improve the performance so it does not cause any major difference. If the parameters are selected as $R = E$, $\mu_1 = K_e$, $\mu_2 = K_f$ and $\mu_3 = 0$, and the hold filter $H(s)$ is removed, then the two diagrams are exactly the same. In other words, this droop controller structurally resembles an enhanced phase-locked loop. It behaves as an enhanced phase-locked loop when there is no power exchanged with the terminal and it functions as a droop controller when it exchanges power with the terminal. As a result, the same droop controller can be utilized as a synchronization unit to achieve pre-synchronization and then as a droop controller to regulate the power flow.

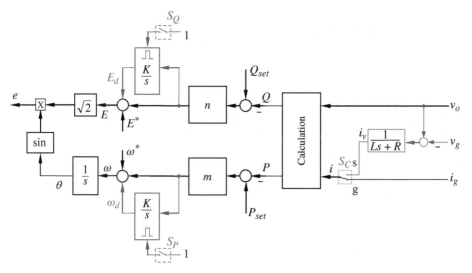

Figure 15.8 The conventional droop controller shown in Figure 15.4(a) after adding two integrators and a virtual impedance.

When the amplitude channel is not considered, the frequency channel is more or less the same as the basic PLL shown in Figure 15.1. In other words, the frequency droop control structurally resembles the basic PLL.

15.4 Operation of a Droop Controller as a Synchronization Unit

In order to demonstrate the above findings, the droop controller shown in Figure 15.4(a) for inverters with inductive output impedance is slightly changed so that the synchronization function can be explicitly demonstrated. As shown in Figure 15.8, two integrators are added to the voltage channel and the frequency channel, one each, to make the hidden integral effect explicit. A virtual impedance $sL + R$ is added to generate the virtual current i_v according to the voltage difference $v_o - v_g$. The current feeding into the power calculation block can be the grid current i_g or the virtual current i_v. The two integrators can be enabled or disabled by switches S_P and S_Q, respectively. This allows the droop controller to work in the synchronization mode or the set mode (sending P_{set} and Q_{set} to the grid), in addition to the normal droop mode (changing the real power and reactive power according to the grid frequency and voltage). In the synchronization mode, the virtual current i_v is used because the inverter is not connected to the grid and the grid current i_g is 0. After the inverter is synchronized with the grid, the circuit breaker can be turned ON. When the circuit breaker is ON, the switch S_C should be turned to Position g so that the grid current i_g is fed into the power calculation block, which operates the inverter in the set mode. Then, if needed, the switches S_P and S_Q can be turned ON to operate the inverter in the droop mode. The operation modes are summarized as shown in Table 15.1. Note that the switches S_P and S_Q

Table 15.1 Operation modes.

Mode	Switch S_C	Switch S_P	Switch S_Q
Synchronization mode	s	OFF	OFF
Set mode	g	OFF	OFF
Droop mode	g	ON	ON

Table 15.2 Parameters of the inverter.

Parameter	Value
Grid voltage (RMS)	110 V
Line frequency f	50 Hz
Switching frequency f_s	19 kHz
DC-bus voltage V_{DC}	200 V
Rated apparent power S	300 VA
Inductance L_s	2.2 mH
Resistance R_s	0.2 Ω
Inductance L_g	2.2 mH
Resistance R_g	0.2 Ω
Capacitance C	10 μF

can be operated independently when the switch S_C is at position g so it is possible to operate the real power and the reactive power in the set mode or the droop mode independently.

15.5 Experimental Results

A single-phase inverter controlled by the controller shown in Figure 15.8 was built and tested. The parameters of the system are shown in Table 15.2. The control circuit of the system was constructed based on TMS320F28335 DSP, with the sampling frequency of 4 kHz. The droop coefficients used in the experiments are calculated as $n = \frac{0.1K_e E^*}{S}$ and $m = \frac{0.01\omega^*}{S}$, where S is the rated apparent power of the inverter, according to (Zhong 2013c), so that 10% increase in the voltage E results in 100% decrease in the reactive power Q and 1% increase in the frequency f results in 100% decrease in the real power P.

15.5.1 Synchronization with the Grid

The time needed for synchronization is different for different voltage v_g. Here, two typical cases with $v_g = 0$ and $v_g = V_g$ are considered. The corresponding results are shown in Figures 15.9(a) and (b), respectively. For the case with $v_g = 0$ when the synchronization was

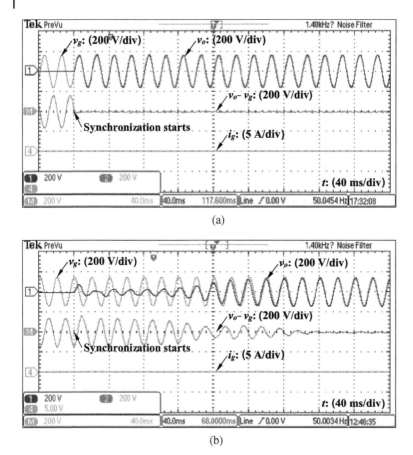

(a)

(b)

Figure 15.9 The synchronization capability of the droop controller shown in Figure 15.8. (a) When v_g crosses 0. (b) When v_g is at the peak value V_g.

started, as shown in Figure 15.9(a), the voltage difference between the output voltage and the grid voltage, i.e. $v_o - v_g$, quickly became very small. It took less than one cycle for the whole synchronization process. For the case with $v_g = V_g$ when the synchronization was started, as shown in Figure 15.9(b), the synchronization took longer, about 12 cycles. This is still acceptable for grid-connected inverters. This shows indeed the droop controller can be applied to achieve synchronization before connecting the inverter to the grid, without the need of a dedicated synchronization unit.

15.5.2 Connection to the Grid

After the synchronization process is finished, the inverter is ready to be connected to the grid. This involves turning the relay ON and the switch S_C to the position g, which shifts the current used for calculating P and Q from the virtual current i_v to the grid current i_g. As shown in Figure 15.10, the grid current i_g was well maintained around zero without large spikes, as expected because $P_{set} = 0$ and $Q_{set} = 0$.

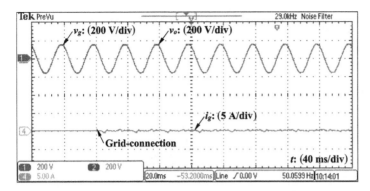

Figure 15.10   Connection of the droop controlled inverter to the grid.

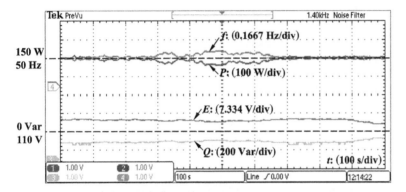

Figure 15.11   Regulation of the grid frequency and voltage in the droop mode.

15.5.3   Operation in the Droop Mode

In order to test the droop function of regulating P and Q corresponding to the variations of f and E, the inverter was continuously operated in the droop mode while being connected to the public grid. The results are shown in Figure 15.11. The real power P is almost symmetrical to the grid frequency f while the reactive power Q is symmetrical to the voltage E, as expected. When the frequency is higher (lower) than the rated frequency, the real power is automatically reduced (increased) proportionally. Similar regulation capability can be seen from the reactive power against the voltage. It is worthy emphasizing that the inverter kept in synchronization with the grid after being connected to the grid, without a dedicated synchronization unit. The synchronization is achieved by the droop controller itself. The voltage did not change much during the experiment but the trend is very clear.

15.5.4   Robustness of Synchronization

In order to test the robustness of the synchronization, the DC-bus voltage V_{DC} was changed when the system was operated in the set mode with $P_{set} = 150$ W and $Q_{set} = 150$ *Var*. At first, V_{DC} was suddenly dropped from 200 V to 180 V. As shown in Figure 15.12(a), there was no problem with the synchronization. The grid current i_g dropped initially because of

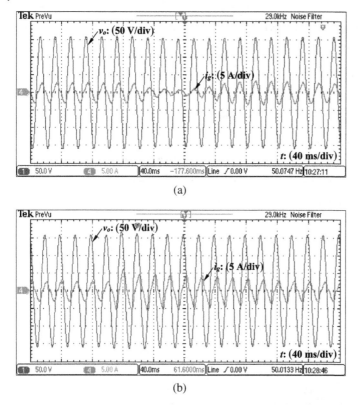

(a)

(b)

Figure 15.12 Robustness of synchronization against DC-bus voltage changes. (a) When the DC-bus voltage V_{DC} was changed from 200 V to 180 V. (b) When the DC-bus voltage V_{DC} was changed from 180 V to 200 V.

the dropped V_{DC} but, after about 10 cycles, it recovered to the original value before the voltage change to maintain the real power and reactive power sent to the grid. Then, V_{DC} was suddenly increased from 180 V to 200 V. Again, there was no problem with the synchronization, as shown in Figure 15.12(b). The grid current i_g increased initially but after about 10 cycles it recovered to the original value before the voltage change.

15.5.5 Change in the Operation Mode

The frequency, voltage, real power, and reactive power of the system when the mode of the droop controller was changed in the sequence of the synchronization mode, connection to the grid, the set mode. and the droop mode are shown in Figure 15.13. At $t = 0$ s, the synchronization was enabled. As shown in Figure 15.13, both the real power and the reactive power were controlled around zero. When the inverter was connected to the grid at 3 s, there was not much transient and both the real and reactive power were maintained around zero. The real and reactive power references were changed at $t = 6$ s and $t = 9$ s to 150 W and 150 Var, respectively. The real power and reactive power responded quickly, with some coupling effect. There was some dynamics in the frequency but it settled down. The voltage E increased because of the increased real power and then the increased reactive

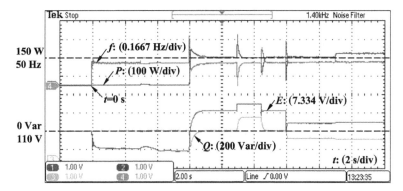

Figure 15.13 System response when the operation mode was changed.

power output. At $t = 10.5$ s, Q_{set} was changed back to 0. At $t = 12$ s, the droop mode was enabled for the reactive power. The reactive power started changing according to the voltage, nearly symmetrically. At $t = 15$ s, the droop mode was enabled for the real power. The real power started changing according to the frequency, nearly symmetrically as well. This has demonstrated that the droop controller with the changes shown in Figure 15.8 can function properly without a dedicated synchronization unit.

15.6 Summary

Based on (Zhong and Boroyevich 2013, 2016), it has been shown in this chapter that a droop controller structurally resembles an enhanced phase-locked loop. This establishes a link between the droop control community and the PLL community, and offers fundamental understanding about the operation of power systems dominated by droop-controlled renewable energy sources interfaced by inverters. As a result, there is no need to have a dedicated synchronization unit in addition to the droop controller for synchronization. This avoids the problems brought by the PLL to grid-tied inverters, e.g. competition with each other, difficulties in tuning PLLs, performance degradation, etc. Moreover, droop control offers a natural technical route to implement SYNDEM smart grids.

The link between droop control and PLLs is shown for the case when the impedance is resistive, but can be easily extended to investigate the cases when the impedance is inductive or capacitive to find the equivalent structure of PLLs. Indeed, the case with inductive output impedance has been demonstrated by extensive experimental results.

16

Robust Droop Control

In the previous Chapter, it was shown that a droop controlled converter structurally resembles a PLL. In this chapter, it will be shown at first that the conventional droop control scheme has some fundamental limitations. The conventional droop control scheme requires all the converters in parallel operation: (1) to have the same per-unit impedance; (2) to generate the same voltage set point for the inverters, in order for the converters to share the load accurately in proportion to their power ratings. Both conditions are difficult to meet in practice, which results in errors in proportional load sharing. Moreover, there is a fundamental trade-off between the power sharing accuracy and the voltage regulation capability. A robust droop controller is then presented to achieve accurate proportional load sharing without these limitations for R-, L-, and C-inverters. The load voltage can be maintained within the desired range around the rated value. The strategy is robust against numerical errors, disturbances, noises, feeder impedance, parameter drifts, component mismatches, etc. The only sharing error comes from the error in measuring the load voltage. When there are errors in the voltage measured, a trade-off between the voltage drop and the sharing accuracy appears. It is also explained that, in order to avoid errors in power sharing, the global settings of the rated voltage and frequency should be accurate. The technology is patented.

16.1 Control of Inverter Output Impedance

It is well known (Zhong and Hornik 2013) that the filter capacitor of an inverter can be regarded as a part of the load and, as a result, the output impedance of the inverter is inductive. Such inverters are referred to as L-inverters. The output impedance of an inverter can be controlled to be resistive and capacitive, respectively. The corresponding inverters are referred to as R-inverters and C-inverters.

16.1.1 Inverters with Inductive Output Impedances (L-inverters)

The circuit of a single-phase inverter under consideration is shown in Figure 16.1(a). It consists of a single-phase H-bridge inverter powered by a DC source, and an LC filter. The inverter is connected to the AC-bus via a circuit breaker (CB) and the load is assumed to be connected to the AC-bus. The control signal u is converted to a PWM signal to drive

Power Electronics-Enabled Autonomous Power Systems: Next Generation Smart Grids,
First Edition. Qing-Chang Zhong.
© 2020 John Wiley & Sons Ltd. Published 2020 by John Wiley & Sons Ltd.

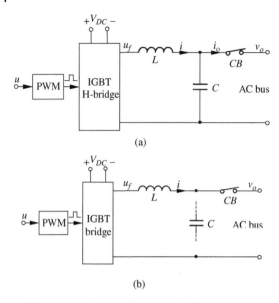

(a)

(b)

the H-bridge so that the average of u_f over a switching period is the same as u, i.e. $u \approx u_f$. Hence, the PWM block and the H-bridge can be ignored in the controller design (Zhong 2013c).

Although the inverter consists of an LC filter, the capacitor (C) can be regarded as a part of the load instead of a part of the inverter. This reduces the control plant to an H-bridge and an inductor, as shown in Figure 16.1(b). The advantages of this are: (1) it reduces the order of the control plant to be 1; (2) it considerably simplifies the design and analysis of the controller, which facilitates the understanding of the nature of inverter control.

Since the average of u_f over a switching period is the same as u, which is the same as the reference voltage v_r, there is approximately

$$u = v_r \approx u_f = v_o + sLi,$$

which gives

$$v_o = v_r - Z_o(s) \cdot i$$

with

$$Z_o(s) = sL.$$

That is, the output impedance $Z_o(s)$ is inductive. Inverters with inductive output impedances are called L-inverters.

16.1.2 Inverters with Resistive Output Impedances (R-inverters)

The inductor current i can be measured to construct a proportional controller K_i, as shown in Figure 16.2, so that the output impedance of the inverter is forced to be resistive and dominates the impedance between the inverter and the AC-bus. This is also widely referred to as a virtual resistor (Dahono et al. 2001; Dahono 2003, 2004; Li 2009) to dampen the resonance of the LC filter.

The following two equations hold for the closed-loop system consisting of Figure 16.1(b) and Figure 16.2:

$$u = v_r - K_i i,$$

$$u_f = sLi + v_o.$$

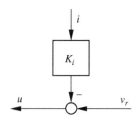

Since the average of u_f over a switching period is the same as u, there is approximately

$$v_r - K_i i = sLi + v_o,$$

Figure 16.2 Controller to achieve a resistive output impedance.

which gives

$$v_o = v_r - Z_o(s) \cdot i$$

with

$$Z_o(s) = sL + K_i.$$

If the gain K_i is chosen big enough, the effect of the inductance is not significant and the output impedance can be made nearly purely resistive over a wide range of frequencies. Then, the output impedance is roughly

$$Z_o(s) \approx K_i,$$

which is independent of the inductance. However, a big K_i causes considerable harmonic components in the output voltage for nonlinear loads and, hence, small K_i is preferred in order to achieve low THD in the output voltage. How to reduce the THD of the output voltage while using large K_i can be found in (Zhong and Hornik 2013). Note that when $K_i = 0$ the output impedance becomes inductive and the inverter becomes an L-inverter.

With the above control strategy, the inverter can be approximated as a controlled ideal voltage supply v_r cascaded with a resistive output impedance R_o described as

$$v_o = v_r - R_o i \tag{16.1}$$

with

$$R_o = K_i.$$

Such inverters are called R-inverters. Note that $v_o \approx u = v_r$ if no load is connected.

16.1.3 Inverters with Capacitive Output Impedances (C-inverters)

The inductor current i can be measured to construct an integral controller $\frac{1}{sC_o}$, as shown in Figure 16.3, so that the output impedance of the inverter is forced to be capacitive and dominates the impedance between the inverter and the AC-bus (Zhong and Zeng 2011, 2014). This is equivalent to having a virtual capacitor C_o connected in series with the filter inductor L, as will be shown later.

The following two equations hold for the closed-loop system consisting of Figure 16.1(b) and Figure 16.3:

$$u = v_r - \frac{1}{sC_o}i \quad \text{and} \quad u_f = sLi + v_o.$$

Since the average of u_f over a switching period is the same as u, there is approximately

$$v_r - \frac{1}{sC_o}i = sLi + v_o,$$

which leads to

$$v_o = v_r - Z_o(s) \cdot i,$$

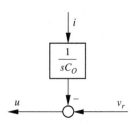

Figure 16.3 Controller to achieve a capacitive output impedance.

with the output impedance $Z_o(s)$ of the inverter given as

$$Z_o(s) = sL + \frac{1}{sC_o}. \tag{16.2}$$

If the capacitor C_o is chosen small enough, the effect of the inductor is not significant and the output impedance can be made nearly purely capacitive at the fundamental frequency, i.e. roughly

$$Z_o(s) \approx \frac{1}{sC_o}.$$

Such inverters are called C-inverters.

Figure 16.4 illustrates the typical output impedances of R-, L-, and C-inverters. Around the fundamental frequency, the impedances are resistive, inductive and capacitive. At very high frequencies, the impedances are all capacitive because of the capacitor in the LC filter. There is also a resonant peak between the L and the C of the LC filter. The resonant peak of the R-inverter is well damped.

16.2 Inherent Limitations of Conventional Droop Control

16.2.1 Basic Principle

As explained in the previous chapter, the conventional droop control depends on the type of the inverter output impedance. In order to facilitate the presentation, the case with two R-inverters in parallel operation as shown in Figure 16.5 is taken as an example.

In this case, the real and reactive power of each inverter injected into the bus are

$$P_i = \frac{E_i V_o \cos \delta_i - V_o^2}{R_{oi}}, \tag{16.3}$$

$$Q_i = -\frac{E_i V_o}{R_{oi}} \sin \delta_i \tag{16.4}$$

for $i = 1, 2$. The corresponding conventional droop controller is

$$E_i = E^* - n_i P_i, \tag{16.5}$$

$$\omega_i = \omega^* + m_i Q_i, \tag{16.6}$$

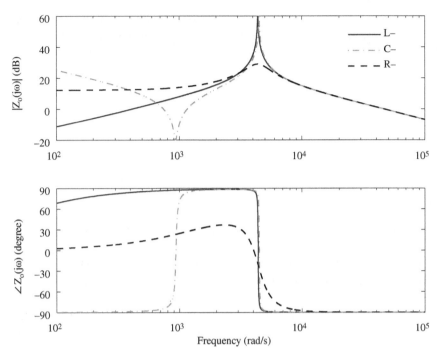

Figure 16.4 Typical output impedances of L-, R-, and C-inverters.

Figure 16.5 Two R-inverters operated in parallel.

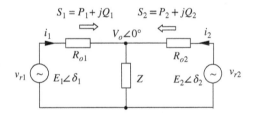

Figure 16.6 Conventional droop control scheme for R-inverters.

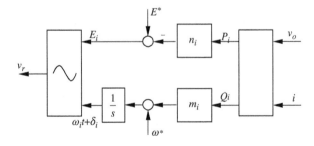

as shown in Figure 16.6. The droop coefficients n_i and m_i are normally determined by the desired voltage drop ratio $\frac{n_i P_i^*}{E^*}$ and frequency boost ratio $\frac{m_i Q_i^*}{\omega^*}$, respectively, at the rated real power P^* and reactive power Q^*. The frequency ω_i is integrated to form the phase of the voltage reference v_{ri}.

In order for the inverters to share the load in proportion to their power ratings, the droop coefficients of the inverters should be in inverse proportion to their power ratings (Tuladhar et al. 1997), i.e. n_i and m_i should be chosen to satisfy

$$n_1 S_1^* = n_2 S_2^*,\tag{16.7}$$

$$m_1 S_1^* = m_2 S_2^*.\tag{16.8}$$

It is easy to see that n_i and m_i also satisfy

$$\frac{n_1}{m_1} = \frac{n_2}{m_2}.$$

16.2.2 Experimental Phenomena

Figure 16.7 shows the experimental results when the conventional droop controller was applied to a system with two R-inverters in parallel operation. The DC bus of the two inverters were provided by two 42 V DC power supplies separately. The values of the inductors and capacitors of the inverters were 2.35 mH and 22 μF, respectively. The switching frequency was 7.5 kHz and the frequency of the system was 50 Hz. The nominal output voltage was 12 V RMS and the load was about 9 Ω. The virtual resistance K_i was chosen as 4 Ω for both inverters and the droop coefficients were: $n_1 = 0.4$ and $n_2 = 0.8$; $m_1 = 0.1$ and $m_2 = 0.2$. It was expected that $P_1 = 2P_2$ and $Q_1 = 2Q_2$.

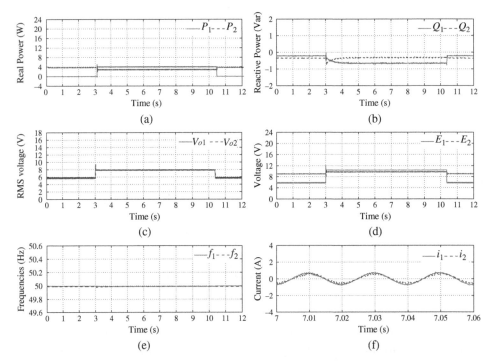

Figure 16.7 Experimental results: two R-inverters in parallel with conventional droop control. (a) Real power. (b) Reactive power. (c) Load voltage. (d) Voltage set point. (e) Frequency. (f) Currents.

A linear load of about 9 Ω was supplied by inverter 2 initially. Inverter 1 was connected to the system at around $t = 3$ s and was then disconnected at around $t = 10.5$ s. The reactive power was shared accurately (in the ratio of 2 : 1) but the real power was not. The load voltage was only about 8 V, far from the rated voltage 12 V.

16.2.3 Real Power Sharing

Substituting (16.5) into (16.3), then the real power of the two inverters can be obtained as

$$P_i = \frac{E^* \cos \delta_i - V_o}{n_i \cos \delta_i + R_{oi}/V_o}. \tag{16.9}$$

Substituting (16.9) into (16.5), the voltage amplitude deviation of the two inverters is

$$\Delta E = E_2 - E_1 = \frac{E^* \cos \delta_1 - V_o}{\cos \delta_1 + \frac{R_{o1}}{n_1 V_o}} - \frac{E^* \cos \delta_2 - V_o}{\cos \delta_2 + \frac{R_{o2}}{n_2 V_o}}. \tag{16.10}$$

It is known from (Tuladhar et al. 1997) that the voltage deviation of the two units leads to considerable errors in load sharing. Indeed, in order to achieve proportional sharing, i.e. for

$$n_1 P_1 = n_2 P_2 \quad \text{or} \quad \frac{P_1}{S_1^*} = \frac{P_2}{S_2^*}$$

to hold, the voltage deviation ΔE should be 0 according to (16.5). This is a very strict condition because there are always numerical computational errors, disturbances, parameter drifts, and component mismatches. This condition is satisfied if

$$\frac{n_1}{R_{o1}} = \frac{n_2}{R_{o2}} \tag{16.11}$$

and

$$\delta_1 = \delta_2, \tag{16.12}$$

according to (16.10). In other words, n_i should be chosen to be proportional to its output impedance R_{oi}.

Taking (16.7) into account, in order to achieve accurate sharing of real power, the (resistive) output impedance should be designed to satisfy

$$R_{o1} S_1^* = R_{o2} S_2^*. \tag{16.13}$$

Since the per-unit output impedance of inverter i is

$$\gamma_i = \frac{R_{oi}}{E^*/I_i^*} = \frac{R_{oi} S_i^*}{(E^*)^2},$$

the condition (16.13) is equivalent to

$$\gamma_1 = \gamma_2.$$

This simply means that the per-unit output impedances of all inverters operated in parallel should be the same in order to achieve accurate proportional real power sharing for the conventional droop control scheme. If this is not met, then the voltage set points E_i are not the same and errors appear in real power sharing.

According to (16.5), the real power deviation ΔP_i due to the voltage set-point deviation ΔE_i is

$$\Delta P_i = -\frac{1}{n_i}\Delta E_i.$$

For two inverters operated in parallel with real power consumption of P_1 and P_2, the relative sharing error is defined (Zhong 2013c) as

$$e_p\% = \left(\frac{P_1}{n_2} - \frac{P_2}{n_1}\right)\frac{n_1 + n_2}{P_1 + P_2}.$$

When $P_1 + P_2 = P_1^* + P_2^*$, the relative real power sharing error due to the voltage set point deviation $\Delta E = E_2 - E_1 = \Delta E_2 - \Delta E_1$ is

$$e_p\% = \frac{P_1}{P_1^*} - \frac{P_2}{P_2^*} = \frac{\Delta P_1}{P_1^*} - \frac{\Delta P_2}{P_2^*} = \frac{E^*}{n_i P_i^*}\frac{\Delta E}{E^*},$$

where $\frac{E^*}{n_i P_i^*}$ is the inverse of the voltage drop ratio at the rated power for inverter i. The smaller the droop coefficient (or the voltage drop ratio), the bigger the sharing error; the bigger the voltage set point deviation ΔE, the bigger the sharing error. In other words, there is a fundamental trade-off between the power sharing and the voltage regulation for the conventional droop control.

16.2.4 Reactive Power Sharing

When the system is in the steady state, the two inverters work under the same frequency, i.e. $\omega_1 = \omega_2$. It is well known that this guarantees the accuracy of reactive power sharing for R-inverters (or the accuracy of real power sharing for L-inverters); see e.g. (Li and Kao 2009). Indeed, from (16.6), there is

$$m_1 Q_1 = m_2 Q_2$$

when the frequencies ω_i are the same. Since the coefficients m_i are chosen to satisfy (16.8), reactive power sharing proportional to their power ratings is (always) achieved, i.e.

$$\frac{Q_1}{S_1^*} = \frac{Q_2}{S_2^*}.$$

16.3 Robust Droop Control of R-inverters

16.3.1 Control Strategy

The conventional droop controller can be improved to avoid the limitations. As a matter of fact, the voltage droop (16.5) can be re-written as

$$\Delta E_i = E_i - E^* = -n_i P_i,$$

and the voltage E_i can be implemented via integrating ΔE_i, that is,

$$E_i = \int_0^t \Delta E_i dt.$$

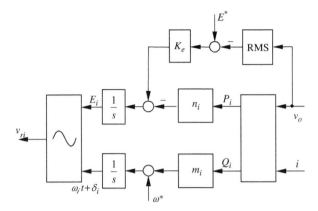

Figure 16.8 Robust droop controller for R-inverters.

This introduces an integrator into the voltage amplitude channel.

At the same time, in order to improve the voltage regulation capability, the voltage drop $E^* - V_o$ can be fed back and added to ΔE_i via an amplifier K_e, as shown in Figure 16.8. This actually leads to a robust droop control strategy that is able to achieve accurate power sharing and good voltage regulation even when there are computational errors, noises, and disturbances.

In the steady state, the input to the integrator should be 0. Hence,

$$n_i P_i = K_e(E^* - V_o). \tag{16.14}$$

The right-hand side of the above equation is always the same for all inverters operated in parallel as long as K_e is chosen the same, which can be easily met. Hence,

$$n_i P_i = \text{constant},$$

which guarantees accurate real power sharing without having the same E_i. This is more natural than the case with the conventional droop controller. The accuracy of real power sharing no longer depends on the inverter output impedances (including the feeder impedance) and is also immune to numerical computational errors and disturbances.

16.3.2 Error due to Inaccurate Voltage Measurements

The only possible error in the real power sharing comes from the error in measuring the RMS value of the load voltage. From (16.14), the real power deviation ΔP_i due to the error ΔV_{oi} in the measurement of the RMS voltage is

$$\Delta P_i = -\frac{K_e}{n_i} \Delta V_{oi}.$$

For two inverters operated in parallel with $P_1 + P_2 = P_1^* + P_2^*$, the relative real power sharing error due to the error in the measurement of the RMS voltage $\Delta V_o = \Delta V_{o2} - \Delta V_{o1}$ is

$$e_P\% = \frac{P_1}{P_1^*} - \frac{P_2}{P_2^*} = \frac{\Delta P_1}{P_1^*} - \frac{\Delta P_2}{P_2^*} = \frac{K_e E^*}{n_i P_i^*} \frac{\Delta V_o}{E^*}.$$

This characterizes the percentage error $e_P\%$ of the real power sharing with respect to the percentage error $\frac{\Delta V_o}{E^*}$ of the RMS voltage measurement. The term $\frac{K_e E^*}{n_i P_i^*}$ is the inverse of the voltage drop ratio with respect to the rated voltage at the rated power. If all inverters measure the voltage at the same point accurately, then the error ΔV_o can be made zero and accurate proportional sharing can be achieved.

16.3.3 Voltage Regulation

The strategy also improves the voltage regulation capability. From (16.14), the load voltage is

$$V_o = E^* - \frac{n_i}{K_e} P_i = E^* - \frac{n_i P_i}{K_e E^*} E^*,$$

where $\frac{n_i P_i}{K_e E^*}$ is the voltage drop ratio. Note that the voltage drop ratio here is the overall effective voltage drop ratio and can be designed, unlike the case of the conventional droop controller. The voltage drop here is no longer determined by the output impedance originally designed but by the parameters n_i, K_e and the actual power P_i. It can be considerably reduced by using a large K_e. If there are errors in the RMS voltage measurement, then the trade-off between the voltage drop and the accuracy of power sharing has to be made because the voltage drop is proportional to $\frac{n_i}{K_e}$ but the sharing error is inverse proportional to $\frac{n_i}{K_e}$.

Here is a calculation example. Assume that the voltage drop ratio at the rated power is $\frac{n_i P_i^*}{K_e E^*} = 10\%$ and the error in the RMS voltage measurement is $\frac{\Delta V_o}{E^*} = 0.5\%$, whether because the local voltages of inverters are measured or because the sensors are not accurate. Then, the error in the real power sharing is $\frac{K_e E^*}{n_i S_i^*} \frac{\Delta V_o}{E^*} = \frac{0.5\%}{10\%} = 5\%$, which is reasonable.

It is worth noting that the robust droop control still contains the voltage droop function. What is different from the conventional droop control is that the voltage droop is applied to the output voltage V_o but not to the voltage set point E. This small change is able to improve the performance significantly.

16.3.4 Error due to the Global Settings for E^* and ω^*

This subsection is devoted to the sensitivity analysis of the error in the global settings E^* and ω^* for the robust droop controller.

Any small error $\Delta \omega_i$ in ω_i^* would lead to the reactive power deviation (if still stable) of

$$\Delta Q_i = -\frac{1}{m_i} \Delta \omega_i,$$

according to (16.6). For two inverters operated in parallel with $Q_1 + Q_2 = Q_1^* + Q_2^*$, the relative reactive power sharing error due to the error $\Delta \omega = \omega_2^* - \omega_1^* = \Delta \omega_2 - \Delta \omega_1$ is

$$e_Q\% = \frac{Q_1}{Q_1^*} - \frac{Q_2}{Q_2^*} = \frac{\Delta Q_1}{Q_1^*} - \frac{\Delta Q_2}{Q_2^*} = \frac{\omega^*}{m_i Q_i^*} \frac{\Delta \omega}{\omega^*},$$

where $\frac{\omega^*}{m_i Q_i^*}$ is the inverse of the frequency boost ratio at the rated reactive power. The smaller the frequency boost ratio, the bigger the reactive power sharing error; the bigger the error $\Delta\omega$, the bigger the sharing error. For example, for a typical frequency boost ratio of $\frac{m_i Q_i^*}{\omega^*} = 1\%$, the error of $\frac{\Delta\omega}{\omega^*} = 1\%$ in the frequency setting would lead to 100% error in the reactive power sharing! Hence, the accuracy of reactive power sharing is very sensitive to the accuracy of the global setting for ω^*, which should be made very accurate.

Similarly, according to (16.14), the real power deviation ΔP_i due to the error ΔE_i^* in E_i^* is

$$\Delta P_i = \frac{K_e}{n_i} \Delta E_i^*.$$

For two inverters operated in parallel with $P_1 + P_2 = P_1^* + P_2^*$, the relative real power sharing error due to the error $\Delta E^* = E_2^* - E_1^* = \Delta E_2^* - \Delta E_1^*$ in the global settings of E^* is

$$e_P\% = \frac{P_1}{P_1^*} - \frac{P_2}{P_2^*} = \frac{\Delta P_1}{P_1^*} - \frac{\Delta P_2}{P_2^*} = -\frac{K_e E^*}{n_i P_i^*} \frac{\Delta E^*}{E^*}.$$

For a typical voltage drop ratio at the rated power of $\frac{n_i P_i^*}{K_e E^*} = 10\%$, a 10% error in $\frac{\Delta E^*}{E^*}$ would lead to a -100% error in the real power sharing. Although the error in E^* is less sensitive than the error in ω^*, it is still quite significant. Hence, in practice, it is very important to make sure that the global settings are accurate and the same for all players.

16.3.5 Experimental Results

The robust droop control strategy is verified in a laboratory setup consisting of two single-phase inverters controlled by dSPACE kits and powered by separate 42 V DC voltage supplies. The inverters are connected to the AC-bus via a circuit breaker (CB) and the load is assumed to be connected to the AC-bus. The values of the inductors and capacitors are 2.35 mH and 22 μF, respectively. The switching frequency is 7.5 kHz and the frequency of the system is 50 Hz. The rated voltage is 12 V RMS and $K_e = 10$. The droop coefficients are: $n_1 = 0.4$ and $n_2 = 0.8$; $m_1 = 0.1$ and $m_2 = 0.2$. Hence, it is expected that $P_1 = 2P_2$ and $Q_1 = 2Q_2$. Due to the configuration of the hardware setup, the voltage for inverter 2 was measured by the controller of inverter 1 and then sent out via a DAC channel, which was then sampled by the controller of inverter 2. This brought some latency into the system but the effect was not noticeable.

16.3.5.1 Inverters having Different Per-unit Output Impedances with a Linear Load

Both K_i were chosen as 4 for the two inverters to intentionally make the per-unit output impedances of the two inverters significantly different. In reality, this could be due to different feeder impedances or component mismatches.

A linear load of about 9 Ω was connected to inverter 2 initially. Inverter 1 was connected to the system at around $t = 3$ s and was then disconnected at around $t = 10.5$ s. The relevant curves from the experiment with the robust droop controller are shown in the left column of Figure 16.9 and the relevant curves from the experiment with the conventional droop controller are shown in the right column of Figure 16.9. In both cases, the reactive power was shared accurately (in the ratio of 2 : 1), although the actual values were different (because the voltages were different). Inverter 1 was able to pick up the load, gradually in the case of the robust droop controller and very quickly in the case of the conventional droop controller. The robust droop controller could share the real power very accurately but

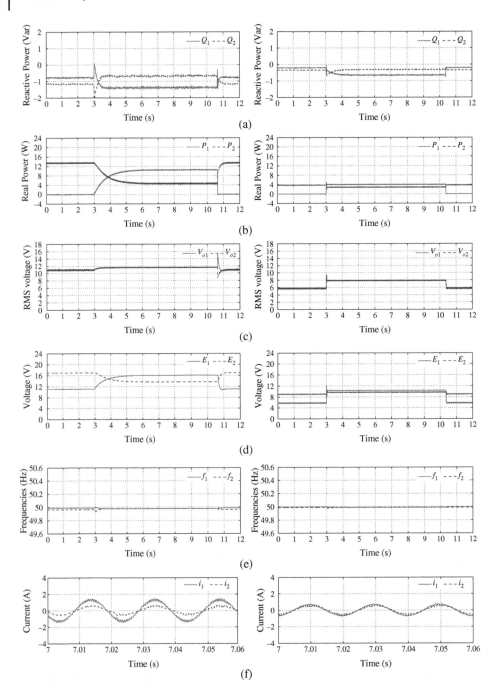

Figure 16.9 Experimental results for the case with a linear load when inverters have different per-unit output impedance: with the robust droop controller (left column) and with the conventional droop controller (right column). (a) Reactive power. (b) Real power. (c) RMS value of the load voltage. (d) Voltage set point. (e) Frequency. (f) Current in the steady state.

the conventional one could not. The robust droop controller has considerably relaxed the trade-off between the sharing accuracy and the voltage drop. The voltage from the inverters equipped with the robust droop controller is very close to the rated voltage but the voltage from the inverters equipped with the conventional controller is only two-thirds of the rated voltage. The voltage set point of the conventional droop controller has to be lower than the rated voltage due to the droop effect. The bigger the voltage drop ratio, the lower the voltage set point. It can also be clearly seen that $E_1 \neq E_2$ because the per-unit output impedances are different and also there are numerical errors and component mismatches, etc. Because of the reduced deviation in the voltage, the reactive power becomes bigger. This leads to a slightly bigger deviation in the frequency but it is expected because of the $Q - \omega$ droop. The current sharing reflects the power sharing well. It is worth noting that there was no need to change the operation mode of inverter 2 when connecting or disconnecting inverter 1.

16.3.5.2 Inverters having the Same Per-unit Output Impedance with a Linear Load

The current feedback gains were chosen as $K_{i1} = 2$ and $K_{i2} = 4$ so that the output impedances are consistent with the power sharing ratio 2 : 1. The results from the robust droop controller with $K_e = 10$ are shown in the left column of Figure 16.10 and the results from the conventional droop controller are shown in the right column of Figure 16.10. The robust droop controller was able to share the load according to the sharing ratio and considerably outperformed the conventional droop controller in terms of sharing accuracy and voltage drop. The difference between the voltage set points can be clearly seen. This indicates the effect of numerical errors, parameter drifts and component mismatches, etc. because the voltage set points were supposed to be identical without these uncertain factors. Comparing the left columns of Figures 16.9 and 16.10, there were no noticeable changes in the performance for the robust droop controller but the difference in the voltage set points was decreased. Comparing the right columns of Figures 16.9 and 16.10, the sharing accuracy and the voltage drop were improved slightly and the voltage set points became closer to each other when the per-unit output impedances were the same.

Another experiment was carried out with $K_e = 1$ to demonstrate the role of K_e and the results are shown in the right column of Figure 16.11. The results with $K_e = 10$ shown in the left column of Figure 16.10 are shown in the left column of Figure 16.11 again for comparison. It can be seen that a large K_e helps speed up the response and reduce the voltage drop. However, a large K_e causes large ripples in the current.

16.3.5.3 With a Nonlinear Load

A nonlinear load, consisting of a rectifier loaded with an LC filter and the same rheostat used in the previous experiments, was connected to inverter 2 initially. A similar procedure to connect/disconnect inverter 1 was followed in the experiment. The relevant curves of the experiment are shown in the left column of Figure 16.12 for the case with the robust droop controller and in the right column of Figure 16.12 for the case with the conventional droop controller. It can be seen that the two inverters with the robust droop controllers were still able to share the load very accurately in the ratio of 2 : 1, although $E_1 \neq E_2$. The dynamic performance did not change much either. The circulating current is very small and does not contain noticeable fundamental component. It should be emphasized that the active power sharing is still very accurate although the output impedances of the inverters are not

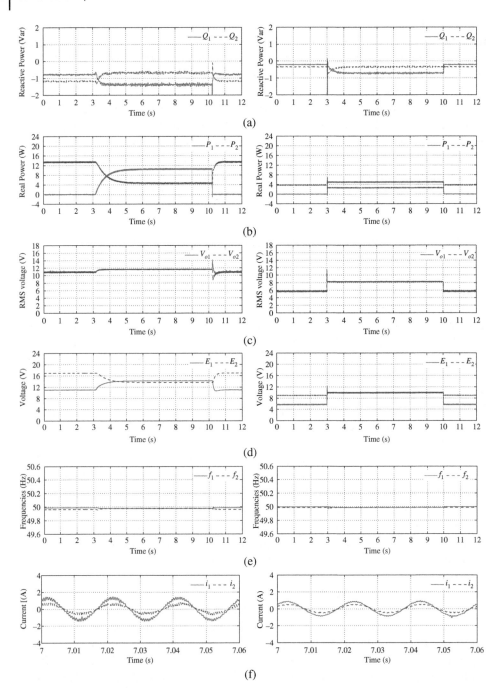

Figure 16.10 Experimental results for the case with a linear load when inverters have the same per-unit impedance: with the robust droop controller (left column) and with the conventional droop controller (right column). (a) Reactive power. (b) Real power. (c) RMS value of the load voltage. (d) Voltage set point. (e) Frequency. (f) Current in the steady state.

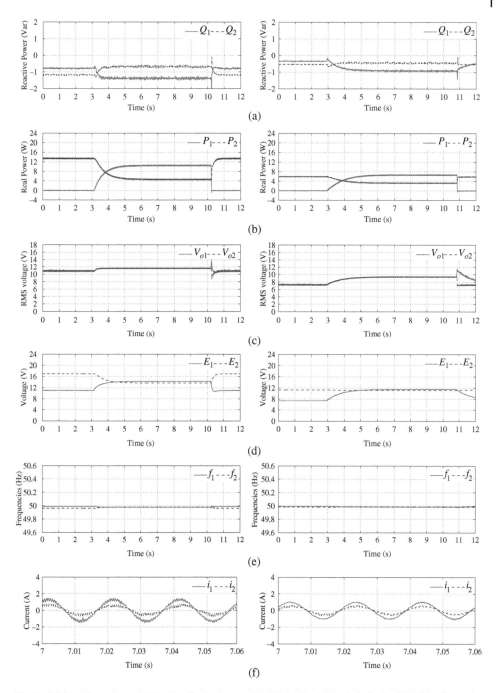

Figure 16.11 Experimental results for the case with the same per-unit impedance using the robust droop controller: with $K_e = 10$ (left column) and $K_e = 1$ (right column). (a) Reactive power. (b) Real power. (c) RMS value of the load voltage. (d) Voltage set point. (e) Frequency. (f) Current in the steady state.

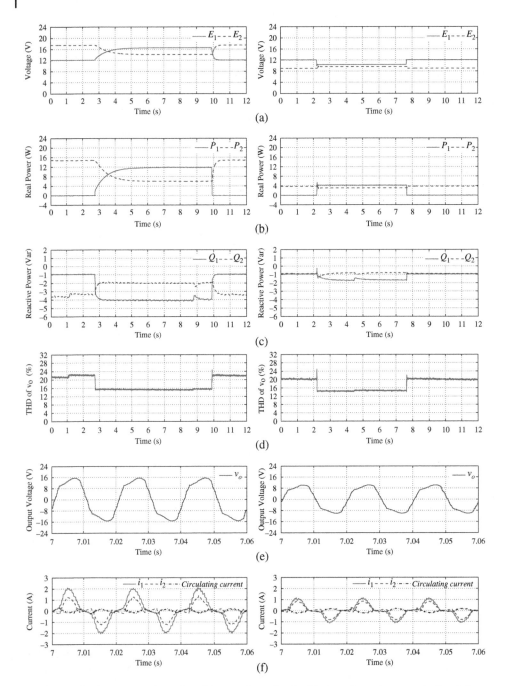

Figure 16.12 Experimental results with a nonlinear load: with the robust droop controller (left column) and with the conventional droop controller (right column). (a) Voltage set point. (b) Real power. (c) Reactive power. (d) THD of the output voltage. (e) Output voltage in the steady state. (f) Currents in the steady state.

resistive over a wide enough frequency range and there is a significant number of harmonic current components.

The only drawback is that the THD of the output voltage v_0 is not satisfactory (22% for one inverter and 16% for two inverters in parallel). However, this is expected because $R_{oi} = K_i = 4$ was used, which dominated the harmonic voltage drop on the output impedance and increased the THD. It can be improved by using smaller K_i while maintaining the output impedance resistive. Various strategies to improve the output-voltage THD can be found in (Zhong and Hornik 2013).

16.4 Robust Droop Control of C-inverters

16.4.1 Control Strategy

The voltage regulator bolted onto the conventional droop controllers for R-inverters can also be bolted onto the droop controller for C-inverters. The integrator in the voltage channel can be added as well. This results in the robust droop controller for C-inverters shown in Figure 16.13. It is able to share both real power and reactive power accurately even if the per-unit output impedances are not the same and/or there are numerical errors, disturbances, and noises because, at the steady state, there is

$$n_i Q_i + K_e(E^* - V_o) = 0. \tag{16.15}$$

This means

$$n_i Q_i = \text{constant},$$

as long as K_e is the same for all inverters. This guarantees accurate sharing of reactive power in proportion to their ratings. As long as the system is stable, which leads to the same frequency, the real power can be guaranteed as well (Zhong 2013c).

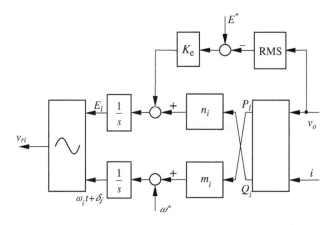

Figure 16.13 Robust droop controller for C-inverters.

According to (16.15), the output voltage is

$$V_0 = E^* + \frac{n_i}{K_e}Q_i = E^* + \frac{n_iQ_i}{K_eE^*}E^*,$$

which can be maintained within the desired range via choosing a big K_e. Hence, the control strategy has very good capability of voltage regulation, in addition to accurate power sharing.

The droop coefficients n_i and m_i can be determined as usual by the desired voltage drop ratio $\frac{n_iQ^*}{K_eE^*}$ and the frequency boost ratio $\frac{m_iP^*}{\omega^*}$, respectively, at the rated reactive power Q^* and real power P^*.

16.4.2 Experimental Results

Experiments were carried out on a test rig that consists of two single-phase inverters powered by two separate 42 V DC voltage supplies. The parameters of the system are the same as those given in the previous section. Experiments were carried out for C-inverters with $C_o = 479\ \mu F$ and R-inverters with $K_i = 4\ \Omega$.

16.4.2.1 With a Linear Load

Experiments were carried out for a linear load with $R_L = 9\ \Omega$. The results from C-inverters and R-inverters are shown in the left and right columns of Figure 16.14, respectively, for comparison. The system worked very well for both cases: with accurate sharing of real power and reactive power, good regulation of output voltage and low THD. In comparison to R-inverters, the voltage regulation performance is slightly better because this is related to the reactive power of the load, which is small, and the frequency variation is slightly higher because this is related to the real power.

16.4.2.2 With a Nonlinear Load

The same experiments were carried out for a full-bridge rectifier load with an LC filter $L = 150\ \mu H$, $C = 1000\ \mu F$ and $R_L = 9\ \Omega$. The results for C-inverters and R-inverters are shown in the left and right columns of Figure 16.15, respectively. Again, the system worked very well for both cases with accurate sharing of real power and reactive power and good capability of voltage regulation. The C-inverters demonstrated much better THD than the R-inverters. Moreover, when another inverter was added into parallel operation, the THD of the output voltage dropped much more when the output impedances of the inverters are capacitive than when the output impedances are resistive.

16.5 Robust Droop Control of L-inverters

16.5.1 Control Strategy

Similar to the cases of R-inverters and C-inverters, an additional loop to regulate the output voltage can be bolted onto the conventional droop control to "robustify" the strategy and to improve the capability of voltage regulation. The resulting control scheme is shown in Figure 16.16. It is able to share both real power and reactive power accurately even if

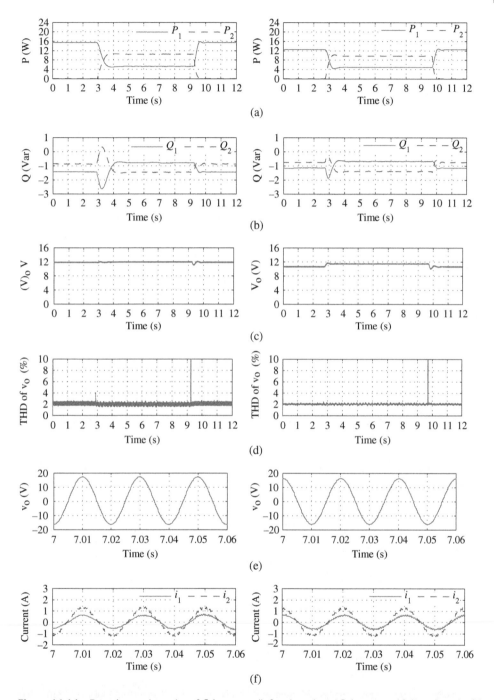

Figure 16.14 Experimental results of C-inverters (left column) and R-inverters (right column) with a linear load $R_L = 9\ \Omega$. (a) Real power. (b) Reactive power. (c) RMS output voltage. (d) THD of the output voltage. (e) Output voltage at the steady state. (f) Current at the steady state.

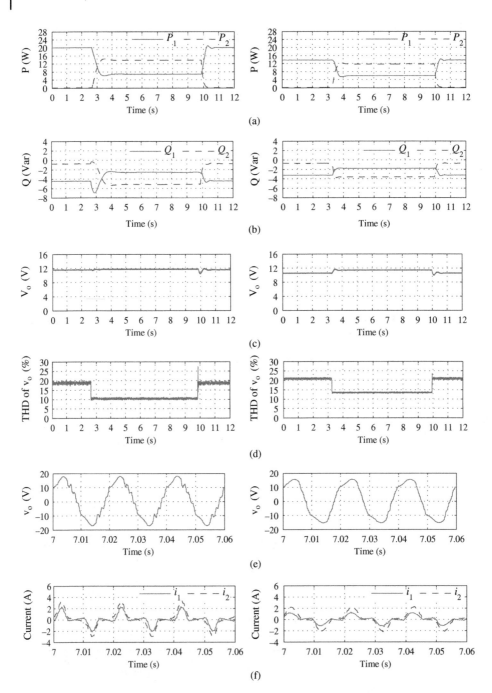

Figure 16.15 Experimental results of C-inverters (left column) and R-inverters (right column) with a nonlinear load. (a) Real power. (b) Reactive power. (c) RMS output voltage. (d) THD of the output voltage. (e) Output voltage at the steady state. (f) Current at the steady state.

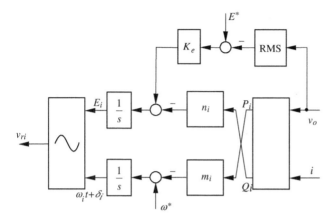

Figure 16.16 Robust droop controller for L-inverters.

the per-unit output impedances are not the same and/or there are numerical errors, disturbances, and noises because, at the steady state, there is

$$-n_i Q_i + K_e(E^* - V_o) = 0. \tag{16.16}$$

This means

$$n_i Q_i = K_e(E^* - V_o) = \text{constant},$$

as long as K_e is the same for all inverters. This guarantees the accurate sharing of reactive
power in proportion to their ratings. As long as the system is stable, which leads to the same
frequency, the real power can be guaranteed as well (Zhong 2013c).

According to (16.16), the output voltage is

$$V_o = E^* - \frac{n_i}{K_e} Q_i = E^* - \frac{n_i Q_i}{K_e E^*} E^*,$$

which can be maintained within the desired range with a large K_e. Thus, the control strategy has very good capability of voltage regulation as well, in addition to accurate power
sharing.

The droop coefficients n_i and m_i can then be determined as usual by the desired voltage
drop ratio $\frac{n_i Q_i^*}{K_e E^*}$ and the frequency drop ratio $\frac{m_i P_i^*}{\omega^*}$, respectively, at the rated reactive power
Q^* and real power P^*.

16.5.2 Experimental Results

The above strategy was verified in a laboratory setup consisting of two single-phase inverters
powered by separate 42 V DC voltage supplies. The inverters were connected to the AC-bus
via a circuit breaker (CB) and the load was connected to the AC-bus. The values of the
inductors and capacitors are 2.35 mH and 22 μF, respectively. The switching frequency is
7.5 kHz and the frequency of the system is 50 Hz. The rated voltage is 12 V RMS and $K_e =$
10. The droop coefficients were $n_1 = 0.4$ and $n_2 = 0.8$, $m_1 = 0.1$ and $m_2 = 0.2$. Hence, it is
expected that $P_1 = 2P_2$ and $Q_1 = 2Q_2$. Note that the droop coefficients and K_e are different
from those in the simulations, which resulted in slower responses.

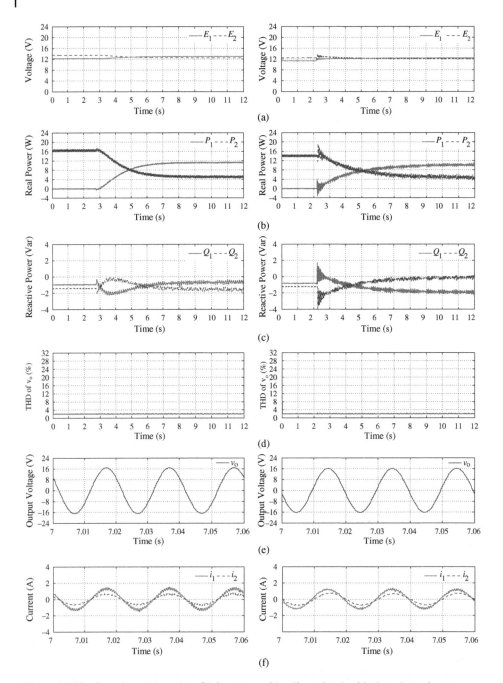

Figure 16.17 Experimental results of L-inverters with a linear load: with the robust droop controller (left column) and the conventional droop controller (right column). (a) Voltage set point. (b) Real power. (c) Reactive power. (d) THD of the output voltage. (e) Output voltage in the steady state. (f) Currents in the steady state.

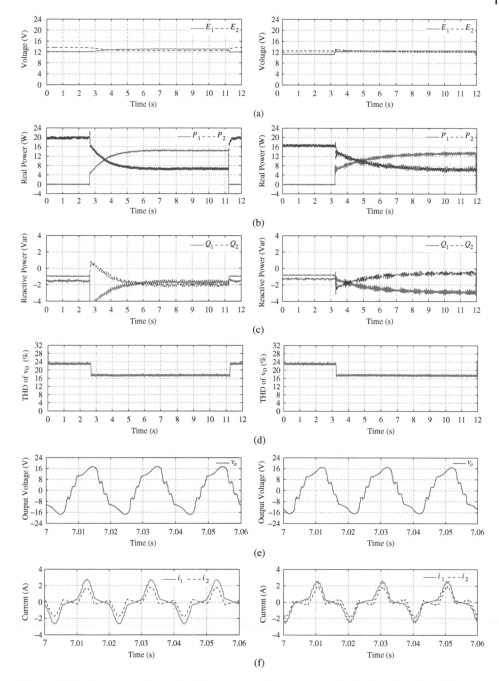

Figure 16.18 Experimental results of L-inverters with a nonlinear load: with the robust droop controller (left column) and with the conventional droop controller (right column). (a) Voltage set point. (b) Real power. (c) Reactive power. (d) THD of the output voltage. (e) Output voltage in the steady state. (f) Currents in the steady state.

16.5.2.1 With a Linear Load

A linear load of about 9 Ω was connected to inverter 2 initially. Inverter 1 was connected to the system at around $t = 2.5$ s. The experimental results are shown in the left column of Figure 16.17 when the robust droop controller was adopted and in the right column of Figure 16.17 when the conventional droop controller was adopted. For the robust droop controller, both the real power and the reactive power were shared accurately. However, for the conventional droop controller, only the real power was shared accurately. The reactive power was not shared properly at all and the two inverter currents were not in phase. The voltage drop in both cases was not bad because the reactive power was small.

16.5.2.2 With a Nonlinear Load

The same experiments were carried out for a full-bridge rectifier load with an LC filter $L = 150 \mu$ H, $C = 1000 \mu$F and $R_L = 9$ Ω. The experimental results are shown in the left column of Figure 16.18 when the robust droop controller was adopted and in the right column of Figure 16.18 when the conventional droop controller was adopted. The real power was shared very accurately in both cases. The robust droop controller improved the accuracy of the reactive power significantly but the sharing was not accurate enough because of the harmonics in the load current. In both cases, the THD of the voltage was very high because there was no mechanism embedded to improve the voltage quality.

16.6 Summary

Based on (Zhong 2013c; Zhong and Zeng, 2011, 2014), the fundamental limitations of the conventional droop control scheme are analyzed and a robust droop control strategy is presented for R-, L-, and C-inverters in different forms. This strategy allows the inverters with the same type of output impedance to share the real power and the reactive power accurately while regulating the output voltage within a given range, even when there are component mismatches, computational errors, noises, etc. Extensive experimental results are presented.

17

Universal Droop Control

As shown in the previous chapter, the droop control strategy has different forms for inverters with different types of output impedance. In this chapter, it is shown that there exists a universal droop control principle for impedance having a phase angle between $-\frac{\pi}{2}$ rad and $\frac{\pi}{2}$ rad. It takes the form of the droop control for R-inverters. In other words, the robust droop control for R-inverters presented in the previous chapter is universal and can be applied to inverters with impedance having a phase angle from $-\frac{\pi}{2}$ rad to $\frac{\pi}{2}$ rad. Both real-time simulation results and experimental results from a test rig consisting of an R-inverter, an L-inverter, and a C-inverter operated in parallel are presented.

17.1 Introduction

Unlike a synchronous machine, of which the impedance is inductive, the impedance of an inverter around the fundamental frequency can be inductive (L-inverters), resistive (R-inverters) (Guerrero et al. 2005; Zhong 2013c), capacitive (C-inverters) (Zhong and Zeng, 2011, 2014), resistive-inductive (R_L-inverters) or resistive-capacitive (R_C-inverters), etc. Compared with L-inverters, R-inverters can enhance system damping and C-inverters can improve power quality and eliminate DC current component sent to the grid. As shown in the previous two chapters, the conventional droop control principle has different forms for inverters with different types of output impedance. This makes it impossible to operate inverters with different types of output impedance in one system, leading to problems for large-scale utilization of distributed generations and renewable energy sources.

In the literature, there have been some attempts to find a droop controller that works for more general cases (Bevrani and Shokoohi 2013; Brabandere et al. 2007; Karimi-Ghartemani 2015; Khan et al. 2013; Sun et al. 2014; Yao et al. 2011). An orthogonal linear rotational transformation matrix was adopted to modify the real power and the reactive power so that, for L-, R-, and R_L-inverters, the power angle could be controlled by the modified real power and the inverter voltage could be controlled by the modified reactive power (Brabandere et al. 2007). However, the ratio of R/X needs to be known, where R and X are the resistance and inductance of the inverter output impedance, respectively. In (Yao et al. 2011), a different droop control method with a virtual complex impedance added to make the angle of the new output impedance around $\pi/4$, leading to a fixed droop control form. However, the virtual complex impedance needs to be carefully designed.

Power Electronics-Enabled Autonomous Power Systems: Next Generation Smart Grids,
First Edition. Qing-Chang Zhong.
© 2020 John Wiley & Sons Ltd. Published 2020 by John Wiley & Sons Ltd.

A generalized droop controller based on an adaptive neuro-fuzzy interface system was developed in (Bevrani and Shokoohi 2013) to handle a wide range of load change scenarios for L-, R-, and R_L-inverters, but resulted in a complex control structure. Additionally, a real power and reactive power flow controller, which took into account all cases of the R–L relationship, was presented for three-phase PWM voltage source inverters (Khan et al. 2013) but the phase shift needs to be obtained for its power transformation. Moreover, an adaptive droop control method was presented based on the online evaluation of a power decouple matrix (Sun et al. 2014), which was obtained by the ratio of the variations of the real power and the reactive power under a small perturbation on the voltage magnitude. Recently, an integrated synchronization and control was presented to operate single-phase inverters in both grid-connected and stand-alone modes (Karimi-Ghartemani 2015). However, the controller, called the R_L-controller to facilitate the presentation in the sequel, only works for L-, R-, and R_L-inverters but not for C- or R_C-inverters.

A droop controller for C-, R-, and R_C-inverters, called the R_C-controller, is presented at first in this chapter. Then, a universal droop control principle is presented for inverters with output impedance having a phase angle between $-\frac{\pi}{2}$ rad and $\frac{\pi}{2}$ rad, which covers any practical L-, R-, C-, R_L-, and R_C-inverters. This universal droop control principle takes the form of the droop control principle for R-inverters, which paves the way for designing universal droop controllers with different methods. In this chapter, the robust droop controller presented in (Zhong 2013c) is adopted for implementation.

17.2 Further Insights into Droop Control

In this section, the widely adopted droop control strategy is reviewed, with many new insights provided.

An inverter can be modeled as a voltage source v_r in series with the output impedance $Z_o \angle \theta$, as shown in Figure 17.1, where E is the amplitude (or RMS value) of the source voltage and δ, called the power angle, is the phase difference between v_r and v_o. The real power and reactive power delivered from the voltage source v_r to the terminal v_o through the impedance $Z_o \angle \theta$ are

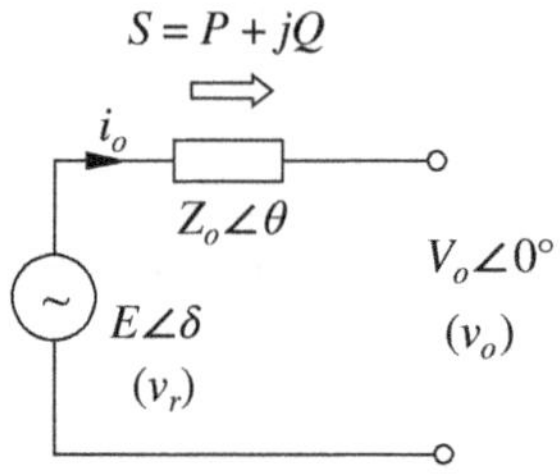

Figure 17.1 The model of a single-phase inverter.

$$P = \left(\frac{EV_o}{Z_o} \cos \delta - \frac{V_o^2}{Z_o} \right) \cos \theta + \frac{EV_o}{Z_o} \sin \delta \sin \theta,$$

$$(17.1)$$

$$Q = \left(\frac{EV_o}{Z_o} \cos \delta - \frac{V_o^2}{Z_o} \right) \sin \theta - \frac{EV_o}{Z_o} \sin \delta \cos \theta. \qquad (17.2)$$

This characterizes a two-input–two-output control plant from the amplitude E and the phase δ of the source v_r to the real power P and the reactive power Q, as shown in the upper part of Figure 17.2. The function of a droop control strategy is to generate appropriate amplitude E and phase δ for the inverter according to the measured P and Q, that is to close the loop, as shown in Figure 17.2. This certainly helps understand the essence

Figure 17.2 The closed-loop system consisting of the power flow model of an inverter and a droop controller.

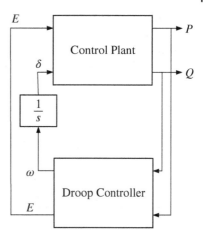

of droop control and motivates the design of other droop control strategies. Indeed, so far, the majority of droop controllers have been static rather than dynamic (Zhong 2013c) and dynamic droop controllers could be developed to improve system performance.

17.2.1 Parallel Operation of Inverters with the Same Type of Impedance

In practice, it is often assumed that δ is small. In this case,

$$P \approx (\frac{EV_0}{Z_0} - \frac{V_0^2}{Z_0}) \cos\theta + \frac{EV_0}{Z_0}\delta\sin\theta, \tag{17.3}$$

$$Q \approx (\frac{EV_0}{Z_0} - \frac{V_0^2}{Z_0}) \sin\theta - \frac{EV_0}{Z_0}\delta\cos\theta. \tag{17.4}$$

This leads to decoupled relationships between the inputs and the outputs, which change with the impedance angle θ. For example, when the output impedance is inductive ($\theta = \frac{\pi}{2}$ rad), P is roughly proportional to δ, noted as $P \sim \delta$, and Q is roughly proportional to E, noted as $Q \sim E$. According to this, the well known droop control strategy, that is to droop the frequency when the real power increases and to droop the voltage when the reactive power increases, can be adopted. The cases when the output impedance is resistive ($\theta = 0$ rad) and capacitive ($\theta = -\frac{\pi}{2}$ rad) can be analyzed similarly, which results in different droop control strategies, as illustrated in Figure 15.5. The cases when the impedance is inductive (L-inverter), capacitive (C-inverter), resistive (R-inverter), resistive-capacitive (R_C-inverter), and resistive-inductive (R_L-inverter) are summarized in Table 17.1 for convenience. Apparently, the input-output relationships are different and so are the droop controllers. This holds true for the conventional droop controller as well as the robust droop controller (Zhong 2013c) that is robust against variations of output impedance, component mismatches, parameter drifts, disturbances, etc.

Since the droop control strategies change the form when the output impedance θ changes, it is difficult to operate inverters with different types of output impedance in parallel. In particular, the droop control strategies for L-inverters and C-inverters act in the opposite way and the parallel operation of a C-inverter with an L-inverter certainly does not work if these droop control strategies are employed.

Table 17.1 Droop controllers for L-, R-, C-, R_L-, and R_C-inverters.

Inverter type	θ	Input–output relationship	Droop controller
L-	$\frac{\pi}{2}$	$P \sim \delta, Q \sim E$	$E = E^* - nQ,$ $\omega = \omega^* - mP$
R-	$0°$	$P \sim E, Q \sim -\delta$	$E = E^* - nP,$ $\omega = \omega^* + mQ$
C-	$-\frac{\pi}{2}$	$P \sim -\delta, Q \sim -E$	$E = E^* + nQ,$ $\omega = \omega^* + mP$
R_C-	$(-\frac{\pi}{2}, 0)$	Coupled	Depends on θ
R_L-	$(0, \frac{\pi}{2})$	Coupled	Depends on θ

17.2.2 Parallel Operation of L-, R-, and R_L-inverters

Some works (Bevrani and Shokoohi 2013; Brabandere et al. 2007; Yao et al. 2011) have been reported in the literature to investigate the parallel operation of inverters with different types of output impedance, although they are limited to the parallel operation of L-, R-, and R_L-inverters. This involves the introduction of the orthogonal transformation matrix

$$T_L = \begin{bmatrix} \sin\theta & -\cos\theta \\ \cos\theta & \sin\theta \end{bmatrix} \tag{17.5}$$

to convert the real power and the reactive power when $\theta \in (0, \frac{\pi}{2}]$ into

$$\begin{bmatrix} P_L \\ Q_L \end{bmatrix} = T_L \begin{bmatrix} P \\ Q \end{bmatrix} = \begin{bmatrix} \frac{EV_o}{Z_o} \sin\delta \\ \frac{EV_o}{Z_o} \cos\delta - \frac{V_o^2}{Z_o} \end{bmatrix}. \tag{17.6}$$

If δ is assumed small, roughly

$$P_L \sim \delta \quad \text{and} \quad Q_L \sim E, \tag{17.7}$$

which results in the droop controller of the form

$$E = E^* - nQ_L \tag{17.8}$$

$$\omega = \omega^* - mP_L. \tag{17.9}$$

This is called the R_L-controller in order to facilitate the presentation in the sequel. Here, n and m are called the droop coefficients. This controller has the same form as the droop controller for L-inverters but the impedance angle θ needs to be known in order to obtain the transformed power P_L and Q_L from (17.6), as introduced in (Bevrani and Shokoohi 2013; Brabandere et al. 2007; Yao et al. 2011).

As a matter of fact, the eigenvalues of T_L in (17.5) are $\sin\theta \pm j\cos\theta$, of which the real part $\sin\theta$ is positive for impedance with $\theta \in (0, \frac{\pi}{2}]$. According to the properties of the linear transformation (Poole 2011) and the mapping described by (17.6), it can be seen that P and Q have positive correlations with P_L and Q_L, respectively. This can be described as

$$P \sim P_L \quad \text{and} \quad Q \sim Q_L. \tag{17.10}$$

So the relationship shown in (17.7) can be passed onto P and Q as

$$P \sim P_{\rm L} \sim \delta \quad \text{and} \quad Q \sim Q_{\rm L} \sim E. \tag{17.11}$$

In other words, for output impedance with $\theta \in (0, \frac{\pi}{2}]$, the real power P always has positive correlation with the power angle δ and the reactive power Q always has positive correlation with the voltage E. Hence, the $R_{\rm L}$-controller can also be designed as

$$E = E^* - nQ, \tag{17.12}$$

$$\omega = \omega^* - mP, \tag{17.13}$$

which is directly related to the real power P and the reactive power Q, regardless of the impedance angle θ. In other words, there is no need to know the impedance angle θ as long as it satisfies $\theta \in (0, \frac{\pi}{2}]$.

In order to better understand the transformation matrix (17.5), the transformation (17.6) can actually be rewritten as

$$P_{\rm L} + jQ_{\rm L} = P\sin\theta - Q\cos\theta + j(P\cos\theta + Q\sin\theta)$$
$$= e^{j(\frac{\pi}{2}-\theta)}(P + jQ),$$

where $j = \sqrt{-1}$. In other words, the transformation (17.5) rotates the power vector $P + jQ$ by $\frac{\pi}{2} - \theta$ rad onto the axis aligned with the L-inverter, as shown in Figure 17.3(a), which leads to the droop controller shown in (17.12)–(17.13).

17.2.3 Parallel Operation of R_C-, R-, and C-inverters

Along the same line of thinking, the transformation matrix

$$T_C = \begin{bmatrix} -\sin\theta & \cos\theta \\ -\cos\theta & -\sin\theta \end{bmatrix} \tag{17.14}$$

can be introduced for C-, R- or R_C-inverters with $\theta \in [-\frac{\pi}{2}, 0)$ to convert the real power and the reactive power into

$$\begin{bmatrix} P_C \\ Q_C \end{bmatrix} = T_C \begin{bmatrix} P \\ Q \end{bmatrix} = \begin{bmatrix} -\frac{EV_o}{Z_o}\sin\delta \\ -\frac{EV_o}{Z_o}\cos\delta + \frac{V_o^2}{Z_o} \end{bmatrix}. \tag{17.15}$$

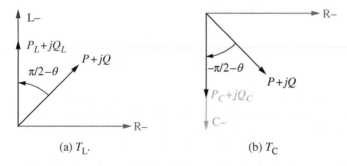

(a) $T_{\rm L}$. (b) T_C

Figure 17.3 Interpretation of transformation matrices $T_{\rm L}$ and T_C. (a) $T_{\rm L}$. (b) T_C.

In this case, for a small δ, roughly

$$P_C \sim -\delta \quad \text{and} \quad Q_C \sim -E, \tag{17.16}$$

which results in the droop controller of the form

$$E = E^* + nQ_C \tag{17.17}$$

$$\omega = \omega^* + mP_C. \tag{17.18}$$

This is called the R_C-controller and it has the same form as the droop controller for C-inverters proposed in (Zhong and Zeng, 2011, 2014). Again, the impedance angle θ needs to be known in order to obtain the transformed active power P_C and reactive power Q_C from (17.15). Apparently, this controller does not work for L- or R_L-inverters because of the negative signs in (17.8)–(17.9).

Similarly, for the R_C-controller, the eigenvalues of T_C in (17.14) are $-\sin\theta \pm j\cos\theta$, of which the real part $-\sin\theta$ is positive for any output impedance with $\theta \in [-\frac{\pi}{2}, 0)$. Hence, according to the mapping described by (17.15), P and Q have positive correlations with P_C and Q_C, respectively. This can be described as

$$P \sim P_C \quad \text{and} \quad Q \sim Q_C. \tag{17.19}$$

So the relationship shown in (17.16) can be passed onto P and Q as

$$P \sim P_C \sim -\delta \quad \text{and} \quad Q \sim Q_C \sim -E. \tag{17.20}$$

In other words, for impedance with $\theta \in [-\frac{\pi}{2}, 0)$, the real power P always has negative correlation with the power angle δ and the reactive power Q always has negative correlation with the voltage E. Then, the R_C-controller can also be designed as

$$E = E^* + nQ, \tag{17.21}$$

$$\omega = \omega^* + mP, \tag{17.22}$$

which is also directly related to the real power P and the reactive power Q. It does not require the knowledge of the impedance angle θ, either, as long as it satisfies $\theta \in [-\frac{\pi}{2}, 0)$.

The transformation (17.15) can also be rewritten as

$$\begin{aligned} P_C + jQ_C &= -P\sin\theta + Q\cos\theta + j(-P\cos\theta - Q\sin\theta) \\ &= e^{j(-\frac{\pi}{2}-\theta)}(P + jQ). \end{aligned}$$

In other words, the transformation (17.14) actually rotates the power vector $P + jQ$ by $-\frac{\pi}{2} - \theta$ rad onto the axis aligned with the C-inverter, as shown in Figure 17.3(b), which leads to the droop controller shown in (17.21)–(17.22).

In summary, the R_L-controller (17.12)–(17.13) can be applied to inverters with the output impedance satisfying $\theta \in (0, \frac{\pi}{2}]$ and the R_C-controller can be applied to inverters with the output impedance satisfying $\theta \in [-\frac{\pi}{2}, 0)$. This widens the application range of the L-controller and the C-controller. However, the R_L-controller cannot be applied to C- or R_C-inverters, and the R_C-controller cannot be applied to L- or R_L-inverters either. There is still a need to develop a controller that is applicable to L-, R-, C-, R_L-, and R_C-inverters.

17.3 Universal Droop Controller

17.3.1 Basic Principle

Following the above analysis, consider the transformation matrix

$$T = \begin{bmatrix} \cos\theta & \sin\theta \\ -\sin\theta & \cos\theta \end{bmatrix}.$$

(17.23)

It transforms the real power P and the reactive power Q to

$$\begin{bmatrix} P_R \\ Q_R \end{bmatrix} = T \begin{bmatrix} P \\ Q \end{bmatrix} = \begin{bmatrix} \frac{EV_o}{Z_o}\cos\delta - \frac{V_o^2}{Z_o} \\ -\frac{EV_o}{Z_o}\sin\delta \end{bmatrix},$$

(17.24)

which can be rewritten as

$$P_R + jQ_R = P\cos\theta + Q\sin\theta + j(-P\sin\theta + Q\cos\theta)$$
$$= e^{-j\theta}(P + jQ).$$

As shown in Figure 17.4, this transformation rotates the power vector $P + jQ$ by $-\theta$ onto the axis aligned with the R-inverter, i.e. clockwise when $\theta \in [0, \frac{\pi}{2})$ and counter-clockwise when $\theta \in (-\frac{\pi}{2}, 0]$. Indeed, the eigenvalues of T in (17.23) are $\cos\theta \pm j\sin\theta$, of which the real part $\cos\theta$ is positive for any output impedance with $\theta \in (-\frac{\pi}{2}, \frac{\pi}{2})$. According to the properties of the linear transformation (Poole 2011) and the mapping described by (17.24), P and Q are proven to have positive correlations with P_R and Q_R, respectively. This can be described as

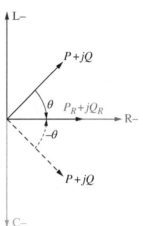

Figure 17.4 Interpretation of the universal transformation matrix T.

$$P \sim P_R \quad \text{and} \quad Q \sim Q_R.$$

(17.25)

According to (17.24), for a small δ, there are

$$P_R \sim E \quad \text{and} \quad Q_R \sim -\delta.$$

(17.26)

Combining these two, there are

$$P \sim P_R \sim E \quad \text{and} \quad Q \sim Q_R \sim -\delta$$

(17.27)

for any $\theta \in (-\frac{\pi}{2}, \frac{\pi}{2})$. This basically indicates that the real power P always has positive correlation with the voltage E and the reactive power Q always has negative correlation with the power angle δ for any impedance angle $\theta \in (-\frac{\pi}{2}, \frac{\pi}{2})$. This results in the following conventional universal droop controller

$$E = E^* - nP,$$

(17.28)

$$\omega = \omega^* + mQ,$$

(17.29)

which is applicable to inverters with output impedance satisfying $\theta \in (-\frac{\pi}{2}, \frac{\pi}{2})$. Note that this droop controller (17.28)–(17.29) takes the form of the droop controller for R-inverters.

Theoretically, when the impedance is purely inductive ($\theta = \frac{\pi}{2}$ rad) or capacitive ($\theta = -\frac{\pi}{2}$ rad), this relationship does not hold but, in practice, there is always an equivalent series resistance (ESR) in series with the filter inductor so the controller (17.28)–(17.29) is actually applicable to all practical L-, R-, C-, R_L-, and R_C-inverters.

17.3.2 Implementation

There are many ways to implement the universal droop control principle revealed in the previous subsection. The most natural way is to take the robust droop controller presented in (Zhong 2013c; Zhong and Hornik 2013) and Chapter 16, which is re-drawn as shown in Figure 17.5 for convenience. This controller can be described as:

$$\dot{E} = K_e(E^* - V_o) - nP, \tag{17.30}$$

$$\omega = \omega^* + mQ, \tag{17.31}$$

In the steady state, there is

$$nP = K_e(E^* - V_o), \tag{17.32}$$

which means the output voltage is

$$V_o = E^* - \frac{nP}{K_e E^*} E^*. \tag{17.33}$$

Here, $\frac{nP}{K_e E^*}$ is the voltage drop ratio, which can be maintained within the desired range via choosing K_e. Moreover, as long as K_e is chosen the same for all inverters in parallel, the right-hand side of (17.32) is the same, which guarantees accurate real power sharing.

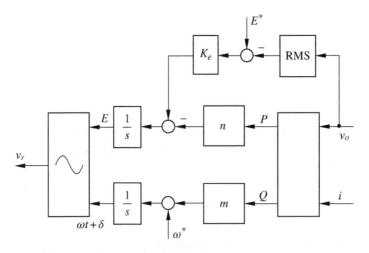

Figure 17.5 Universal droop controller.

17.4 Real-time Simulation Results

Three single-phase inverters were operated together to power a 20 Ω linear load. The capacities of inverters 1 (L-inverter), 2 (C-inverter) and 3 (R-inverter with a virtual 4 Ω resistor) were 1 kVA, 2 kVA and 3 kVA, respectively. It was expected that $P_2 = 2P_1$, $Q_2 = 2Q_1$, $P_3 = 3P_1$ and $Q_3 = 3Q_1$. The switching frequency was 10 kHz and the system frequency was 50 Hz. The rated output voltage was 230 V and $K_e = 10$. The filter inductor was $L = 0.55$ mH with a parasitic resistance of 0.3 Ω and the filter capacitor C was 20 μF. The desired voltage drop ratio $\frac{n_i S_i^*}{K_e E^*}$ was chosen as 0.25% and the frequency boost ratio $\frac{m_i S_i^*}{\omega^*}$ was 0.1% so the droop coefficients are $n_1 = 0.0057$, $n_2 = 0.0029$, $n_3 = 0.0019$, $m_1 = 3.1416 \times 10^{-4}$, $m_2 = 1.5708 \times 10^{-4}$ and $m_3 = 1.0472 \times 10^{-4}$.

The real-time simulation results are shown in Figure 17.6. At $t = 0$ s, the three inverters were operated separately with the load connected to the R-inverter only. Then, at $t = 10$ s, the C-inverter was connected in parallel with the R-inverter and the two inverters shared the real power and reactive power accurately in the ratio of 2:3. At $t = 30$ s, the L-inverter was put into parallel operation. The three inverters shared the real power and reactive power accurately in the ratio of 1:2:3. Then the R-inverter was disconnected at $t = 60$ s and the C-inverter and the L-inverter shared the power accurately in the ratio of 2:1. Finally, the L-inverter was disconnected at $t = 80$ s and the load was powered by the C-inverter only. The frequency and the voltage were regulated to be very close to the rated values, respectively, as can be seen from Figures 17.6(c) and (d).

The waveforms of the load voltage and the inductor currents of the three inverters after taking away the switching ripples with a hold filter when the three inverters were in parallel operation are shown in Figures 17.6(e) and (f). It can be seen that indeed the three inverters shared the load accurately in the ratio of 1:2:3.

17.5 Experimental Results

Experiments were carried out on a system consisting of three inverters operated in parallel, as shown in Figure 17.7. Each single-phase inverter is powered by a 30 V DC voltage supply and loaded with a 3.8 Ω resistor in series with two 2.2 mH inductors. The on-board filter is not optimized. The filter inductor is $L = 7$ mH with a parasitic resistance of 1 Ω and the filter capacitor is $C = 1$ μF. The PWM switching frequency is 10 kHz; the rated system frequency is 50 Hz and the cut-off frequency ω_f of the measuring filter is 10 rad s^{-1}. The rated output voltage is 12 V and $K_e = 20$. The desired voltage drop ratio $\frac{n_i S_i^*}{K_e E^*}$ is chosen as 10% and the frequency boost ratio $\frac{m_i S_i^*}{\omega^*}$ is chosen as 0.5%. Here the subscript i is the inverter index. These inverters are operated as an R-inverter with a virtual 8 Ω resistor (Zhong 2013c), a C-inverter with a virtual 161 μF capacitor in series with a virtual 2.5 Ω resistor (Zhong and Zeng, 2011, 2014), and an original L-inverter, respectively.

17.5.1 Case I: Parallel Operation of L- and C-inverters

In this case, the L-inverter and the C-inverter are designed to have the power ratio of 1:2, with $P_2 = 2P_1$ and $Q_2 = 2Q_1$. The droop coefficients are $n_1 = 0.96$, $n_2 = 0.48$, $m_1 = 0.06$ and

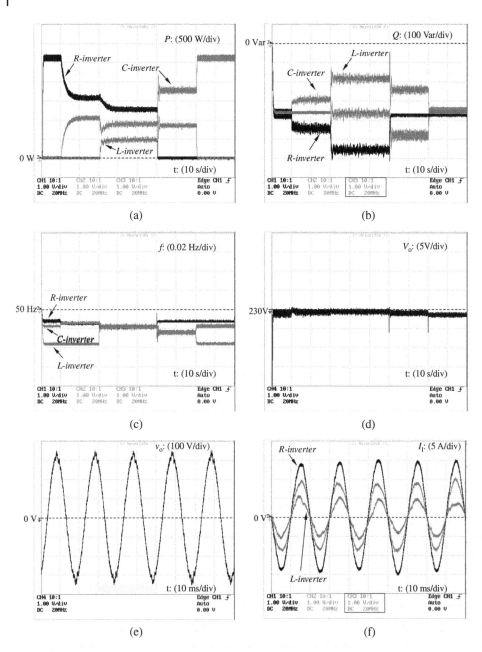

Figure 17.6 Real-time simulation results of three inverters with different types of output impedance operated in parallel. (a) Real power. (b) Reactive power. (c) Frequencies. (d) RMS output voltage. (e) Output voltage. (f) Inductor currents.

Figure 17.7   Experimental set-up consisting of an L-inverter, an R-inverter, and a C-inverter.

$m_2 = 0.03$. The experimental results are shown in the left column of Figure 17.8. At $t = 3$ s, the C-inverter was started to take the load. Then, at about $t = 6$ s, the L-inverter was started to synchronize with the C-inverter. At about $t = 12$ s, the L-inverter was paralleled with the C-inverter. They shared the power with a ratio of 1:2. The inverter output voltage and inductor currents were regulated well and the currents were shared accurately with a ratio of 1:2. Note that the spikes in the frequency before the connection were caused by the phase resetting (zero crossing) applied for synchronization.

17.5.2   Case II: Parallel Operation of L-, C-, and R-inverters

In this case, the L-inverter, the C-inverter, and the R-inverter were designed to have a power capacity ratio of 1:2:3, with $P_3 = 1.5P_2 = 3P_1$ and $Q_3 = 1.5Q_2 = 3Q_1$. The droop coefficients are $n_1 = 1.44, n_2 = 0.72, n_3 = 0.48, m_1 = 0.09, m_2 = 0.045$, and $m_3 = 0.03$. The parallel operation of the three inverters is tested, and the experimental results are shown in the right column of Figure 17.8.

At $t = 3$ s, the R-inverter was started to supply the load. Then, at about $t = 6$ s, the C-inverter was started and began to synchronize with the R-inverter. The RMS output voltage of the C-inverter stepped up to be almost the same as that of the R-inverter and the frequency of the C-inverter stepped up to be around 50 Hz. At about $t = 12$ s, the C-inverter was connected to the load and thus in parallel with the R-inverter. After a short transient, the R-inverter and the C-inverter shared the real power and the reactive power with the ratio of 3:2, as designed. The RMS value of the output voltage and the frequency of both inverters became the same. The inverter output voltage RMS value slightly increased and the R-inverter frequency decreased a little bit. Then, at about $t = 15$ s, the L-inverter was started to synchronize with the terminal voltage established by the R-inverter and the C-inverter. The RMS output voltage of the L-inverter stepped up to be almost the same as

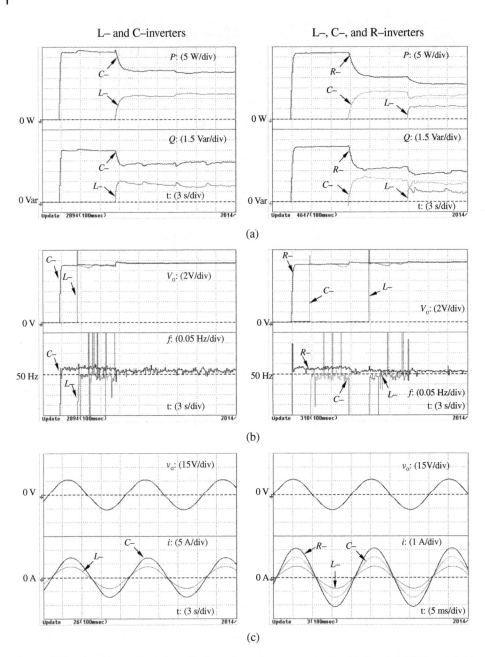

Figure 17.8 Experimental results with the universal droop controller. (a) *P* and *Q*. (b) V_o and *f*. (c) v_o and *i*.

Table 17.2 Steady-state performance of the three inverters in parallel operation.

Variable	R-, L-, and C-inverters
Apparent power 1	6.07 + 1.54jVA
Apparent power 2	11.62+2.83jVA
Apparent power 3	16.60+3.97jVA
Output voltage	11.55V (rms)
Inductor current 1	0.54A (rms)
Inductor current 2	1.03A (rms)
Inductor current 3	1.48A (rms)
Frequency f	50.016 Hz
Current sharing error $\frac{I_3 - 3I_1}{4I_3} \times 100\%$	−2.4%
Voltage drop $\frac{E^* - V_o}{E^*} \times 100\%$	3.8%
Frequency drop $\frac{f^* - f}{f^*} \times 100\%$	0.03%

that of the load and the frequency of the L-inverter stepped up to be around 50 Hz. After that, at about $t = 21$ s, the L-inverter was connected to the load and thus in parallel with the R-inverter and the C-inverter. The L-inverter, the C-inverter, and the R-inverter shared the real power and the reactive power with the designed ratio of 1:2:3, as expected. The RMS value of the output voltage and the frequency of these three inverters became the same. The RMS voltage of the load slightly increased and the frequency decreased a little bit. The load voltage was regulated well and the inverter currents were shared accurately with the ratio of 1:2:3 in the steady state.

The measured steady-state performance is summarized and shown in Table 17.2. The current sharing error is just −2.4%, which is very low taking into account the fact that the inverters were not optimized. The performance for voltage regulation and frequency regulation is very good too.

17.6 Summary

Based on (Zhong and Zeng 2016), a universal droop control principle is presented for inverters with output impedance having an impedance angle between $-\frac{\pi}{2}$ rad and $\frac{\pi}{2}$ rad. Coincidentally, it takes the form of the the robust droop controller for R-inverters. Both simulation and experimental results have demonstrated the effectiveness of the universal droop controller for the parallel operation of inverters with different types of output impedance, achieving accurate proportional power sharing, tight voltage and frequency regulation.

18

Self-synchronized Universal Droop Controller

In this chapter, a self-synchronization mechanism is embedded into the universal droop controller presented in the previous chapter. Both the voltage loop and the frequency loop of the universal droop controller are modified to facilitate the standalone and grid-connected operation of inverters. Importantly, the dedicated PLL that is often needed for grid-connected or parallel-operated converters is removed. The inverter is able to achieve synchronization before and after connection without the need of a dedicated synchronization unit. Since the original structure of the universal droop controller is kept, its properties, such as accurate power sharing and tight output voltage regulation, are well maintained. Extensive experimental results are presented to demonstrate the performance of the strategy. In order to further validate the strategy, real-time simulation results from multiple inverters having different types of output impedance connected in parallel to a weak grid are presented. The technology is patented.

18.1 Description of the Controller

In order to synchronize the output voltage of an inverter with the grid voltage, a droop controller, including the universal droop controller, often needs a dedicated synchronization unit, such as a PLL. However, it is well known that the parameters of PLLs are usually difficult and time-consuming to tune. As discussed in Chapter 15, a droop controller structurally resembles a PLL so it is possible to adopt the "hidden" synchronization mechanism in the universal droop controller to achieve synchronization without a dedicated synchronization unit. This results in a self-synchronized universal droop controller, which could automatically synchronize the output voltage with the grid voltage by itself so that the inverter can be smoothly connected to the grid without any noticeable inrush currents. After the inverter is connected to the grid, the controller should accurately regulate real and reactive power between the inverter and the grid while maintaining synchronization with the grid.

The resulting self-synchronized universal droop controller is shown in Figure 18.1. Compared to the universal droop controller given in Figure 17.5, the major changes made are: (1) because the original universal droop controller is designed to operate under the droop mode only, in order to operate the controller in the set mode and to facilitate the self-synchronization process, summation blocks are added after the power calculation block to compare the measured real power P and reactive power Q with the power

Power Electronics-Enabled Autonomous Power Systems: Next Generation Smart Grids,
First Edition. Qing-Chang Zhong.
© 2020 John Wiley & Sons Ltd. Published 2020 by John Wiley & Sons Ltd.

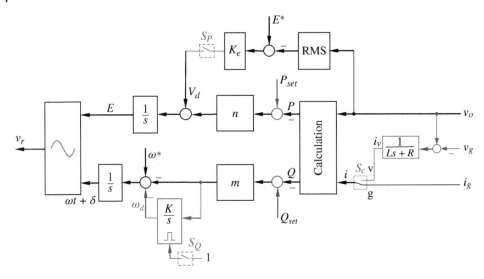

Figure 18.1 The self-synchronized universal droop controller.

Table 18.1 Operation modes of the self-synchronized universal droop controller.

Mode	Switch S_C	Switch S_P	Switch S_Q
Self-synchronization mode	s	OFF	OFF
P-mode, Q-mode	g	OFF	OFF
P_D-mode, Q-mode	g	ON	OFF
P-mode, Q_D-mode	g	OFF	ON
P_D-mode, Q_D-mode	g	ON	ON

references P_{set} and Q_{set}, respectively. This is equivalent to setting the rated operational point of the inverter at $P = P_{set}$ and $Q = Q_{set}$ when $V_o = E^*$ and $\omega = \omega^*$; (2) a virtual current i_v is generated via passing the voltage error $v_o - v_g$ through a virtual impedance $Ls + R$, which consists of a virtual inductance L in series with a resistance R; (3) a switch S_C is added so that the current sent to the controller can be switched between the virtual current i_v and the grid current i_g; (4) a switch S_P is added to enable or disable the addition of the term $K_e(E^* - V_o)$ from the controller; (5) an integrator is added to regulate $Q_{set} - Q$ to zero, with the reset function to enable or disable it via turning the switch S_Q ON or OFF. The operation modes of the controller are summarized in Table 18.1, which will be discussed in detail later.

As can be seen from Figure 18.1, the controller can be described as

$$\dot{E} = V_d + n(P_{set} - P) \tag{18.1}$$

$$\omega = \omega^* + \omega_d - m(Q_{set} - Q) \tag{18.2}$$

with

$$V_d = \begin{cases} 0, & (S_P = \text{OFF}) \\ K_e(E^* - V_o), & (S_P = \text{ON}) \end{cases}$$

$$\omega_d = \begin{cases} \frac{mK}{s}(Q - Q_{\text{set}}), & (S_Q = \text{OFF}) \\ 0, & (S_Q = \text{ON}) \end{cases}$$

where K is a positive gain. When the switch S_P is turned OFF, the addition of the term $K_e(E^* - V_o)$ is disabled; when the switch S_P is turned ON, the addition of the term $K_e(E^* - V_o)$ is enabled. Moreover, when the switch S_Q is turned ON, the reset function of the integrator $\frac{K}{s}$ is enabled; when the switch S_Q is turned OFF, the reset function is disabled and the integrator is added into the controller.

The real power P and reactive power Q are calculated from v_o and i. Note that i can be switched between the grid current i_g and the virtual current

$$i_v = \frac{v_o - v_g}{Ls + R}. \tag{18.3}$$

When the switch S_C is set at position v, the virtual current i_v is sent to the power calculation block; when the switch S_C is set at position g, the grid current i_g is sent instead.

18.2 Operation of the Controller

18.2.1 Self-synchronization Mode

Before the inverter is connected to the grid, the terminal voltage v_o should be synchronized with the grid voltage v_g, which means $V_o = V_g$ and $\omega = \omega_g$. According to (18.1) and (18.2), when S_P is OFF and S_Q is OFF, at the steady state, the real power P and the reactive power Q sent to the grid are controlled to be around their reference values P_{set} and Q_{set}, respectively, which are set at zero in the self-synchronization mode. However, both P and Q are always zero before the inverter is connected to the grid. In order to force the controller to start synchronizing with the grid, a virtual impedance $Ls + R$ is introduced to generate a virtual current i_v described by (18.3) according to the voltage difference $v_o - v_g$. For this purpose, the switch S_C is set at position v to route i_v for power calculation. In this mode, the controller becomes

$$\dot{E} = n(P_{\text{set}} - P) \tag{18.4}$$

$$\omega = \omega^* + \frac{mK}{s}(Q - Q_{\text{set}}) - m(Q_{\text{set}} - Q) \tag{18.5}$$

with

$$i = i_v = \frac{v_o - v_g}{Ls + R}. \tag{18.6}$$

Under this condition, if P_{set} and Q_{set} are both set at zero, then, at the steady state, the current carried by the virtual impedance is regulated to zero and, hence, the terminal voltage v_o is synchronized with the grid voltage v_g. In order to enable this mode, both the switch S_P and the switch S_Q should be turned OFF with the switch S_C set at position v. After the synchronization is achieved, the relay in the inverter can be turned ON to connect the inverter to the grid. At the same time, the switch S_C should be turned to position g so that the real grid current i_g can be fed into the controller for the power calculation.

The parameters L and R can be chosen smaller than the inductance and resistance of the filter inductor to speed up the synchronization process. In order to filter out the effect of the harmonics on the synchronization, the ratio $\frac{L}{R}$ can be chosen larger than the fundamental system period (Zhong et al. 2014).

18.2.2 Set Mode (*P*-mode and *Q*-mode)

After the inverter is connected to the grid, the switch S_C is at position g and the real grid current i_g is fed into the controller for the power calculation. When S_P is OFF, there is

$$\dot{E} = n(P_{set} - P).$$

The voltage magnitude E settles down at a constant value in the steady state with

$$P = P_{set}.$$

The desired real power P_{set} is sent to the grid. Similarly, when switch S_C is at position g and S_Q is turned OFF, there is

$$\omega = \omega^* + \frac{mK}{s}(Q - Q_{set}) - m(Q_{set} - Q)$$

and the frequency settles down at a certain value in the steady state with

$$Q = Q_{set}.$$

The desired reactive power Q_{set} is sent to the grid.

This mode is called the set mode, and the set mode for the real power is called the *P*-mode and the set mode for the reactive power is called the *Q*-mode.

18.2.3 Droop Mode (*P*$_D$-mode and *Q*$_D$-mode)

When the switch S_C is at position g and the switch S_P is ON, there is

$$\dot{E} = n(P_{set} - P) + K_e(E^* - V_o)$$

and the voltage magnitude settles down at a constant value in the steady state with

$$P = P_{set} + \frac{K_e}{n}(E^* - V_o). \tag{18.7}$$

This is the droop function of the real power with respect to the voltage. Similarly, when switch S_C is at position g and S_Q is ON, there is

$$\omega = \omega^* - m(Q_{set} - Q)$$

and the frequency settles down at a certain value with

$$Q = Q_{set} + \frac{\omega - \omega^*}{m}. \tag{18.8}$$

This is the droop function of the reactive power with respect to the frequency. As can be seen from (18.7) and (18.8), the actual real power and reactive power sent to the grid are changed automatically according to the voltage V_o and the frequency ω, respectively.

Table 18.2 Parameters of the inverter.

Parameter	Value
Grid voltage (RMS)	110 V
Line frequency f	50 Hz
Switching frequency f_s	19 kHz
DC-bus voltage V_{DC}	200 V
Rated apparent power S^*	300 VA
Inductance L_s	2.2 mH
Resistance R_s	0.2 Ω
Inductance L_g	2.2 mH
Resistance R_g	0.2 Ω
Capacitance C	10 μF

18.3 Experimental Results

Intensive experiments were conducted on a single-phase grid-connected inverter to validate the control strategy. Here, the results from two cases with the inverter operated as an R-inverter and an L-inverter with the self-synchronized universal droop controller are presented. Since the currently dominant droop control is for inductive impedance, the experimental results from an L-inverter with the robust droop controller equipped with the self-synchronization mechanism are also presented. The virtual resistor used for the R-inverter is 4 Ω. The inverter has an LCL filter to filter out the high-frequency components in the output voltage and the grid current. The parameters of the system are summarized in Table 18.2. The control circuit of the system was constructed based on TMS320F28335 DSP, with the sampling frequency of 4 kHz. A DC power supply was used to provide the DC-bus voltage at 200 V.

The droop coefficients are set in such a way that 100% increase of real power P results in 10% decrease of voltage E and 100% increase of reactive power Q results in 1% increase of the frequency f. Then, the droop coefficients can be calculated as $n = \frac{0.1K_eE^*}{S^*}$ and $m = \frac{0.01\omega^*}{S^*}$, according to Chapter 16, where S^* is the rated apparent power of the inverter.

The experiments were conducted in the following sequence of actions:

1. Starting the self-synchronization mode (S_C: position v; S_P: OFF; and S_Q: OFF) with $P_{set} = 0$ W, $Q_{set} = 0$ Var at $t = 0$ s
2. Turning the relay ON and switching S_C to the position g at $t = 3$ s
3. Applying $P_{set} = 150$ W at $t = 6$ s
4. Applying $Q_{set} = 150$ Var at $t = 9$ s
5. Switching S_P ON to enable the P_D-mode at $t = 12$ s
6. Switching S_Q ON to enable the Q_D-mode at $t = 15$ s
7. Stopping data acquisition at about $t = 18$ s.

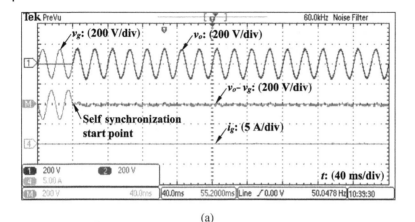

(a)

(b)

Figure 18.2 Experimental results of self-synchronization with the R-inverter. (a) When started at $v_g = 0$. (b) When started at $v_g = V_g$.

18.3.1 R-inverter with Self-synchronized Universal Droop Control

18.3.1.1 Self-synchronization

Two cases are shown in Figure 18.2 with the synchronization started at $v_g = 0$ (at $0°$) and $v_g = V_g$ (at $90°$). The time it takes to synchronize with the grid depends on the moment the synchronization started. For the case with $v_g = 0$, the voltage difference between the output voltage and the grid voltage, i.e. $v_o - v_g$, quickly became very small, as shown in Figure 18.2(a). It took less than one cycle for the whole self-synchronization process. For the case with $v_g = V_g$, as shown in Figure 18.2(b), it took about six cycles for the synchronization.

18.3.1.2 Connection to the Grid

After the synchronization process was finished, the inverter was ready to be connected to the grid. At $t = 3$ s, the relay was turned ON and S_C was turned to position g, which changed the current used for calculating P and Q from the virtual current i_v to the real grid current

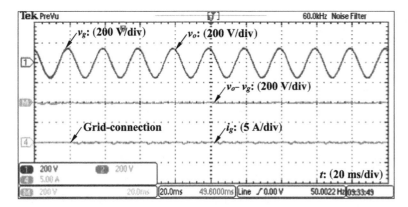

Figure 18.3 Experimental results when connecting the R-inverter to the grid.

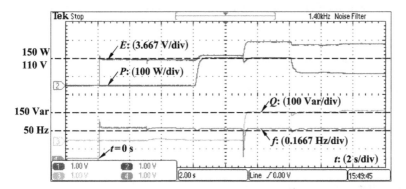

Figure 18.4 Experimental results with the R-inverter: performance during the whole experimental process.

i_g. As shown in Figure 18.3, the connection was seamless and the grid current i_g was well maintained around zero without any spikes, as expected, because $P_{set} = 0$ and $Q_{set} = 0$.

18.3.1.3 Regulation of Real and Reactive Power

The real power, reactive power, frequency and voltage during the whole process are shown in Figure 18.4. After the self-synchronization was enabled at $t = 0$ s, both the real power and the reactive power were controlled around zero. When the inverter was connected to the grid at 3 s, there was not much transient and both the real and reactive power were maintained around zero. After that, the system responded quickly to the step changes of the real and reactive power demands at $t = 6$ s and $t = 9$ s, respectively, without any static error. After the P_D-mode was enabled at $t = 12$ s, the voltage E was 112.93 V, which is about 2.67% higher than the nominal value 110 V. In this case, according to the given droop coefficients, i.e. 10% increase in E results in 100% decrease in P, the real power is expected to drop by $\frac{2.67\%}{10\%} \times 300$ W ≈ 80 W. Indeed, as shown in Figure 18.4, the real power P dropped by about 80 W from 150 W to 70 W. On the other hand, the reactive power increased after the Q_D-mode was enabled at $t = 15$ s because the frequency f was 50.03 Hz, which is 0.06% higher than the nominal value. According to the droop coefficients, i.e.

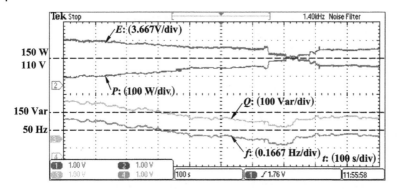

Figure 18.5 Experimental results with the R-inverter: regulation of system frequency and voltage in the droop mode.

1% increase in f results in 100% increase in Q, the reactive power is expected to increase by $\frac{0.06\%}{1\%} \times 300$ Var ≈ 20 Var. As shown in Figure 18.4, Q indeed increased by 20 Var from 150 Var to about 170 Var.

In order to test the regulation of the real power P and reactive power Q with respect to the variations of voltage E and frequency f, the inverter was kept running continuously in the P_D-mode and the Q_D-mode. The results are shown in Figure 18.5. The real power P is symmetrical to the changing voltage E and the reactive power Q follows the trend of the frequency f very well. It is worthy highlighting that this was achieved without a dedicated synchronization unit and the inverter maintained in synchronization with the grid all the time, even when the frequency and voltage changed.

18.3.1.4 Impact of Change in the DC-bus Voltage V_{DC}

In order to further test the robustness of the system, the DC-bus voltage V_{DC} was changed from 200 V to 180 V and then from 180 V to 200 V when the system was operated in the P-mode and Q-mode with $P_{set} = 150$ W and $Q_{set} = 150$ Var. When the V_{DC} was suddenly dropped from 200 V to 180 V, as shown in Figure 18.6, the grid current i_g dropped because of the lowered V_{DC}. However, it only took about five cycles for the grid current i_g to recover in order to maintain the real power and reactive power sent to the grid at the reference values. When V_{DC} was suddenly increased from 180 V to 200 V, the grid current increased and it took about five cycles for the grid current to recover to its value before the voltage change.

18.3.2 L-inverter with Self-synchronized Universal Droop Control

18.3.2.1 Self-synchronization

Again, two cases with $v_g = 0$ and $v_g = V_g$ when the synchronization was started are considered here, with the results shown in Figure 18.7. When $v_g = 0$, it took less than one cycle to synchronize with the grid. When $v_g = V_g$, it took about six cycles to synchronize.

18.3.2.2 Connection to the Grid

At $t = 3$ s, the relay was turned ON and the S_C in the controller was turned to position g to connect the inverter to the grid. As shown in Figure 18.8, the grid current i_g was smoothly maintained at around zero without any noticeable spikes.

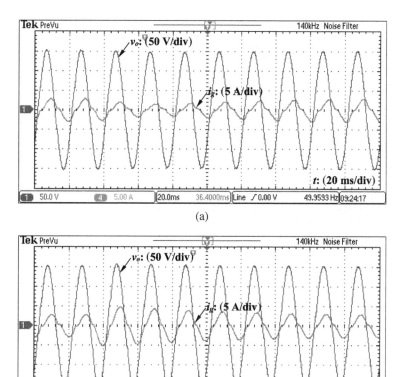

Figure 18.6 Experimental results with the R-inverter: change in the DC-bus voltage V_{DC}. (a) From 200 V to 180 V. (b) From 180 V to 200 V.

18.3.2.3 Regulation of Real and Reactive Power

The real power, reactive power, frequency and voltage during the whole process are shown in Figure 18.9. After the self-synchronization was enabled at $t = 0$ s, both the real power and the reactive power were controlled around zero. At $t = 3$ s, the inverter was connected to the grid without much transient, and both the real and reactive power were maintained around zero. At $t = 6$ s and $t = 9$ s, the inverter quickly responded to the real and reactive power demands, respectively. After the P_D-mode was enabled at $t = 12$ s, the voltage E was 114.4 V, which is about 4% higher than the nominal value 110 V. In this case, according to the set droop coefficients, i.e. 10% increase in the E results in 100% decrease in the P, the real power is expected to drop by $\frac{4\%}{10\%} \times 300$ W ≈ 120 W. Indeed, as shown in Figure 18.9, the real power P dropped by about 120 W from 150 W to about 30 W. On the other hand, the reactive power increased after the Q_D-mode was enabled at $t = 15$ s, because f is about 50.0667 Hz, which is 0.133% higher than the nominal value. According to the droop coefficients, i.e. 1% increase in f results in 100% increase in Q, the reactive power is expected to increase by $\frac{0.133\%}{1\%} \times 300$ Var ≈ 40 Var. As shown in Figure 18.9, the reactive power indeed increased by 40 Var from 150 Var to about 190 Var.

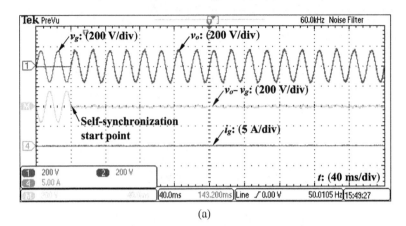

(a)

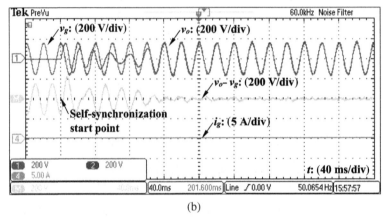

(b)

Figure 18.7 Experimental results of self-synchronization with the L-inverter. (a) When started at $v_g = 0$. (b) When started at $v_g = V_g$.

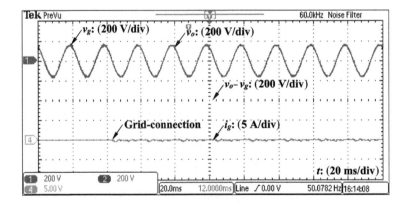

Figure 18.8 Experimental results with the L-inverter: connection to the grid.

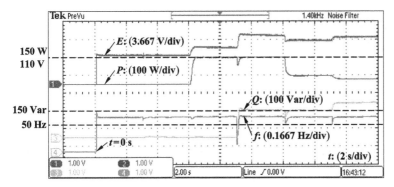

Figure 18.9 Experimental results with the L-inverter: performance during the whole experimental process.

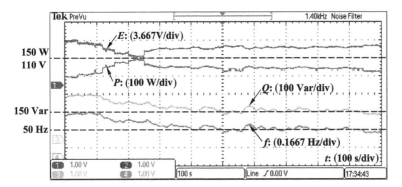

Figure 18.10 Experimental results with the L-inverter: regulation of system frequency and voltage in the droop mode.

In order to test the regulation of P and Q with respect to the variations of E and f, the inverter was kept running continuously in the droop mode (P_D- and Q_D-modes). The results are shown in Figure 18.10. The real power P is symmetrical to the varying voltage E and the reactive power Q closely follows the trend of the varying frequency f. Note again that this was achieved without a dedicated synchronization unit and the inverter maintained in synchronization with the grid all the time, even when the frequency and voltage changed.

18.3.2.4 Impact of the Change in the DC-bus Voltage
In order to further test the robustness of the system, the DC-bus voltage V_{DC} was changed from 200 V to 180 V and then from 180 V to 200 V when the system was operated in the P-mode and Q-mode with the real power reference $P_{set} = 150$ W and the reactive power reference $Q_{set} = 150$ Var. The results are shown in Figure 18.11. The performance of the L-inverter during the change in the V_{DC} is very similar to that of the R-inverter. In both cases, it took about five cycles to recover.

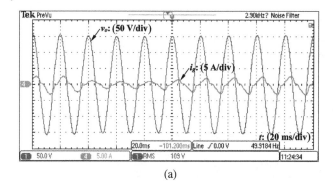

(a)

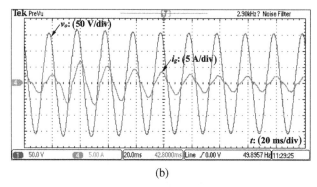

(b)

Figure 18.11 Experimental results with the L-inverter: change in the DC-bus voltage V_{DC}. (a) From 200 V to 180 V. (b) From 180 V to 200 V.

18.3.3 L-inverter with Self-synchronized Robust Droop Control

Since the $P - f$ and $Q - E$ droop control principle is adopted in the current power systems, the self-synchronization mechanism is added into the robust droop controller for L-inverters to demonstrate its performance. The droop coefficients are set in such a way that 10% increase in the voltage E results in 100% decrease in the reactive power Q and 1% increase in the frequency f results in 100% decrease in the real power P. In other words, the droop coefficients are calculated as $n = \frac{0.1K_eE^*}{S^*}$ and $m = \frac{0.01\omega^*}{S^*}$.

The experiments were conducted in the following sequence of actions:

1. Starting the self-synchronization mode (S_C: position v; S_P: OFF; and S_Q: OFF) with $P_{set} = 0$ W, $Q_{set} = 0$ Var at $t = 0$ s
2. Turning the relay on and switching S_C to the position g at $t = 3$ s
3. Applying $P_{set} = 150$ W at $t = 6$ s
4. Applying $Q_{set} = 150$ Var at $t = 9$ s
5. Switching S_Q ON to enable the Q_D-mode at $t = 12$ s
6. Switching S_P ON to enable the P_D-mode at $t = 15$ s
7. Stop of acquisition of data at about $t = 18$ s.

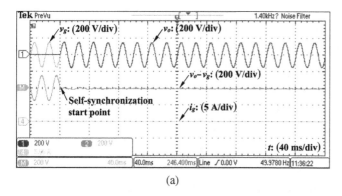

(a)

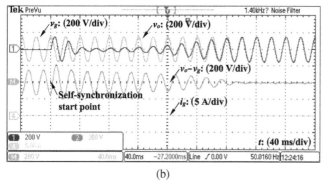

(b)

Figure 18.12 Experimental results of self-synchronization with the L-inverter with the robust droop controller. (a) When started at $v_g = 0$. (b) When started at $v_g = V_g$.

18.3.3.1 Self-synchronization

Again, two cases with $v_g = 0$ and $v_g = V_g$ when the synchronization was started are considered here, with the results shown in Figure 18.12. For the case with $v_g = 0$, it only took about one cycle for the whole self-synchronization process. On the other hand, for the case with $v_g = V_g$ at $t = 0$, it took about 14 cycles for the synchronization process, which is much longer than the cases with the universal droop controller but is still acceptable.

18.3.3.2 Connection to the Grid

After the synchronization process was finished, the inverter was ready to be connected to the grid. At $t = 3$ s, the relay was turned ON and the S_C was switched to the position g, which changed the current used for calculating P and Q from the virtual current i_v to the real grid current i_g. As shown in Figure 18.13, the grid current i_g was well maintained around zero, as expected because $P_{set} = 0$ and $Q_{set} = 0$, without any noticeable spikes.

18.3.3.3 Regulation of Real and Reactive Power

At $t = 0$ s, the self-synchronization was enabled. As shown in Figure 18.14, both the real power and the reactive power were controlled around zero. When the inverter was

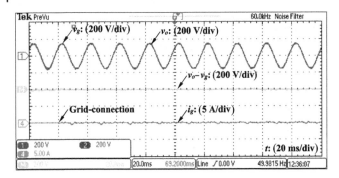

Figure 18.13 Experimental results from the L-inverter with the robust droop controller: connection to the grid.

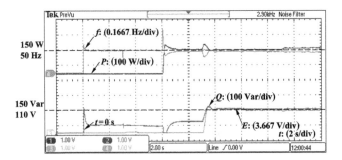

Figure 18.14 Experimental results from the L-inverter with the robust droop controller: performance during the whole experimental process.

connected to the grid at 3 s, there was not much transient and both the real and reactive power were maintained around zero. After that, the system responded quickly to the step change of the real and reactive power demand at $t = 6$ s and $t = 9$ s, respectively. After the Q_D-mode was enabled at $t = 12$ s, the reactive power was still maintained around 150 *Var*, because E was very close to its nominal value 110 V.

At $t = 15$ s, the P_D-mode was enabled. The frequency f was 49.967 Hz, which is about 0.067% lower than the nominal value 50 Hz. In this case, according to the set droop coefficients, i.e. 1% increase in f results in 100% decrease in P, the real power is expected to increase by $\frac{0.067\%}{1\%} \times 300$ W ≈ 20 W. Indeed, as shown in Figure 18.4, the real power P is increased by about 20 W from 150 W to 170 W.

In order to test the regulation of P and Q corresponding to the variations in f and E, the inverter was continuously operated in the P_D-mode and Q_D-mode and the results are shown in Figure 18.15; the real power P is symmetrical to the frequency f and the reactive power Q is symmetrical to the voltage E, as expected.

18.3.3.4 Impact of the Change in the DC-bus Voltage
Two step changes were applied to V_{DC} when the system was operated in the P-mode and Q-mode with $P_{set} = 150$ W and $Q_{set} = 150$ Var and the results are shown in Figure 18.16. At first, the V_{DC} was suddenly dropped from 200 V to 180 V. The grid current i_g dropped because of the lowered V_{DC} and recovered after about 10 cycles to maintain the real power

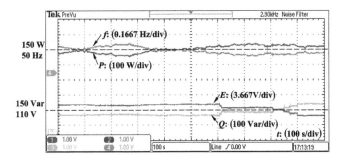

Figure 18.15 Experimental results from the L-inverter with the robust droop controller: regulation of system frequency and voltage in the droop mode.

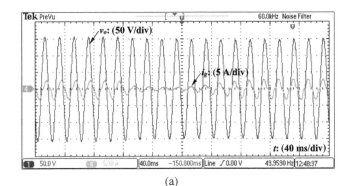

(a)

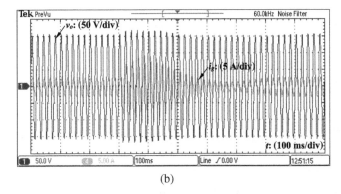

(b)

Figure 18.16 Experimental results with the L-inverter under robust droop control: change in the DC-bus voltage V_{DC}. (a) From 200 V to 180 V. (b) From 180 V to 200 V.

and reactive power at the reference values. Then the V_{DC} was suddenly increased from 180 V to 200 V. The grid current increased and recovered after about 25 cycles.

18.4 Real-time Simulation Results from a Microgrid

Real-time simulation results from a microgrid consisting of three inverters that are connected in parallel to a weak grid are described here. As shown in Figure 18.17, the microgrid

Table 18.3 Parameters of the microgrid.

	Weak grid	R-inverter	L-inverter	C-inverter
Rated apparent power S^*	30 kVA	15 kVA	10 kVA	5 kVA
Inductance L_s	0.35 mH	0.7 mH	1.05 mH	2.1 mH
Resistance R_s	0.02 Ω	0.04 Ω	0.06 Ω	0.12 Ω
Inductance L_g	0.035 mH	0.07 mH	0.105 mH	0.21 mH
Resistance R_g	0.002 Ω	0.004 Ω	0.006 Ω	0.012 Ω
Capacitance C	10 μF	10 μF	10 μF	10 μF

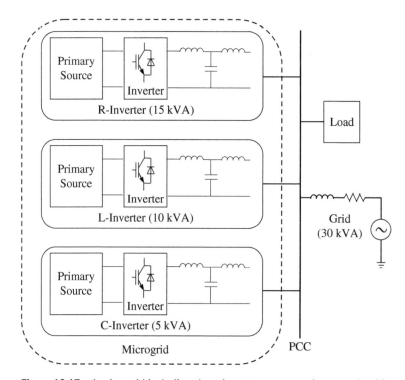

Figure 18.17 A microgrid including three inverters connected to a weak grid.

consists of an R-inverter, an L-inverter, and a C-inverter, with capacities of 15 kVA, 10 kVA, and 5 kVA, respectively. The weak grid is implemented by an R-inverter with a capacity of 30 kVA, which is controlled by the universal droop controller. The line impedance between the microgrid and the grid is modeled as a 0.24 mH inductor in series with a 0.1 Ω resistor (Lee et al. 2013). The rated RMS system voltage is 230 V, and the rated system frequency is 50 Hz. All the inverters are controlled with the self-synchronized universal droop controller. The LCL filter parameters are given in Table 18.3.

The droop coefficients are chosen in such a way that 100% increase in real power P results in 10% decrease in voltage E and 100% increase of reactive power Q results in 1% increase

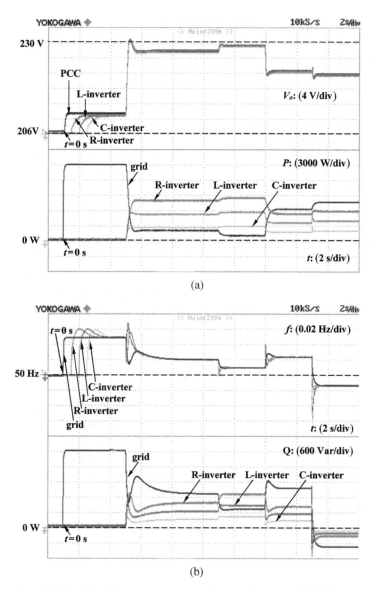

(a)

(b)

Figure 18.18 Real-time simulation results from the microgrid. (a) Real power and RMS voltage. (b) Reactive power and frequency.

in the frequency f. The inductance of the virtual impedance used in the controller for the three inverters to generate the virtual current for synchronization is 1 mH for all inverters with resistance of 2 Ω, 3 Ω, and 6 Ω for the R-, L- and C-inverters, respectively.

A series of actions were taken during the real-time simulation. The results are shown in Figure 18.18. After the grid was started at $t = 0$ s, all the power consumed by the load (3 Ω in series with 1.8 mH) was provided by the grid. The grid voltage was 208.4 V, which was about 9.4% lower than the nominal value 230 V because of the load. The system frequency was

50.05 Hz, which was about 0.1% higher than the nominal value. This is mainly caused by the reactive power 3 kVar. After the self-synchronization mode was enabled, the R-inverter, L-inverter, and C-inverter all synchronized with the PCC voltage. At $t = 3$ s, the inverters were connected to the grid at the PCC and the connection was very smooth. The real power P and reactive power Q of the inverters were maintained around zero without any spike. At $t = 4$ s, P_{set} was changed to 7.5 kW, 5 kW and 2.5 kW, and Q_{set} to 0.9 kW, 0.6 kW and 0.3 kW for the R-inverter, L-inverter, and C-inverter, respectively. The PCC voltage increased to 227.6 V, which was about 1% lower than the nominal value 230 V, because all the inverters shared the real power with the grid. The frequency decreased to 50.02 Hz, which was about 0.04% higher than the nominal value, because all the inverters shared the reactive power with the grid. As can be seen, all the inverters injected the required real power and reactive power to the grid and the grid picked up the rest of the load. At $t = 10$ s, the P_D-mode and Q_D-mode were enabled, which forced the R-inverter, L-inverter, and C-inverter to take more load (proportionally). As a result, the power provided by the grid decreased, which brought both the grid voltage and grid frequency closer to the nominal values, respectively. At $t = 13$ s, P_{set} and Q_{set} were set to 0 and the PCC voltage decreased to 222 V because the inverters sent less real power to the grid. In this case, for the R-inverter, according to the droop coefficients, i.e. 10% decrease in voltage E results in 100% increase in real power P; the real power was expected to be $\frac{3.4\%}{10\%} \times 15$ kW ≈ 5.1 kW. Indeed, as shown in Figure 18.18, the real power P of the R-inverter was around 5.1 kW. On the other hand, the frequency was 50.024 Hz, which was 0.048% higher than the nominal value 50 Hz. For the R-inverter, according to the set droop coefficients, i.e. 1% increase in f results in 100% increase in Q; the reactive power was expected to be $\frac{0.048\%}{1\%} \times 15$ kVar ≈ 0.72 kVar. As shown in Figure 18.18, the reactive power was indeed around 0.72 kVar. Similarly, the L-inverter and C-inverter supplied the real power P and reactive power Q proportionally and accurately. At $t = 16$ s, an RC load (20 Ω in parallel with 300 μF) was connected at the PCC, the real power was further increased but the reactive power was decreased to a negative value. This brought the voltage even lower and the frequency below the nominal value. In the whole process, all the inverters responded well, although their output impedances have different types and values and there was no dedicated synchronization unit.

18.5 Summary

Based on (Zhong et al. 2016), a self-synchronized universal droop controller is presented. It does not require a dedicated synchronization unit, which reduces the complexity and computational burden of the controller. The output voltage of inverters can be synchronized with the grid before and after the connection by the controller itself. Additionally, the controller can be easily configured to operate in the set mode (P- and Q-mode) and the droop mode (P_D- and Q_D-modes). As a result, the inverter equipped with the controller can fully take part in the regulation of the voltage and the frequency. Experimental and real-time simulation results have validated the system operation under different modes.

19

Droop-Controlled Loads for Continuous Demand Response

As discussed in Chapter 5, it is possible to operate a rectifier-fed load as a VSM to take part in grid regulation with the synchronverter technology. In this chapter, a general framework based on the universal droop control is presented for a rectifier-fed load to continuously take part in the regulation of grid voltage and frequency. As a result, such a load can provide primary frequency response, exceeding the FERC requirement on newly integrated generators to provide primary frequency response. It can automatically change the power consumed to support the grid, without affecting the normal operation of the load. It has a built-in storage port, in addition to normal AC and DC ports. The flexibility required by the AC port to support the grid is provided by the storage port without affecting the DC port. The grid support of the AC port is achieved through the universal droop controller. A DC-bus voltage controller is cascaded to the droop controller to regulate the DC-bus voltage of the storage port within a wide range to provide the flexibility needed while the DC-port voltage is maintained constant. An illustrative example is presented with the patented θ-converter, which consists of three ports with only four switches. Experimental results are provided to verify the capability of the rectifier to regulate the grid voltage and frequency without affecting the load. The technology is patented.

19.1   Introduction

Nowadays, many different types of loads are being connected to power grids through power electronic rectifiers. Motors, lighting, and internet devices that consume more than 80% electricity are expected to be equipped with rectifiers at the front end for various reasons (Zhong, 2017e). These rectifiers mostly consume constant power from the grid and do not take part in the regulation of the grid. A lot of attempts are being made to enable loads to take part in the grid regulation at the system level, called demand-side management (DSM), with the objective of improving the system stability from the demand side (Costanzo et al. 2012; Mohsenian-Rad et al. 2010; Palensky and Dietrich, 2011). These methods can remotely control the energy consumption of certain appliances or enable users to actively respond to the price information and take part in the power system regulation. However, these DSM methods rely on the communication infrastructure among utilities and users, and often require human beings at the demand side to make the final decision. Moreover, the demand-side regulation available nowadays is done on an ON/OFF basis. It would

Power Electronics-Enabled Autonomous Power Systems: Next Generation Smart Grids,
First Edition. Qing-Chang Zhong.
© 2020 John Wiley & Sons Ltd. Published 2020 by John Wiley & Sons Ltd.

bring enormous benefits if loads could take part in the regulation in a continuous way like generators, which is called continuous demand response in this book. Take the Texas ERCOT grid as an example. Its capacity is about 70 GW and the capacity of its largest interconnection is 650 MW. If all loads in ERCOT are able to contribute 1% when needed, then the total contribution is 700 MW. This means there is no need to shed any load even if its largest interconnection is tripped off, significantly improving its stability and reliability.

In this chapter, it will be shown that droop control can be applied to operate a rectifier as a VSM to continuously take part in the grid regulation. With the universal droop control, the power drawn by the rectifier changes according to the voltage and frequency deviations of the power grid. However, the power consumption of the load is often fixed without much flexibility. Hence, an additional energy storage unit is often added to meet the power difference. As is well known, most rectifiers have DC-bus capacitors to reduce voltage ripples and maintain the DC-bus voltage constant. The objective of this chapter is to show that, with proper design and control, these capacitors can also function as energy storage devices to provide the flexibility needed for grid support. This requires the rectifier to have three ports: one AC port, one DC port, and one storage port. In other words, this requires a three-port DC/DC/AC converter. To obtain a DC/DC/AC converter, the simplest way is to connect a DC/DC converter and a DC/AC converter in series, which increases the complexity of the topology and the controller. There are some simple topologies with three ports, e.g. the DC/DC/AC converter in (Cai et al. 2015), the θ-converter (Zhong and Ming, 2016), and the Beijing converter (Zhong et al. 2017). In this chapter, the θ-converter is taken as an example to illustrate the concept of rectifiers that can continuously take part in grid regulation.

The θ-converter was originally presented to reduce the usage of bulky electrolytic capacitors needed to buffer the ripple energy that exists in most single-phase converters (Zhong and Ming, 2016). Similar to the conventional bridge converters, a θ-converter has two legs with only four switches. However, the two legs are independently controlled, which is enabled by another filter consisting of one inductor and an auxiliary capacitor. The voltage of the auxiliary capacitor can be maintained constant while the DC-bus voltage can be allowed to vary within a wide range. Because of this, the DC-bus capacitor can also be adopted to cover the power difference between the DC and AC ports, which makes it possible for the rectifier to take part in the voltage and frequency regulation of the grid as well. In this way, the DC-bus capacitor needs to be designed to reflect the amount of grid support the converter needs to provide, in addition to storing the ripple energy.

19.2 Control Framework with a Three-port Converter

19.2.1 Generation of the Real Power Reference

Figure 19.1 illustrates a general three-port converter, where v_{DC} is the instantaneous DC-bus voltage of the storage port, P_{DC} is the real power exchanged with the DC-bus capacitor (the storage) and P_L is the real power consumed by the load connected to the DC port. P and Q are the real and reactive powers exchanged with the grid at the AC port. Note that in Figure 19.1 the power flow of the system can be bidirectional, although this chapter focuses on the unidirectional power flow from the AC port to the DC port. Because of the additional storage port, the DC-bus voltage v_{DC} is designed to vary in a wide range

Figure 19.1 A general three-port converter with an AC port, a DC port, and a storage port.

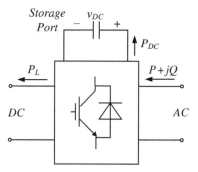

while the voltage of the DC port remains constant. When neglecting the power loss P_{loss}, the power P_{DC} exchanged with the DC-bus capacitor is equal to the power P drawn by the rectifier minus the power P_L consumed by the load, i.e.

$$P_{DC} = P - P_L. \tag{19.1}$$

Normally, there is $P = P_L$ in the steady state so $P_{DC} = 0$. However, the DC-bus capacitor acts as an energy storage unit here to meet the power difference so $P_{DC} \neq 0$. This corresponds to a varying DC-bus voltage by design. In order to ensure the proper operation of the converter, the range of the DC-bus voltage $[V_{DC}^* - V_t, V_{DC}^* + V_t]$ can be designed according to the voltage rating of the circuit components and the amount of the grid support provided. It is recommended to have a V_t as high as possible to provide enough grid support. However, when designing V_t, $V_{DC}^* + V_t$ should not exceed the voltage rating of the DC-bus capacitor. $V_{DC}^* - V_t$ should also meet any requirement to operate the converter properly.

Accordingly, there is a need to equip the rectifier with a DC-bus voltage controller to make sure that the DC-bus voltage does not exceed the specified range, in addition to a controller that regulates the DC-port voltage for the load and a controller that interacts with the grid.

Figure 19.2 illustrates such a controller. It generates the real power reference P_{set} for the inner-loop controller that regulates the power exchanged with the grid at the AC port. It consists of two parts: the power consumed by the load P_L and the power needed to regulate v_{DC}. Because in a single-phase rectifier v_{DC} often contains second-order ripples, a hold filter $H(s)$ is adopted to filter the second-order ripples and extract the average value V_{DC} of v_{DC}. In the following, the notation v_{DC} means the instantaneous value and V_{DC} means the average value. When V_{DC} is above the specified range, no more power should be pushed into the DC-bus capacitor, which means P_{set} should be reduced; when V_{DC} is below the specified range, no more power should be drawn from the DC-bus capacitor, which means P_{set} should be increased; when V_{DC} is within the specified range, no action should be taken to change

Figure 19.2 DC-bus voltage controller to generate the real power reference.

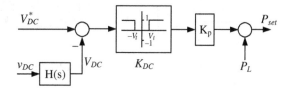

P_{set}. This can be achieved by the control block K_{DC} shown in Figure 19.2, e.g., with

$$K_{\text{DC}} = \begin{cases} 1 & (|V_{\text{DC}} - V^*_{\text{DC}}| \geq V_{\text{t}}), \\ 0 & (|V_{\text{DC}} - V^*_{\text{DC}}| < V_{\text{t}}), \end{cases}$$

where V^*_{DC} is the rated DC bus voltage determined by the design of the converter and V_{t} is the designed threshold voltage. Hence, the desired voltage range is $[V^*_{\text{DC}} - V_{\text{t}}, V^*_{\text{DC}} + V_{\text{t}}]$. When V_{DC} is within the range, $K_{\text{DC}} = 0$, the rectifier is said to work in the grid-support mode, denoted "GS". When V_{DC} is outside of the range, $K_{\text{DC}} = 1$ and the rectifier is said to operate in the no-support mode, denoted "NS". This leads to the following real power reference

$$P_{\text{set}} = \begin{cases} P_{\text{L}} + K_{\text{P}}(V^*_{\text{DC}} - V_{\text{DC}}) & \text{(no-support mode)}, \\ P_{\text{L}} & \text{(grid-support mode)}, \end{cases} \tag{19.2}$$

where K_{P} is a positive proportional gain.

19.2.2 Regulation of the Power Drawn from the Grid

There are different ways to regulate the power drawn from the grid; see e.g. (Escobar et al. 2003; Katiraei and Iravani, 2006; Rodriguez et al. 2007b; Shan et al. 2010). In order for the rectifier to take part in the regulation, the universal droop controller (Zhong and Zeng, 2016) discussed in Chapter 17 is adopted and re-drawn as shown in Figure 19.3 with the positive sign of the current changed to flowing into the rectifier, which is opposite to the case of an inverter. It is constructed according to the following droop relationship

$$E = E^* + nP, \tag{19.3}$$

$$\omega = \omega^* - mQ. \tag{19.4}$$

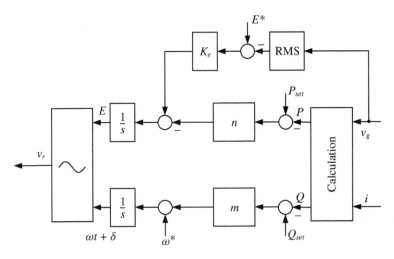

Figure 19.3 The universal droop controller when the positive direction of the current is taken as flowing into the converter.

Here, E^* is the rated RMS value of the system voltage, ω^* is the rated system frequency, and n and m are the droop coefficients. P and Q are the real and reactive power at the AC port. Note that the positive direction of the current is taken as flowing into the converter. This is reflected in Figure 19.3 and in (19.3) and (19.4), where the signs before P and Q are opposite to those for inverters (Zhong and Zeng, 2016). It is worth mentioning that the conventional droop relationship can still be used as long as there are no converters with capacitive impedance.

The universal droop controller shown in Figure 19.3 allows setting the real power and the reactive power at references P_{set} and Q_{set}, respectively. This is equivalent to setting the rated operational point of the rectifier at $P = P_{set}$ and $Q = Q_{set}$ when $V_g = E^*$ and $\omega = \omega^*$. Here, Q_{set} is set as 0 and P_{set} is generated by the DC-bus voltage controller in Figure 19.2.

According to Figure 19.3, the real power P drawn by the rectifier satisfies

$$P = P_{set} + \frac{\dot{E}}{n} - \frac{K_e}{n}(E^* - V_g),$$

where K_e is an amplifying gain that helps improve voltage regulation. The actual power P drawn by the rectifier can be different from the real power reference P_{set} generated by the DC-bus voltage controller in Figure 19.2. Hence, the rectifier is able to provide the essential grid support while meeting the needs of the DC load and regulating the DC-bus voltage of the storage port. Normally, the real power loop is designed to be much slower than the voltage loop, which means $\dot{E}$ can be regarded as 0 when investigating the change of P. As a result, the real power drawn by the rectifier is

$$P \approx P_{set} - \frac{K_e}{n}(E^* - V_g) = P_{set} + \frac{K_e}{n}\Delta V_g, \tag{19.5}$$

where

$$\Delta V_g = V_g - E^*$$

represents the deviation of the grid RMS voltage V_g from the rated value E^*. Apparently, the power P drawn by the rectifier is in proportion to the voltage variation ΔV_g. The higher the grid voltage, the more the power P drawn; the lower the grid voltage, the less the power P drawn. It indeed offers the desired continuous demand response.

According to Figure 19.3, the reactive power satisfies

$$Q = Q_{set} + \frac{\omega^* - \omega}{m}. \tag{19.6}$$

In order to achieve unity power factor at the rated operational point, Q_{set} can be set as 0. It can also be set at other values to provide reactive power compensation if needed, which enables the rectifier to provide reactive power support as well. Note that the reactive power changes according to the grid frequency deviation in the droop mode so the reactive power is not zero when the frequency is not at the rated value. If a unity power factor is desired, the $Q \sim \omega$ droop can be designed to be soft so that the change of the reactive power is small with respect to the change in the frequency.

19.2.3 Analysis of the Operation Modes

As discussed above, the rectifier can operate under the no-support mode and the grid-support mode. There are two no-support modes that correspond to the cases with the

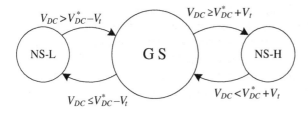

Figure 19.4 Finite state machine of the droop-controlled rectifier.

DC-bus voltage at the low voltage limit and at the high voltage limit, respectively. These modes are denoted as GS, NS-L and NS-H, respectively, and the corresponding finite state machine is shown in Figure 19.4. When the rectifier is operated in the NS-L mode, the DC-bus voltage controller tries to increase the power drawn by the rectifier according to (19.2). If the grid voltage becomes higher than the rated voltage, it forces the rectifier to draw more power than P_{set}, according to (19.5). These would push the DC-bus voltage above the low voltage limit. When the rectifier is operated in the NS-H mode, the DC-bus voltage controller tries to reduce the power drawn by the rectifier according to (19.2). If the grid voltage becomes lower than the rated voltage, it forces the rectifier to draw less power than P_{set}, according to (19.5). This would push the DC-bus voltage below the high voltage limit. When the rectifier is operated in the GS mode, the DC-bus voltage changes with the grid voltage to provide grid support. If the DC-bus voltage reaches the high voltage limit, the rectifier enters into the NS-H mode. If the DC-bus voltage reaches the low voltage limit, the rectifier enters into the NS-L mode. Figure 19.5 illustrates the possible scenarios of operation modes:

- During t_0–t_1, it is operated in the NS-L mode
- At t_1, V_{DC} becomes higher than the low voltage limit and the rectifier enters the GS mode
- At t_2, V_{DC} reaches the high voltage limit and the rectifier enters the NS-H mode
- At t_3, V_{DC} becomes lower than the high voltage limit and the rectifier enters the GS mode
- At t_4, V_{DC} reaches the low voltage limit and the rectifier enters the NS-L mode.

It is worth noting that, after the rectifier enters into the GS mode from the NS-L mode, it is possible for it to enter the NS-L mode again. The same is true when it enters the GS mode from the NS-H mode.

19.2.4 Determination of the Capacitance for Grid Support

For the desired DC-bus voltage range $[V_{\text{DC}}^* - V_t, V_{\text{DC}}^* + V_t]$ of the storage port and the given DC-bus capacitance C, the change in the energy stored is

$$\Delta \mathcal{E} = \frac{1}{2} C((V_{\text{DC}}^* + V_t)^2 - (V_{\text{DC}}^* - V_t)^2) = 2CV_{\text{DC}}^* V_t. \tag{19.7}$$

This is the maximum energy that can be stored in the capacitor starting from the NS-L mode or the maximum energy that can be drawn from the capacitor starting from the NS-H mode. Taking the case when the rectifier is operated in the NS-L mode at the beginning, as shown in Figure 19.5, as an example. The rectifier enters the GS mode at $t_s = t_1$ with $V_{\text{DC}}(t_s) = V_{\text{DC}}^* - V_t$ and quits the GS mode at $t_e = t_2$ with $V_{\text{DC}}(t_e) = V_{\text{DC}}^* + V_t$. Substituting

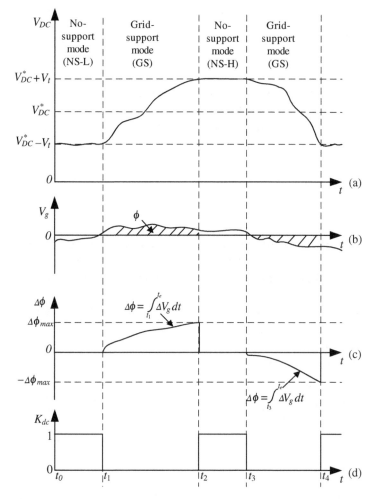

Figure 19.5 Illustration of the operation of the droop-controlled rectifier.

(19.5) and (19.2) into (19.1), the power P_{DC} that enters the storage capacitor is

$$P_{DC} \approx \frac{K_e}{n} \Delta V_g. \tag{19.8}$$

Taking integration on both sides, then the change in the energy stored in the capacitor is

$$\Delta \mathcal{E} = \int_{t_s}^{t_e} P_{DC} dt \approx \frac{K_e}{n} \int_{t_s}^{t_e} \Delta V_g dt = \frac{K_e}{n} \Delta \phi, \tag{19.9}$$

where

$$\Delta \phi = \int_{t_s}^{t_e} \Delta V_g dt,$$

i.e. the integral of grid voltage deviation ΔV_g from time t_s to t_e, is a new metric introduced to characterize the variation of the grid voltage over a period of time. The larger the absolute value of $\Delta \phi$, the larger the grid voltage variation along one direction. A value of $\Delta \phi$ around

0 means that the grid voltage mostly stay around the rated voltage. A similar metric can be introduced for the grid frequency as

$$\Delta\theta = \int_{t_s}^{t_e} \Delta\omega dt,$$

where $\Delta\omega = \omega - \omega^*$, to characterize the variation of the grid frequency over a period of time. These are accumulative metrics to evaluate the grid quality.

Combining (19.7) and (19.9), for a given $\Delta\phi_{max}$, there is

$$2CV_{DC}^* V_t = \frac{K_e}{n} \Delta\phi_{max}.$$

Hence, the required DC-bus capacitance can be found as

$$C = \frac{K_e}{2nV_{DC}^* V_t} \Delta\phi_{max}. \tag{19.10}$$

It can be reduced by increasing the rated DC-bus voltage V_{DC}^*, the range of the DC-bus voltage variation $2V_t$ and the depth of the voltage regulation $\frac{K_e}{n}$. Of course, it can be reduced if the grid voltage variation is small, i.e. to have a small $\Delta\phi_{max}$ over time. These are consistent with normal expectations.

When the power loss P_{loss} is not negligible, the real power entering the DC-bus storage capacitor given in (19.8) is

$$P_{DC} = P - P_L - P_{loss} \approx \frac{K_e}{n} \Delta V_g - P_{loss}.$$

Because of the P_{loss}, the power flows into the DC-bus capacitor is smaller than the ideal value $\frac{K_e}{n} \Delta V_g$. Taking the integration for both sides, the change of the energy stored in the capacitor is

$$\Delta\mathcal{E} = \int_{t_s}^{t_e} P_{DC} dt \approx \frac{K_e}{n} \Delta\phi - \int_{t_s}^{t_e} P_{loss} dt.$$

As a result,

$$\Delta\phi = \frac{n}{K_e} \Delta\mathcal{E} + \frac{n}{K_e} \int_{t_s}^{t_e} P_{loss} dt, \tag{19.11}$$

which means a positive item $\frac{n}{K_e} \int_{t_s}^{t_e} P_{loss} dt$ is added to $\Delta\phi$ in (19.9). In practice, the actual $\Delta\phi$ that can be allowed is then larger than $\Delta\phi_{max}$. In other words, the capacitance C obtained from (19.10) with the power loss ignored is larger than the actual capacitance needed for the given $\Delta\phi_{max}$.

In Figure 19.5, $\Delta\phi$ is reset to 0 when the rectifier is in the NS mode to illustrate the evolution of $\Delta\phi$ during the GS mode.

19.3 An Illustrative Implementation with the θ-converter

Here, the three-port four-switch θ-converter proposed in (Zhong and Ming, 2016) is taken as an example to illustrate the implementation of the control framework for a rectifier to achieve continuous demand response.

19.3.1 Brief Description about the θ-converter

As shown in Figure 19.6, the θ-converter (Zhong and Ming, 2016) has two legs: one conversion leg and one neutral leg. The conversion leg consists of two switches Q_1 and Q_2, one inductor L_g and one DC-bus capacitor C. It is operated as a AC/DC converter. The neutral leg consists of two switches Q_3 and Q_4, one inductor L and one capacitor C_+. It is operated as a DC/DC converter. Note that there are only four switches, the same as conventional single-phase bridge converters. Unlike a conventional full-bridge converter, the load is connected to the DC port with the output voltage V_+ and the conventional DC-bus capacitor is treated as a storage port with the voltage V_{DC}, which is allowed to vary in a wide range. The neutral inductor L is added in order to enable the regulation of the voltage V_+.

A prominent feature of the θ-converter is that the two legs are independently controlled. The conversion leg is responsible for regulating the power exchanged with the grid and the neutral leg is responsible for maintaining the DC-port voltage V_+ and diverting the ripples to the capacitor C. Here, the DC-bus voltage is allowed to vary in a wide range to provide grid support.

The detailed operation principle and control design of the θ-converter can be found in (Zhong and Ming, 2016). Here, only the basic operation is presented. The switches Q_1 and Q_2, and Q_3 and Q_4 are operated so that they complement each other. The duty cycle of the switch Q_4 satisfies

$$d_4 = \frac{V_+}{V_{DC}}. \tag{19.12}$$

Since the switches Q_3 and Q_4 are operated so that they complement each other, the duty cycle of Q_3 is

$$d_3 = 1 - d_4 = 1 - \frac{V_+}{V_{DC}} \tag{19.13}$$

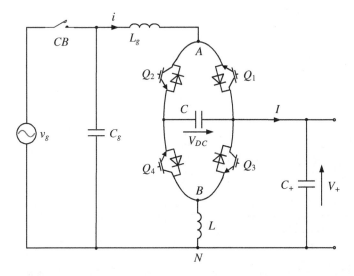

Figure 19.6 The θ-converter.

The conversion leg is a half bridge converter. The duty cycle of the switch Q_2 is

$$d_2 = \frac{V_+}{V_{DC}} - \frac{v_g}{V_{DC}} \tag{19.14}$$

while the duty cycle of the switch Q_1 is

$$d_1 = 1 - d_2 = 1 - \frac{V_+}{V_{DC}} + \frac{v_g}{V_{DC}}. \tag{19.15}$$

The duty cycles of switches Q_1 and Q_2 contain a DC component $\frac{V_+}{V_{DC}}$ and an AC component $\frac{v_g}{V_{DC}}$. The DC component $\frac{V_+}{V_{DC}}$ is needed to eliminate the effect brought from the output voltage V_+. The AC component $\frac{v_g}{V_{DC}}$ represents the conversion ratio between the grid voltage v_g and the DC-bus voltage V_{DC}.

19.3.2 Control of the Neutral Leg

The controller proposed in (Zhong and Ming, 2016) for the neutral leg, as shown in Figure 19.7(a), can be adopted without any change. It adopts a PI controller to regulate the output voltage V_+ and a repetitive controller to remove the low-frequency component in V_+. A band-pass filter

$$B(s) = \frac{10000s}{(s+10)(s+10000)}$$

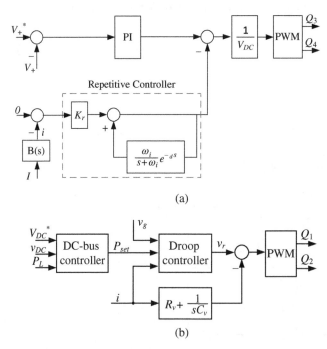

(a)

(b)

Figure 19.7 Control structure for the droop-controlled rectifier. (a) Controller for the neutral leg. (b) Controller for the conversion leg.

is used to extract the low-frequency component in the output current I, which is then put through a repetitive controller

$$K(s) = \frac{K_r}{1 - \frac{\omega_l}{s+\omega_l}e^{-\tau_d s}}.$$

It forces the low-frequency component in I to zero. Therefore, the low-frequency ripples, which are mainly fundamental and second-order components, are forced not to go through the load, maintaining a ripple-free V_+ while the DC-bus voltage is varying.

19.3.3 Control of the Conversion Leg

The role of the conversion leg is to regulate the power exchanged with the grid through the droop controller. One particular feature to note is that the voltage reference v_r generated by the droop controller can only regulate the AC component in the duty cycles d_1 and d_2, which corresponds to $\frac{v_g}{V_{DC}}$ in (19.14). In order to ensure the proper operation of the rectifier, a bias needs to be added to generate the DC component $\frac{V_+}{V_{DC}}$ in the duty cycles d_1 and d_2. This can be achieved by adopting the technique proposed in (Zhong and Zeng, 2014) that forms C-converters, which are converters with capacitive output impedances, i.e. by introducing a local inner current feedback loop through an integrator. The overall controller for the conversion leg is shown in Figure 19.7(b). In order to increase the damping, a virtual resistor R_v is added to the virtual capacitor $\frac{1}{sC_v}$ in the local current feedback loop. The power of the DC load can be obtained as $P_L = V_+ I$.

19.4 Experimental Results

The parameters of an experimental system are given in Table 19.1. It is connected to the grid through a variable transformer so that the grid voltage can be changed continuously. A 400 Ω resistor is connected to the θ-converter as a load. In the θ-converter, L_g is chosen as 2.2 mH and C_g is chosen as 10 μF to filter out the switching frequency ripples. In order to maintain the output voltage and to facilitate the grid support, the capacitors C_+ and C are chosen to be larger than those used in (Zhong and Ming, 2016). The capacitor C_+ is chosen as 1000 μF to reduce the voltage drop of V_+ when the load is connected. This is in line with the capacitance needed for a normal bridge converter at this power level. The controller is implemented on a TMS320F28335 DSP with the sampling frequency of 10 kHz.

19.4.1 Design of the Experimental System

The droop coefficients are set in such a way that 40% increase of the voltage leads to 100% increase of real power and 1% increase of the frequency leads to 100% decrease of reactive power. The gain K_e is chosen as 10. The droop coefficients can be calculated as $n = \frac{0.4K_e E^*}{S^*} = \frac{0.4*10*120}{300} = 1.6$ and $m = \frac{0.01\omega^*}{S^*} = \frac{0.01*2\pi*60}{300} = 0.0126$, where $S^* = 300$ VA is the rated capacity of the converter. The virtual RC impedance in the experiment is chosen as $6 + \frac{30}{s}$ Ω.

In the experiments, the reactive power reference was set as $Q_{set} = 0$, and the real power reference P_{set} was generated by the DC-bus voltage controller. The rated DC-bus voltage

Table 19.1 Parameters of the experimental droop-controlled rectifier.

Parameter	Value
Grid voltage (RMS)	120 V
Grid frequency f	60 Hz
Switching frequency f_s	20 kHz
Inductor L_g	2.2 mH
Inductor L	2.2 mH
DC output voltage V_+^*	200 V
Rated DC-bus voltage V_{DC}^*	500 V
Load R	400 Ω
Capacitor C_+	1000 μF
Capacitor C	500 μF
Capacitor C_g	10 μF
Rated apparent power S	300 VA

was $V_{DC}^* = 500$ V with $V_t = 100$ V so the DC-bus voltage was controlled within 400–600 V. In order to make sure that the rectifier can operate under the grid-support mode as long as possible, the DC-bus capacitor C needs to be large enough. Assume $\Delta\phi_{max} = 4$, meaning that the average grid voltage variation during one second is 4 V, which is significant. Then, according to (19.10), the required capacitance C is

$$C = \frac{K_e}{2nV_{DC}^* V_t} \Delta\phi_{max} = \frac{10 \times 4}{2 \times 1.6 \times 500 \times 50} = 500 \ \mu F.$$

Two 400 V 1000 μF capacitors are connected in series in the experimental system. This may slightly increase the cost of the system but it is worth paying because of the important function of supporting the grid gained.

Compared to a traditional single-phase bridge rectifier, no extra switches are needed. Actually, as shown in (Zhong and Ming, 2016), the θ-converter has higher efficiency than a conventional bridge rectifier. Here, because the DC-bus capacitor C acts as storage, there is real power exchange corresponding to the grid voltage deviation, which may slightly increase the loss.

19.4.2 Steady-state Performance

19.4.2.1 Operation under the Grid-support Mode

The experimental results of the system under the grid-support mode are shown in Figure 19.8. The DC-bus voltage v_{DC} follows the trend of V_g, as clearly shown in Figure 19.8(b), to support the grid. It stays within the specified range from 400 V to 600 V. This justifies that the DC-bus capacitor can act as an energy storage unit to provide the flexibility for grid regulation. The DC-bus voltage controller does not take action,

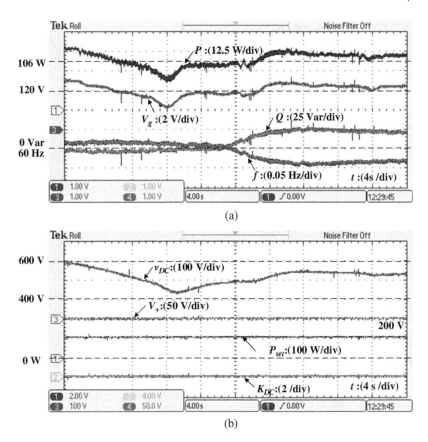

Figure 19.8 Experimental results in the GS mode. (a) Real power P, grid voltage V_g, reactive power Q and frequency f. (b) DC-bus voltage v_{DC}, DC-port voltage V_+, P_{set} and K_{DC}.

with $K_{DC} = 0$, so $P_{set} = P_L = 100$ W. The output voltage V_+ is maintained well at around 200 V. Since the load is 400 Ω, the rated power consumed by the load is about $\frac{200^2}{400} = 100$ W. Because of the inaccuracy of the load resistance and the loss in the converter, the baseline of the real power P is about 106 W. Clearly, the converter is able to take part in the voltage and frequency regulation of the grid according to the P–E and Q–f droop mechanism. As shown in Figure 19.8, the real power P tracks the trend of V_g very well according to the droop coefficient designed. At $t = 12$ s, V_g decreases to about 118 V, which is about 1.67% lower than the nominal value. According to the preset droop coefficients, the real power is expected to drop by $1.67\% \times \frac{100\%}{40\%} \times 300 = 12.5$ W. Indeed, at $t = 12$ s, P is about 95 W, dropping by about $106 - 95 = 11$ W. There exists some error due to the measurement and estimation, but, in general, P can track the changing V_g well according to the preset droop coefficient. Moreover, as clearly shown in Figure 19.8, the reactive power Q is symmetrical to the changing frequency f. At about $t = 28$ s, f is 0.05 Hz below the nominal value. According to the preset droop coefficient, Q is expected to increase by $\frac{0.05}{60} \times \frac{100\%}{1\%} \times 300 = 25$ Var. Indeed, it is about 25 Var at $t = 28$ s.

19.4.2.2 Operation under the No-support Mode

As mentioned above, the NS mode has two modes: the NS-H mode when the V_{DC} reaches the high limit and the NS-L mode when the V_{DC} reaches the low limit.

The experimental results when the converter is operated in the NS-H mode are shown in Figure 19.9. During $t = 0$ to 8 s, v_{DC} increases gradually as V_g and P increase. At about $t = 8$ s, v_{DC} reaches the high voltage limit 600 V and the system enters the NS-H mode. The DC-bus voltage controller takes actions to maintain v_{DC} at 600 V with K_{DC} being zero and one. P_{set} is equal to P_L plus a negative value, which is $(V_{DC}^* - v_{DC}) \times K_{DC} \times K_P = (500 - 600) \times 1 \times 0.2 = -20$ W. This reduces the power drawn by the converter and tries to push the DC-bus voltage back to the preset range. In this case, the converter no longer takes part in the grid regulation. Indeed, as shown in Figure 19.9, P does not track the changing V_g. At $t = 32$ s, v_{DC} is pushed back to the preset range of v_{DC} (below 600 V) and the system enters the GS mode again.

The experimental results when the converter is operated in the NS-L mode are shown in Figure 19.10. The converter operates in the GS mode initially with V_g below 120 V and P less than the baseline power of 106 W. This forces the DC-bus capacitor to provide additional power to the DC-port load, leading to a decrease in v_{DC}. When v_{DC} drops to the low voltage limit of 400 V at about $t = 15.2$ s, the converter enters the NS-L mode. Similar to the NS-H

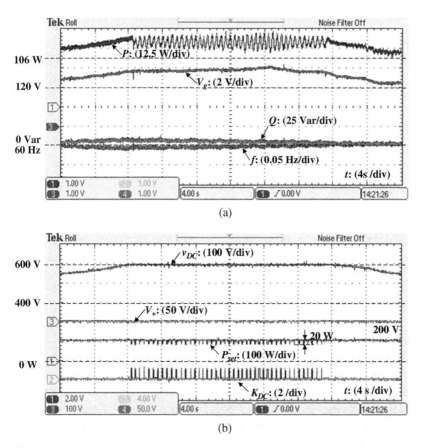

(a)

(b)

Figure 19.9 Experimental results in the NS-H mode. (a) Real power P, grid voltage V_g, reactive power Q and frequency f. (b) DC-bus voltage v_{DC}, DC-port voltage V_+, P_{set} and K_{DC}.

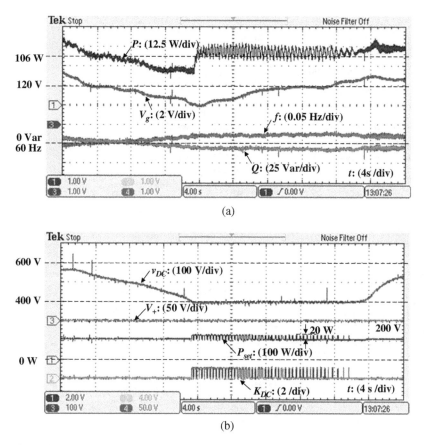

Figure 19.10 Experimental results in the NS-L mode. (a) Real power P, grid voltage V_g, reactive power Q and frequency f. (b) DC-bus voltage v_{DC}, DC-port voltage V_+, P_{set} and K_{DC}.

mode, the DC-bus voltage controller takes action with $K_{DC} = 1$ and increasing the P_{set} by a positive value of 20 W, which is about 20% of P_L. This tries to draw more power into the DC-bus capacitor to push the DC-bus voltage back to the preset range. The grid voltage V_g becomes higher than the rated value at about $t = 31$ s, which pushes even more power into the converter. Eventually, the DC-bus voltage is pushed back to the preset range at about $t = 35.2$ s and the converter enters the GS mode again. Note that, as shown in Figure 19.10, the real power P has an abrupt change to about 106 W when the converter enters the NS-L mode because the DC-bus capacitor can no longer provide additional power to the load and all the power needed has to come from the grid.

19.4.3 Transient Performance

19.4.3.1 System Start-up

The experimental results when the system was started without the load are shown in Figure 19.11. Before the PWM signal is enabled, the converter works as an uncontrolled diode rectifier. v_{DC} is about 340 V and V_+ is half of v_{DC}. The PWM signal is enabled at $t = 0.6$ s (after synchronization). Since, initially, $v_{DC} < 600$ V, the rectifier operates under

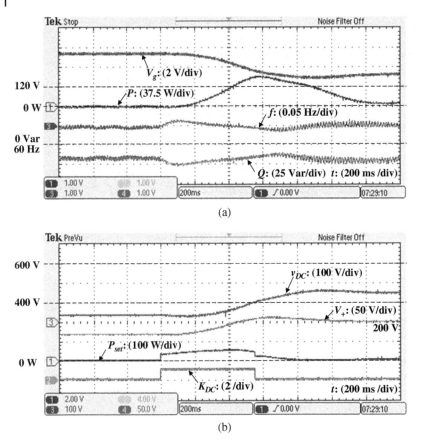

Figure 19.11 Transient response when the system starts up. (a) Real power P, grid voltage V_g, reactive power Q and frequency f. (b) DC-bus voltage v_{DC}, DC-port voltage V_+, P_{set} and K_{DC}.

the NS-L mode. The DC-bus voltage controller takes actions and forces v_{DC} to increase. The grid voltage has dropped a bit because of the start-up, as expected. The power drawn by the converter increases gradually and then settles down to a small value to cover the losses (because no load is connected). At about $t = 1.15$ s, v_{DC} reaches 400 V and the converter enters the GS mode. It takes about 400 ms for V_+ to reach the rated value of 200 V with the overshoot of about 10 V. This is comparable to the transient performance of the θ-converter in (Zhong and Ming, 2016, Figure 9 on p. 8444), in which it takes about 360 ms for V_+ to reach the steady state. It can be seen that the process of starting the system up is very smooth, without excessive power drawn from the grid. The frequency is slightly above the rated value so there is a small negative reactive power Q. Clearly, the frequency and the reactive power are symmetrical, indicating the desired capability of grid regulation.

19.4.3.2 Connection of the Load

The experimental results when the 400 Ω load was connected at $t = 0.32$ s are shown in Figure 19.12. The DC-bus voltage v_{DC} drops from 560 V to 480 V (at about 0.6 s) while V_+ is maintained at 200 V with very little transient, which means C_+ can be chosen much smaller than 1000 μF. The energy stored in the DC-port capacitor C_+ is transferred to the load, which

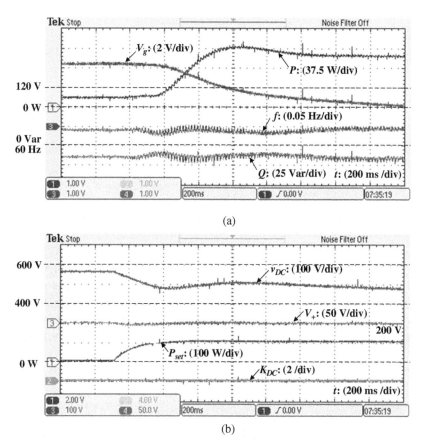

Figure 19.12 Transient response when a load is connected to the system. (a) Real power P, grid voltage V_g, reactive power Q and frequency f. (b) DC-bus voltage v_{DC}, DC-port voltage V_+, P_{set} and K_{DC}.

makes P_{set} rise gradually, and the energy stored in the DC-bus capacitor C is transferred to the load, causing v_{DC} to decrease. Then, the real power P drawn from the grid increases to meet the demand of the load. During the whole process, the converter operates in the GS mode and the v_{DC} stays within the range. Note that the current I does not include the current provided by the capacitor C_+ to the load so the current I, and hence P_{set}, increase gradually rather than sharply. This helps smooth the operation of the system. It can be seen that the process of connecting the load is very smooth, without excessive power drawn from the grid. The frequency is slightly above the rated value so there is a small negative reactive power Q. Clearly, the frequency and the reactive power are symmetrical, indicating the desired capability of grid regulation.

19.4.4 Capacity Potential

Larger grid voltage disturbances were applied to the rectifier with results shown in Figure 19.13 to show its capacity potential. The curves of V_g, v_{DC}, P, and $\Delta\phi$ are placed in the same plot to clearly demonstrate their relationship. The DC-bus voltage followed the grid voltage well. The system operation mode changed from NS-L to NS-H and then back

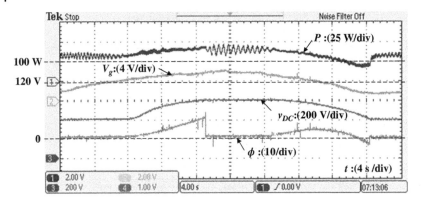

Figure 19.13 Experimental results showing the capacity potential of the rectifier: real power P, grid voltage V_g, DC-bus voltage v_{DC}, and $\Delta\phi$.

to NS-L. The value of $\Delta\phi$ was reset when the system entered into NS-L and NS-H mode. As shown in Figure 19.13, the system operated under the NS-L mode at the beginning with $v_{DC} = 400$ V and $\Delta\phi$ was reset to 0. At around $t = 8$ s. The system entered into the GS mode and $\Delta\phi$ started to increase. When v_{DC} reached the maximum value (600 V) at around $t = 17$ s, the system entered into the NS-H mode, and $\Delta\phi$ reached its maximum value, which is about 12. After that $\Delta\phi$ was reset to zero. $\Delta\phi$ was kept at zero until the system entered into the GS mode again at around $t = 24$ s. $\Delta\phi$ kept increasing until V_g was lower than 120 V. At $t = 36.4$ s, the system entered into the NS-L and at that time $\Delta\phi = -3$. There is some error in $\Delta\phi$ from the designed $\Delta\phi_{max} = 4$ because of the power loss P_{loss} of the system. According to (19.11), a positive item $\frac{n}{K_e}\int_{t_s}^{t_e} P_{loss}dt$ is added to the expression of $\Delta\phi$. Hence, the resulting $\Delta\phi_{max}$ is larger than the designed value. The difference is determined by the total energy loss in the time interval from t_s to t_e. The higher the efficiency, the smaller the difference. The potential of grid regulation can be increased via increasing the capacitor C.

19.4.5 Comparative Study

The controller for the conversion leg in (Zhong and Ming, 2016) is shown in Figure 19.14 with a small change from controlling the minimum value of v_{DC} to its average value V_{DC}.

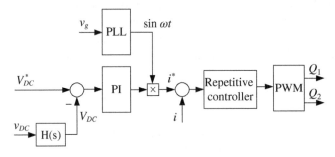

Figure 19.14 Controller for the conversion leg.

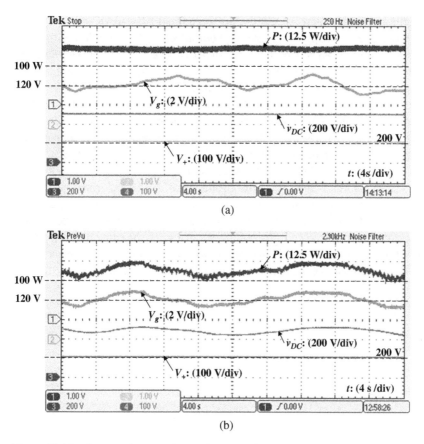

Figure 19.15 Comparative experimental results with a conventional controller. (a) With the controller from (Zhong and Ming, 2016) after a small change. (b) With the droop-controlled converter.

A PI controller is adopted to regulate V_{DC} and to generate the reference input current. A repetitive controller is adopted to regulate the input current.

The results from the same experimental system are shown in Figure 19.15(a). It is clearly shown that v_{DC} is regulated well at 500 V no matter how V_g changes and the power P consumed is about 110 W without much change all the time. Moreover, the DC capacitor cannot act as an energy storage unit to provide grid regulation either. Figure 19.15(b) shows the results from the droop-controlled controller for direct comparison. The function of grid regulation is evident, demonstrating fast transient performance.

19.5 Summary

Based on (Zhong, 2017d; Zhong and Lyu, 2019b), a control framework is presented to operate a rectifier to continuously take part in the regulation of the grid frequency and voltage, i.e. to achieve continuous demand response. This exceeds the FERC requirement on newly integrating generators to provide a primary frequency response (FERC, 2018). The rectifier requires an additional storage port, in addition to the normal AC and DC ports, to provide the flexibility needed to support the grid without affecting the operation of the DC load. As an example, it has been shown that this can actually be achieved by adopting the patented θ-converter, which only needs four switches for single-phase applications, i.e. the same number of switches as conventional bridge rectifiers. The grid support, the DC-port voltage control, and the storage voltage control can all be achieved through controlling the four switches. As a result, the continuous demand response is achieved without much additional hardware cost with comparison to conventional bridge converters. In this chapter, the universal droop control is adopted to achieve the grid support function, but other control strategies can be adopted as well. Moreover, the θ-converter can also be replaced with other three-port converters, e.g. the Beijing converter (Zhong et al. 2017). Experimental results from a θ-converter equipped with the framework are presented to demonstrate the desired function of continuous demand response, as well as the scenarios when the converter works in the no-support mode, when the converter starts up, and when the converter starts supplying a load. While the chapter focuses on single-phase applications, the framework can be easily applied to poly-phase applications after adding more conversion legs.

20

Current-limiting Universal Droop Controller

Power electronic converters are known to be vulnerable to excessive over-currents. In this chapter, a current-limiting universal droop controller is presented to operate a grid-connected inverter under both normal and faulty grid conditions. The controller introduces bounded nonlinear dynamics with a current-limiting property, which is analytically proven by using nonlinear input-to-state stability theory. The controller can be operated in the set mode to accurately send the desired power to the grid or in the droop mode to take part in grid regulation, while maintaining the inverter current below a given value at all times. In contrast to existing current-limiting approaches, the current limitation is achieved without external limiters, additional switches or monitoring devices. Furthermore, this function is independent of grid voltage and frequency variations, achieving desired control performance under grid faults as well. Extensive experimental results are presented to verify the droop function of the controller and its current-limiting capability under normal and faulty grid conditions.

20.1 Introduction

An important issue about grid-connected converters is to maintain the current below a given maximum limit. At the moment, most grid-connected converters are current-controlled so the current-limiting property is not a problem but these converters lack the capability of voltage regulation, which is a crucial function of power-electronics-enabled autonomous power systems. It is important for every voltage-controlled, in particular, droop-controlled, distributed generation unit to possess a current-limiting property in a completely decentralized manner, which should be maintained at all times during both normal and abnormal grid conditions, i.e. during grid faults (Guo et al. 2015; Laaksonen, 2010; Yazdanpanahi et al. 2012). Fault-current-limiting controllers can be applied to achieve the desired current limitation by either triggering properly designed protection circuits (Haj-ahmed and Illindala, 2014; Laaksonen, 2010; Zeineldin et al. 2006) or by using several low-voltage ride-through structures (El Moursi et al. 2013; Yang et al. 2014), which continue injecting power to the grid with a limited current. Several of these methods are based on algorithmic control schemes and lack a rigorous stability proof. Additionally, external limiters and saturation units are often added into the current or voltage control loops to achieve the desired current-limiting property, but these approaches

Power Electronics-Enabled Autonomous Power Systems: Next Generation Smart Grids,
First Edition. Qing-Chang Zhong.
© 2020 John Wiley & Sons Ltd. Published 2020 by John Wiley & Sons Ltd.

can lead to undesired oscillations and instability (Bottrell and Green, 2014; Paquette and Divan, 2015; Xin et al. 2016). A controller for grid-connected inverters that partially overcomes these limitations is reported in (Konstantopoulos et al. 2016a), based on a bounded control structure (Konstantopoulos et al. 2016b), but it can only operate with the unity power factor. Hence, the reactive power could not be controlled and always remained close to zero. This clearly indicates that the droop functions were not implemented.

20.2 System Modeling

Figure 20.1 shows the system under consideration. It consists of a single-phase inverter connected to the grid via an *LCL* filter. The *LCL* filter inductances are denoted as L and L_g with small parasitic resistances in series r and r_g, respectively, and the filter capacitor is given by C with a large parasitic resistance R_c in parallel. The inverter output voltage and current are v and i, respectively. v_c is the capacitor voltage and v_g, i_g are the grid voltage and current, respectively. Initially, the grid is considered stiff with $v_g = \sqrt{2}V_g \sin \omega_g t$, where V_g is the RMS grid voltage and ω_g is the grid angular frequency. Later, the case when the grid voltage and frequency vary from the rated values will be considered.

The dynamic model of the system is given by

$$L\frac{di}{dt} = -ri + v - v_c$$

$$C\frac{dv_c}{dt} = i - \frac{v_c}{R_c} - i_g \tag{20.1}$$

$$L_g\frac{di_g}{dt} = v_c - r_g i_g - v_g.$$

This is a linear dynamic system with state vector $x = \begin{bmatrix} i & v_c & i_g \end{bmatrix}^T$, and v_g represents an uncontrolled external input. The control input is the inverter voltage v.

In order for the inverter to support the grid voltage and frequency regulation, droop control is adopted in the control design, where the control input takes the form of $v = \sqrt{2}E \sin \theta$ with $\dot{\theta} = \omega$ being the angular frequency of the inverter. Although several droop control methods have been presented in the literature, the universal droop controller

$$\omega = \omega^* + m(Q - Q_{set}) \tag{20.2}$$

$$\dot{E} = K_e(E^* - V_c) - n(P - P_{set}) \tag{20.3}$$

is adopted, where ω^* and E^* are the rated angular frequency and voltage, respectively, K_e is a positive constant gain, V_c is the RMS value of the capacitor voltage, P_{set} and Q_{set} correspond

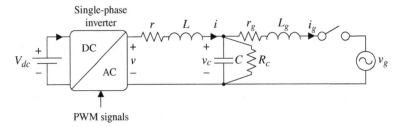

Figure 20.1 A grid-connected single-phase inverter with an *LCL* filter.

to the reference values of the real and the reactive power, and n and m are the droop coefficients. The measured real and reactive power P and Q are usually obtained at the capacitor node as the average values of the instantaneous power expressions over a period T in the form of

$$P = \frac{1}{T} \int_t^{t+T} v_c(\tau) i(\tau) d\tau, \qquad Q = -\frac{1}{T} \int_t^{t+T} v_{cg}(\tau) i(\tau) d\tau, \tag{20.4}$$

where v_{cg} is the ghost signal of the capacitor voltage. It is obvious that the power expressions are nonlinear due to the multiplication of the system states, resulting in a nonlinear closed-loop system for which it is difficult to analyze the stability.

20.3 Control Design

20.3.1 Structure

For grid-connected inverters, the inverter RMS voltage E and the phase θ are often controlled to obtain the inverter voltage $v = \sqrt{2}E \sin \theta$. The output impedance can be controlled to be resistive, which improves the power quality and enhances the stability of the system, especially under grid voltage variations (Vandoorn et al. 2012; Zhong and Hornik, 2013). In this case, the inverter voltage v is subtracted by a term $r_o i$, where r_o is the (virtual) output resistance and i is the inverter current, thus forming an R-inverter. In this section, the focus is on the design of the virtual resistance to obtain the desired current-limiting property. In order to achieve this, the following generic controller for grid-connected inverters is designed:

$$v = v_0 + \Delta V \sin(\omega_o t + \delta) - r_o i, \tag{20.5}$$

which consists of a base voltage v_0, a controllable voltage source $\Delta V \sin(\omega_o t + \delta)$, and a dynamic virtual resistance r_o. Here, v_0, ΔV, ω_o, δ, and r_o should be controlled to achieve different operating modes, realizing the droop control functions (20.2)–(20.3) and the crucial current-limiting property. The equivalent circuit of the controller is shown in Figure 20.2.

20.3.2 Implementation

The controller (20.5) can be designed to have

1) $v_0 = v_c$
2) $\Delta V = \sqrt{2}(1 - w_q)V_g$
3) $\omega_o = \omega_g$
4) $r_o = (1 - w_q)w$.

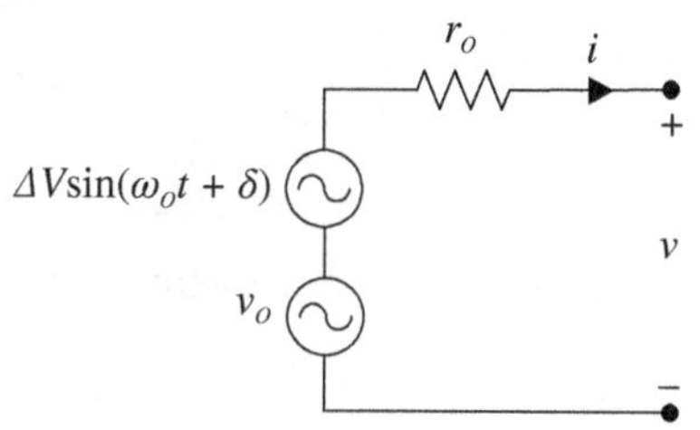

Figure 20.2 The equivalent circuit diagram of the controller.

In other words, the controller is

$$v = v_c + (1 - w_q)(\sqrt{2}V_g \sin(\omega_g t + \delta) - wi). \tag{20.6}$$

Here, the controller states w, w_q, and δ are designed to dynamically change within given bounded sets while introducing the droop functions and the current-limiting property. Since $r_o = (1 - w_q)w$ is a virtual resistance, it has to be positive and larger than a minimum value. Considering the desired droop function expressions in (20.2)–(20.3), the dynamics of the controller states w, w_q, and δ, together with an additional controller state δ_q, are designed according to the bounded integral controller introduced in (Konstantopoulos et al. 2016b) and adopted in Chapter 13 as

$$\dot{w} = -c_w(K_e(E^* - V_c) - n(P - P_{set}))w_q^2 \tag{20.7}$$

$$\dot{w}_q = \frac{c_w(w - w_m)w_q}{\Delta w_m^2}(K_e(E^* - V_c) - n(P - P_{set})) - k_w\left(\frac{(w - w_m)^2}{\Delta w_m^2} + w_q^2 - 1\right)w_q \tag{20.8}$$

$$\dot{\delta} = c_\delta(\omega^* - \omega_g + m(Q - Q_{set}))\delta_q^2 \tag{20.9}$$

$$\dot{\delta}_q = -\frac{c_\delta\delta\delta_q}{\Delta\delta_m}(\omega^* - \omega_g + m(Q - Q_{set})) - k_\delta\left(\frac{\delta^2}{\Delta\delta_m^2} + \delta_q^2 - 1\right)\delta_q, \tag{20.10}$$

with c_w, c_δ, w_m, Δw_m, $\Delta\delta_m$, k_w, and k_δ being positive constants. The initial conditions of w, w_q and δ, δ_q are defined as $w_0 = w_m$, $w_{q0} = 1$, and $\delta_0 = 0$, $\delta_{q0} = 1$, respectively. Note that both V_g and ω_g can be obtained using a traditional PLL. The overall control system is shown in Figure 20.3. Note that the universal droop control principle (20.2)–(20.3) is embedded in the controller dynamics and hence it is a droop controller. Moreover, the virtual resistance r_o is dynamically controlled according to the $P \sim V$ droop and the phase shift δ in the controllable voltage source is dynamically controlled according to the $Q \sim -\omega$ droop. Hence, in order to guarantee system stability, these terms should be proven to remain bounded.

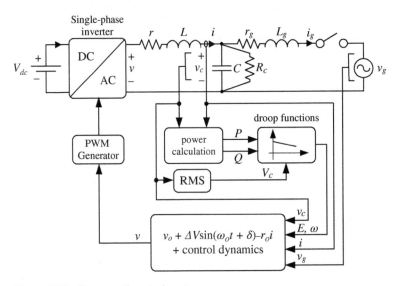

Figure 20.3 The overall control system.

For the control dynamics (20.7)–(20.8), consider the Lyapunov function candidate

$$W = \frac{(w - w_{\mathrm{m}})^2}{\Delta w_{\mathrm{m}}^2} + w_{\mathrm{q}}^2. \tag{20.11}$$

Its time derivative is

$$\dot{W} = \frac{2(w - w_{\mathrm{m}})\dot{w}}{\Delta w_{\mathrm{m}}^2} + 2w_{\mathrm{q}}\dot{w}_{\mathrm{q}}.$$

By substituting $\dot{w}$ and $\dot{w}_{\mathrm{q}}$ from (20.7) and (20.8), respectively, $\dot{W}$ can be found, after some calculations, as

$$\dot{W} = -2k_w \left(\frac{(w - w_{\mathrm{m}})^2}{\Delta w_{\mathrm{m}}^2} + w_{\mathrm{q}}^2 - 1 \right) w_{\mathrm{q}}^2. \tag{20.12}$$

According to the initial conditions w_0 and $w_{\mathrm{q}0}$, both w and w_{q} start and stay at all times on the ellipse

$$W_0 = \left\{ w, w_{\mathrm{q}} \in R : \frac{(w - w_{\mathrm{m}})^2}{\Delta w_{\mathrm{m}}^2} + w_{\mathrm{q}}^2 = 1 \right\},$$

as shown in Figure 20.4(a), because on W_0 there is

$$\dot{W} = 0 \Rightarrow W(t) = W(0) = 1, \ \forall t \geq 0.$$

By choosing $w_{\mathrm{m}} > \Delta w_{\mathrm{m}} > 0$, the ellipse is defined on the right half plane, resulting in $w \in [w_{\min}, w_{\max}] = [w_{\mathrm{m}} - \Delta w_{\mathrm{m}}, w_{\mathrm{m}} + \Delta w_{\mathrm{m}}] > 0$ for all $t \geq 0$. Now, define the transformation

$$w = w_{\mathrm{m}} + \Delta w_{\mathrm{m}} \sin \phi, \quad \text{and} \quad w_{\mathrm{q}} = \cos \phi. \tag{20.13}$$

Note that the term $-k_w \left(\frac{(w-w_{\mathrm{m}})^2}{\Delta w_{\mathrm{m}}^2} + w_{\mathrm{q}}^2 - 1 \right) w_{\mathrm{q}}$ in (20.8) is zero on the ellipse W_0. Then by substituting (20.13) into (20.7) and (20.8), after some calculations, there is

$$\dot{\phi} = \frac{-c_w(K_e(E^* - V_{\mathrm{c}}) - n(P - P_{\mathrm{set}}))w_{\mathrm{q}}}{\Delta w_{\mathrm{m}}}, \tag{20.14}$$

which means that the states w and w_{q} travel on the ellipse W_0 with angular velocity $\dot{\phi}$. When $K_e(E^* - V_{\mathrm{c}}) = n(P - P_{\mathrm{set}})$, the controller settles down in the steady state at the universal droop controller with $\dot{\phi} = 0$. Hence, w and w_{q} converge to some constant values w_e and $w_{\mathrm{q}e}$, respectively, corresponding to the desired equilibrium point.

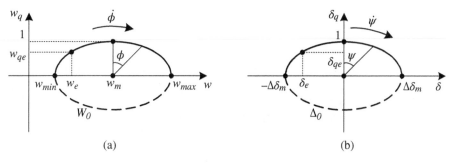

Figure 20.4 Controller states. (a) w and w_{q}. (b) δ and δ_{q}.

Note that by starting from point $(w_m, 1)$ on the w–w_q plane, the controller states w and w_q are restricted only on the upper semi-ellipse of W_0. This is due to the fact that the angular velocity $\dot{\phi}$ depends on w_q from (20.14), since if the states try to reach the horizontal axis, then $w_q \to 0$ and $\dot{\phi} \to 0$ independently of the term $K_e(E^* - V_c) - n(P - P_{set})$. This forces the controller states to slow down and remain on the upper semi-ellipse of W_0, avoiding a continuous oscillation around W_0. Therefore, it holds true that $w_q \in [0, 1]$ for all $t \geq 0$.

Similarly, the controller states δ and δ_q in (20.9)–(20.10) operate exclusively on the upper semi-ellipse of

$$\Delta_0 = \left\{ \delta, \delta_q \in R : \frac{\delta^2}{\Delta\delta_m^2} + \delta_q^2 = 1 \right\}$$

with angular velocity

$$\dot{\psi} = \frac{c_\delta(\omega^* - \omega_g + m(Q - Q_{set}))\delta_q}{\Delta\delta_m} \tag{20.15}$$

and similar properties as described above, resulting in $\delta \in [-\Delta\delta_m, \Delta\delta_m]$ and $\delta_q \in [0, 1]$ for all $t \geq 0$, as shown in Figure 20.4(b). Thus, all controller states w, w_q, δ and δ_q are proven to remain bounded for all $t \geq 0$.

The controller can be operated in different modes. In the steady state, $\dot{w} = 0$ and $\dot{\delta} = 0$, which means the controller achieves the droop function given in (20.7)–(20.10). After removing the term $K_e(E^* - V_c)$ from (20.7)–(20.8) and the term $\omega^* - \omega_g$ from (20.9)–(20.10), the controller can be operated in the set mode to achieve power flow control so that accurate real and reactive power can be sent to the grid. This allows a simple change in the operating modes of the inverter from the droop mode to the set mode at any time. In fact, this is one of the main differences of the controller compared to the one in (Konstantopoulos et al. 2016a). The control structure (20.6), together with the control dynamics (20.7)–(20.10), allow the regulation of both the real power and the reactive power as well as the implementation of droop functions, but the approach in (Konstantopoulos et al. 2016a) can only achieve unity power factor control.

20.4 System Analysis

20.4.1 Current-limiting Property

By applying the controller (20.6) in the original system dynamics (20.1), the inverter current equation becomes

$$L\frac{di}{dt} = -(r + (1 - w_q)w)i + (1 - w_q)\sqrt{2}V_g \sin(\omega_g t + \delta). \tag{20.16}$$

From the above controller analysis, it holds true that $w \in [w_{min}, w_{max}] > 0$ and $w_q \in [0, 1]$ for all $t \geq 0$. Then, for system (20.16), consider the Lyapunov function candidate

$$V = \frac{1}{2}Li^2. \tag{20.17}$$

It represents the energy stored in the inductor L. Its time derivative is

$$\dot{V} = -(r + (1 - w_q)w)i^2 + (1 - w_q)\sqrt{2}V_g i \sin(\omega_g t + \delta)$$

$$\leq -(r + (1 - w_q)w_{min})i^2 + (1 - w_q)\sqrt{2}V_g |i||\sin(\omega_g t + \delta)|.$$

This shows that $\dot{V} < 0$ when $|i| > \frac{(1-w_q)\sqrt{2}V_g|\sin(\omega_g t+\delta)|}{r+(1-w_q)w_{min}}$, proving that (20.16) is input-to-state stable (ISS) (Khalil, 2001). Since $(1 - w_q)\sqrt{2}V_g \sin(\omega_g t + \delta)$ is bounded, then the inverter current i is bounded for all $t \geq 0$. According to the ISS property, it holds true that

$$|i| \leq \frac{(1 - w_q)\sqrt{2}V_g}{r + (1 - w_q)w_{min}}, \quad \forall t \geq 0,$$

if initially $i(0)$ satisfies the previous inequality. By choosing

$$w_{min} = \frac{V_g}{I_{max}} \tag{20.18}$$

then

$$|i| \leq \frac{(1 - w_q)}{r\frac{I_{max}}{V_g} + (1 - w_q)}\sqrt{2}I_{max} < \sqrt{2}I_{max}, \tag{20.19}$$

since $(1 - w_q) \geq 0$ and $r\frac{I_{max}}{V_g} > 0$. The previous inequality holds for any $t \geq 0$ and for any constant positive I_{max}. As a result

$$I < I_{max}, \quad \forall t \geq 0,$$

where I is the RMS value of the inverter current, proving that the controller introduces an inherent current-limiting property independently of the droop function, the nonlinear expressions of P, Q, V_c and the dynamics of δ. This is a crucial property since the inverter is protected at all times by limiting the current, even if a large reference value P_{set} is applied.

It is possible to find a different Lyapunov function candidate, which results in a different limit for the current. However, it is impossible to find a Lyapunov function candidate to obtain a limit lower than I_{max} through the choice of w_{min} from (20.18) because the closed-loop system (20.16) is an RL circuit with a voltage source. Note that the actual limit of I depends on the small parasitic resistance r. As can be seen from (20.19), when the value of r is significant (not zero), the maximum limit of the current is reduced. However, the current-limiting property below I_{max} is still guaranteed and this holds independently from the filter, i.e. without requiring any knowledge of the inductor L and its parasitic resistance r.

20.4.2 Closed-loop Stability

As shown above, the controller states and the filter inductor current always remain bounded. In order to investigate the stability of the rest of the plant states, the dynamics of the capacitor voltage and the grid current in (20.1) can be re-written as

$$\begin{bmatrix} \frac{dv_c}{dt} \\ \frac{di_g}{dt} \end{bmatrix} = \begin{bmatrix} -\frac{1}{R_c C} & -\frac{1}{C} \\ \frac{1}{L_g} & -\frac{r_g}{L_g} \end{bmatrix} \begin{bmatrix} v_c \\ i_g \end{bmatrix} + \begin{bmatrix} \frac{i}{C} \\ -\frac{v_g}{L_g} \end{bmatrix}, \tag{20.20}$$

which can be seen as a linear time-invariant system of the form $\dot{x} = Ax + u$ with state $x = \begin{bmatrix} v_c & i_g \end{bmatrix}^T$ and input $u = \begin{bmatrix} \frac{i}{C} & -\frac{v_g}{L_g} \end{bmatrix}^T$. By choosing

$$P = \begin{bmatrix} C & 0 \\ 0 & L_g \end{bmatrix} > 0,$$

it can be found that

$$PA + A^T P = \begin{bmatrix} -\frac{2}{R_c} & 0 \\ 0 & -2r_g \end{bmatrix} < 0,$$

which proves that A is Hurwitz and (20.20) is a bounded-input bounded-state stable system. Since $v_g = \sqrt{2}V_g \sin \omega_g t$ is bounded and i is bounded from the ISS and the current-limiting properties, both the capacitor voltage v_c and the grid current i_g are proven to remain bounded at all times.

The virtual resistance r_o introduced by the controller, as shown in Figure 20.2, is in series with the filter inductor L, which is equivalent to increasing the parasitic resistance r of the inductor L or adding a resistor in series with the filter inductor. Hence, it is able to enhance the damping of the system.

20.4.3 Selection of Control Parameters

The term $(1 - w_q)w$ represents a dynamic virtual resistance at the output of the inverter and w_{min} corresponds to the maximum current I_{max}. Similarly, the initial value $w_0 = w_m$ can be determined by the given initial current I_m as

$$w_m = \frac{V_g}{I_m}.$$

Note that initially, when the inverter is not connected to the grid, a small amount of current still flows through the LC filter. In particular, since the RMS capacitor voltage is almost at V_g to have a smooth connection ($v_c \approx v_g$), then the inverter current before the connection would be

$$I_m = \omega^* C V_g.$$

As a result, w_m can be chosen as

$$w_m = \frac{1}{\omega^* C}. \tag{20.21}$$

According to the ellipse W_0, the parameter Δw_m is given as

$$\Delta w_m = w_m - w_{min} = \frac{1}{\omega^* C} - \frac{V_g}{I_{max}}. \tag{20.22}$$

The parameter $\Delta\delta_m$ corresponds to the maximum absolute value of δ. According to (20.16), the controller state δ describes the phase shifting applied to the inverter voltage. By neglecting the small phase shifting applied by the filter inductor L, the value of δ corresponds to the reactive power of the inverter. In other words, $\delta = 0$ and $\delta = -\frac{\pi}{2}$ rad approximately correspond to $Q = 0$ and $Q = S_n$, respectively, where S_n is the rated power of the inverter. Therefore, $\Delta\delta_m$ is chosen as $\frac{\pi}{2}$ rad in order to control the reactive power in the range $Q \in [-S_n, S_n]$. In practice, $\Delta\delta_m$ can be chosen slightly smaller to take into account the small inductance L.

Parameters k_w and k_δ should be arbitrary positive constants since they are multiplied with the terms $\frac{(w-w_m)^2}{\Delta w_m^2} + w_q^2 - 1$ and $\frac{\delta^2}{\Delta\delta_m^2} + \delta_q^2 - 1$ in (20.8) and (20.10), which are zero on the ellipses W_0 and Δ_0, respectively. In fact, these terms are used to increase the robustness of the w_q and δ_q dynamics in an actual implementation due to calculation errors or external disturbances.

Parameters c_w and c_δ are found in the angular velocities (20.14) and (20.15) of the controller states, respectively. The selection of c_w is discussed at first. Since w and w_q start from point $(w_m, 1)$, they travel on the ellipse W_0 and can reach the point $(w_{min}, 0)$ at the limit of the current after a settling time t_s, then by considering the worst case scenario where the controller states travel on the arc of W_0 with central angle $\frac{\pi}{2}$ rad and with a maximum angular velocity $\frac{\pi}{2t_s}$ rad s^{-1}, one can calculate a minimum value of c_w. Since tight voltage regulation can be achieved for the capacitor voltage ($V_c \approx E^*$) and $w_q \leq 1$, assuming the real power starts from zero and reaches the maximum real power $P_{set} = S_n$, then from (20.14) it yields

$$\dot{\phi}_{max} = \frac{\pi}{2t_s} = \frac{c_w n S_n}{\Delta w_m},$$

which finally gives

$$c_w = \frac{\pi \Delta w_m}{2t_s n S_n}. \tag{20.23}$$

Similarly, parameter c_δ can be determined as

$$c_\delta = \frac{\pi \Delta \delta_m}{2t_s m S_n}. \tag{20.24}$$

Note that both (20.23) and (20.24) provide some guidance for selecting c_w and c_δ only because they are obtained for the worst case scenario. In practice, larger values can be chosen or equivalently smaller t_s can be used.

20.5 Practical Implementation

Since during the grid-connected operation in most applications L_g and r_g are relatively small, the small phase shifting and the voltage drop across the inductor can be ignored, which gives $v_g \approx v_c$. Hence, the base voltage v_0 can be chosen equal to the grid voltage v_g, resulting in the following controller:

$$v = v_g + (1 - w_q)(\sqrt{2}V_g \sin(\omega_g t + \delta) - wi). \tag{20.25}$$

Since there is $v = v_g$ before the connection with the grid, according to the initial condition of the controller state $w_{q0} = 1$, a smooth connection to the grid can be achieved. After connecting with the grid, the controller can be enabled at any time; thus no pre-synchronization unit is required.

Based on the controller (20.25), the measured signals v_g and i are directly used in the control input v and therefore they represent feed-forwarded terms, which can introduce a small phase shift due to computational delays, PWM modulation, and the inverter filter. To overcome this small phase shift, a phase-lead low-pass filter can be used for the measurements of v_g and i, e.g. $F(s) = \frac{33(0.05s+1)}{(s+300)(0.002s+1)}$ as adopted in (Zhong and Hornik, 2013). Since this phase shift is different in every grid-tied inverter system, the filter gain, poles and zero can be adjusted accordingly to match the requirements of the system. In order to design the

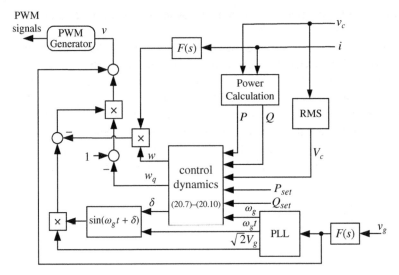

Figure 20.5 Implementation of the current-limiting universal droop controller.

filter $F(s)$ appropriately, the simplest way is to observe the difference $v_c - v_g$ before connecting the inverter to the grid. Hence, one can tune the parameters of $F(s)$ in order to achieve $v_c - v_g = 0$, which can be observed with an oscilloscope. The final implementation of the controller is shown in Figure 20.5.

20.6 Operation under Grid Variations and Faults

Since w_{min} is selected according to (20.18) with $V_g = E^*$, the current-limiting property is guaranteed for any $V_g \leq E^*$ and for any frequency ω_g, i.e. under grid variations (non-stiff grid) and grid faults. This is particularly interesting because, when the grid voltage drops, the current normally increases to high values. The presented controller is able to limit the current without changing its control structure, different from traditional approaches that have to identify the fault and switch to a difference controller.

To further clarify this, assume that a voltage sag occurs in the grid with a percentage $p \times 100\%$, i.e. the grid voltage V_g becomes $(1 - p)V_g$. Then the closed-loop system is given as

$$L\frac{di}{dt} = -(r + (1 - w_q)w)i + (1 - w_q)(1 - p)\sqrt{2}V_g \sin(\omega_g t + \delta), \tag{20.26}$$

which according to the same analysis for the current-limiting property yields

$$I < (1 - p)I_{max}. \tag{20.27}$$

Therefore the RMS voltage of the inverter current I still remains less than I_{max}, satisfying the current-limiting property and protecting the inverter. In particular, the controller state w is forced to reduce to its minimum value w_{min}, corresponding to the maximum allowed current independently from the droop function. When the fault is cleared, the closed-loop system becomes again as the one in (20.16), forcing the current i to return to its original value. Furthermore, during the fault, since $w \to w_{min}$ then $w_q \to 0$, which results from (20.7) in $\dot{w} \to 0$ and the integration automatically slows down, which is an inherent anti-windup property. Hence, the presented controller can overcome wind-up and latch-up problems without additional switches or monitoring devices, which constitute two of the most important challenges in grid-connected inverters operation under grid faults (Bottrell and Green, 2014). Since the current limitation is maintained independently from the phase shift δ, the inverter is protected even if the PLL does not extract accurately the phase of the grid during faults, which is common in fault conditions.

If I_{max} is chosen according to the ratings of the inverter, e.g., given a rated power S_n, then (20.27) is equivalent to

$$S < (1 - p)S_n, \tag{20.28}$$

by ignoring the small voltage drop on the filter. This provides a limit for the apparent power of the inverter at all times, even during faults.

In order to cope with possible over-voltage in the grid, the voltage V_g could be chosen as the maximum possible grid voltage when selecting w_{min}.

It should be noted that the controller can guarantee the current-limiting property independently from the phase shift δ or an error in the phase angle of the grid obtained by the PLL, which is critical, especially under grid faults. The reason is that the current-limiting property has been mathematically proven in Section 20.4.1 to hold independently from the phase shift δ, since for the ISS property the maximum value of $|\sin(\omega_g t + \delta)|$ is required, which is equal to 1. Even if there is an error in the phase angle of the grid obtained from the PLL, this error would appear as an additional component in the sinusoidal term, which does not affect the current-limiting property. This is a significant advantage of the controller compared to existing methods that include a PLL to introduce a current-limiting property.

20.7 Experimental Results

A 220 VA single-phase grid-connected inverter with an *LCL* filter was experimentally tested. The system and controller parameters are shown in Table 20.1. The RMS voltage of the grid V_g is equal to the rated value E^* and the grid frequency ω_g is slightly less than ω^*; see Table 20.1. It is expected that 5% increase in the voltage corresponds to 100% decrease in

Table 20.1 System and controller parameters.

Parameter	Value	Parameter	Value
L, L_g	2.2 mH	ω^*	$2\pi \times 50$ rad s^{-1}
r, r_g	0.5 Ω	ω_g	$2\pi \times 49.97$ rad s^{-1}
C	10 μF	I_{max}	2 A
$V_g = E^*$	110 V	I_m	0.2 A
S_n	220 VA	K_e	150
t_s	0.1 s	k_w, k_δ	1

the real power and 1% increase in the frequency corresponds to 100% increase in the reactive power. Then, the droop coefficients can be calculated as $n = \frac{0.05K_eE^*}{S_n}$ and $m = \frac{0.01\omega^*}{S_n}$ according to (Zhong, 2013c; Zhong and Hornik, 2013), where S_n is the rated power of the inverter. A switching frequency of 15 kHz was used for the inverter operation and the sinusoidal tracking algorithm (STA) (Zhong and Hornik, 2013) was applied to obtain the required V_g and ω_g for the controller design. The controller parameters c_w and c_δ are directly calculated from (20.23) and (20.24), respectively. The controller was implemented using the TMS320F28335 DSP with a sampling frequency of 4 kHz.

20.7.1 Operation under Normal Conditions

Initially, a stiff grid is considered with the system parameters given in Table 20.1. The time response of the inverter is shown in Figure 20.6. It is clear that the grid frequency is constant at 49.97 Hz. The inverter is connected to the grid at $t = 6$ s and the real and reactive power references are operated in the set mode with $P_{set} = 50$ W and $Q_{set} = 0$ Var, respectively. As shown in Figure 20.6(a), both the real power and the reactive power are regulated to their reference values, although the reactive power is slightly positive (less than 5 Var) due to the limitation of the power analyzer that cannot show negative reactive power and introduces small inaccuracies close to zero. At $t = 9$ s the real power reference is changed to 100 W and at $t = 12$ s the reactive power reference is changed to 50 Var. Figure 20.6(a) clearly demonstrates the ability of the controller to regulate the injected power. At $t = 15$ s, the reference P_{set} is changed to 250 W, which exceeds the inverter capacity and forces the inverter current to exceed I_{max}. However, as shown in Figure 20.6(a), the real power is regulated at around 180 W because the current has reached the maximum allowed value. In particular, the RMS value of the current is limited at 1.73 A. It is less than $I_{max} = 2$ A because (i) the parasitic resistance r of the L inductor is not zero and (ii) the controller uses the feed-forwarded voltage term v_g from (20.25) instead of v_c, which is slightly different. Nevertheless, according to the analysis and the results, the current still remains below the maximum value I_{max} as required. In practice, I_{max} can be chosen slightly larger to cover these issues. If the parasitic resistance r is known, one can choose $w_{min} = \frac{V_g}{I_{max}} - r$ instead of using (20.18), and since at the limit $w_q \to 0$, (20.19) will result in $|i| \le \sqrt{2}I_{max}$, achieving a limit for the RMS value at I_{max}. However, even if r is neglected in the control design, the inverter still stays in the safe range below the maximum current value. The transient responses of the inverter and grid currents and voltages are shown in Figure 20.6(b), where the inverter current increases and reaches the maximum allowed value. At $t = 18$ s, P_{set} is changed to 150 W and in order to check the droop functions of the controller, at $t = 21$ s, the $P \sim V$ droop function is enabled and the real power drops in order to bring the capacitor voltage V_c closer to the rated value E^*. This is clearly observed in Figure 20.6(a). Finally, at $t = 23$ s, the $Q \sim -\omega$ droop function is enabled and the reactive power drops since the system frequency is lower than the rated value. This verifies the capability of the controller to operate in both the set mode and the droop mode. The steady-state response of the system is shown in Figure 20.6(c).

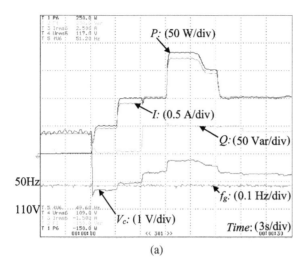

(a)

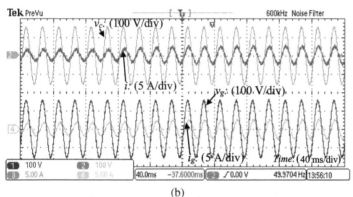

(b)

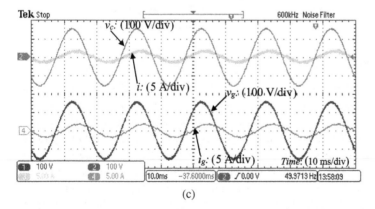

(c)

Figure 20.6 Operation with a normal grid. (a) Real and reactive power, RMS capacitor voltage and inverter current and grid frequency. (b) Transient response at $t = 15$ s (current-limiting property). (c) Steady-state response after 30 s.

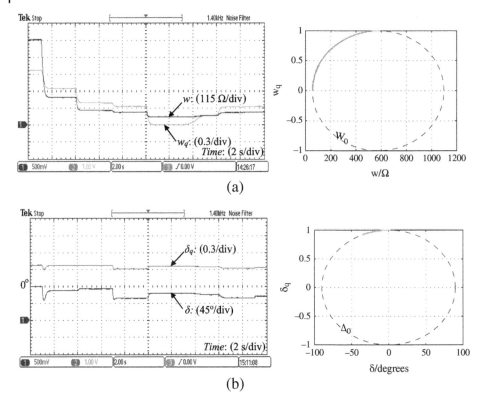

Figure 20.7 Transient response of the controller states with a normal grid. (a) w and w_q. (b) δ and δ_q.

To verify the analysis in Section 20.3, the responses of the controller states w, w_q and δ, δ_q are shown in Figures 20.7(a) and 20.7(b), respectively. The controller states are restricted on the upper semi-ellipse of W_0 and Δ_0, respectively, during the whole operation.

20.7.2 Operation under Grid Faults

To further validate the current-limiting control performance, two different cases of grid voltage sags were investigated while the inverter remains working in the droop mode.

At first, the grid voltage drops rapidly from 110 V to 90 V and the fault is cleared after 9 s. During the fault, the current tries to increase but is eventually limited according to the current-limiting property of the controller, as shown in Figures 20.8(a) and 20.8(b). Based on (20.27) and the percentage of the voltage drop ($p = \frac{110-90}{110} \times 100\% = 18.2\%$), the maximum RMS current would be $0.818I_{max} = 1.64$ A. From Figure 20.8(a), it is clear that the current is limited at around 1.4 A, slightly lower than the limit due to the reasons mentioned in the previous subsection. Figure 20.8(b) indicates the drop of the grid voltage and the capacitor voltage and the smooth increase of the current. When the fault is cleared, the system returns to its initial operation after a short transient as shown in Figures 20.8(a) and 20.8(c). This transient is caused due to the fact that the controller state w, which represents the main part of the virtual resistance, is regulated at w_{min} for the duration of the fault. Then, when the grid voltage rapidly increases, the current will increase due to the small resistance w_{min} but it never violates the limit, as rigorously proven in the theory. This

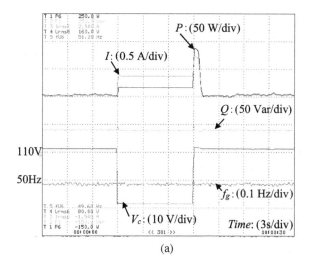

(a)

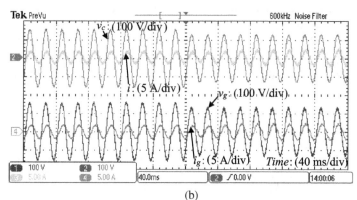

(b)

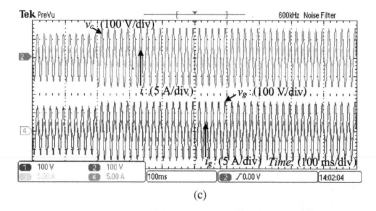

(c)

Figure 20.8 Operation under a grid voltage sag 110 V → 90 V → 110 V for 9 s. (a) Real and reactive power, RMS capacitor voltage and inverter current and grid frequency. (b) Transient response when the fault occurs. (c) Transient response when the fault is cleared.

is clearly shown in the time response of the controller states w and w_q in Figure 20.9(a). Nevertheless, the inverter voltage is immediately restored close to the rated value after a fault clearance, as required by the Grid Code (National Grid, 2016). Since the frequency of the grid remains constant and (20.28) is not violated according to the given $Q_{set} = 50$ Var and the droop control, the reactive power is regulated at the same value during the fault, as shown in Figure 20.8(a), which is achieved from the performance of the controller states δ and δ_q, as shown in Figure 20.9(b). The controller state trajectories remain again on the desired upper ellipses on the w–w_q and δ–δ_q planes.

In the second scenario, the grid voltage drops rapidly from 110 V to 55 V, i.e. a 50% voltage sag, and returns to its original value after 9 s, as shown in Figure 20.10(a). In this case, the inverter current reduces since it is limited below 1 A and the real power drops during the fault to protect the system. This is shown in Figure 20.10(b). As in the previous grid fault scenario, when the fault is cleared, the system returns to its initial operation after a short transient as shown in Figure 20.10(c). Some oscillations, which are caused by the slow response of the PLL and the dynamics of the *LCL* filter, appear in the inverter current and capacitor voltage during the fault and during its clearance since for simplicity the controller (20.25) was used instead of its initial form (20.6). However, this transient only lasts for less than a half cycle with limited amplitude so it is acceptable in practice. A difference between this and the previous fault scenario is that during the fault and the current limitation (20.28), the droop control in the reactive power can no longer be accomplished

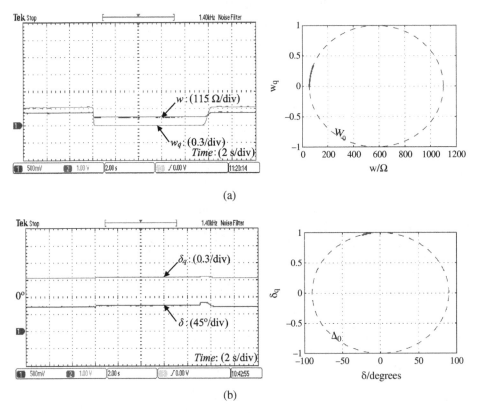

Figure 20.9 Controller states under the grid voltage sag 110 V → 90 V → 110 V for 9 s. (a) w and w_q. (b) δ and δ_q.

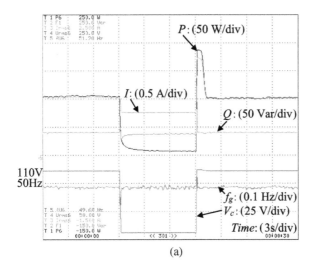

(a)

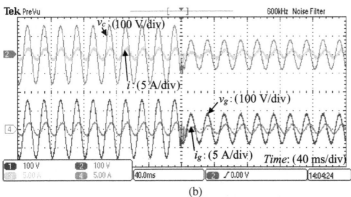

(b)

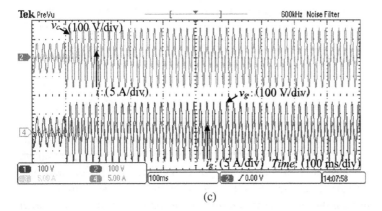

(c)

Figure 20.10 Operation under a grid voltage sag 110 V → 55 V → 110 V for 9 s. (a) Real and reactive power, RMS capacitor voltage, and inverter current and grid frequency. (b) Transient response when the fault occurs. (c) Transient response when the fault is cleared.

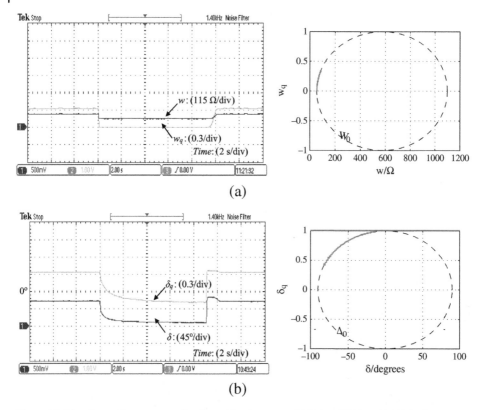

Figure 20.11 Controller states under the grid voltage sag 110 V → 55 V → 110 V for 9 s. (a) w and w_q. (b) δ and δ_q.

with the given $Q_{set} = 50$ Var. This means that the controller states δ and δ_q also decrease and converge to lower values that correspond to a lower reactive power that guarantees (20.28). This is shown in the time response of the controller states in Figure 20.11(b). Finally, both pairs w, w_q and δ, δ_q remain on the ellipses as imposed by the theory.

20.8 Summary

Based on (Zhong and Konstantopoulos, 2017), a nonlinear droop controller with a current-limiting property is presented for single-phase grid-connected inverters. In addition to achieving the desired droop functions with a tight regulation of the output voltage close to the rated value or accurate real and reactive power regulation in the set mode, the controller is able to limit the inverter current under normal or faulty grid conditions. Based on the nonlinear dynamics of the system, the stability of the control system is established by using the input-to-state stability theory. The desired performance of the controller is verified with extensive experimental results. It is noticed that, during the fault, the controller shows some oscillations due to the PLL dynamics and also the current is limited to a value lower than the given maximum value because of the grid voltage drop. These can be improved.

Part IV

3G VSM: Cybersync Machines

21

Cybersync Machines

With the advancement of the first generation of VSMs (synchronveters) and the second generation of VSMs (universal droop controllers) discussed in the previous two parts, this part briefly touches upon the third generation VSMs that are expected to be able to guarantee the stability of a power system with multiple power electronic converters. In this chapter, a generic control framework is presented to render the controller of a power electronic converter passive by using the port-Hamiltonian (PH) system theory and the ghost operator. The controller consists of two symmetric control loops and an engendering block. The two control loops, which are the torque-frequency loop to generate the frequency and the quorte-flux loop to generate the flux that corresponds to the voltage, form a passive block. Here, the new word *quorte* represents the quantity that is dual to torque. The engendering block generates the control signal and the torque and quorte feedback signals for the two control loops. With the critical concepts of the *ghost signal* and the *ghost system* introduced in Chapter 3, the engendering block is augmented as a lossless interconnection between the control block and the plant pair that consists of the original plant and its ghost plant. The whole system is then passive if the plant pair is passive. Moreover, some practical issues, such as controller implementation, power regulation, and self-synchronization without a dedicated synchronization unit, are also discussed. Simulation and experimental results are presented to demonstrate its operation in both grid-connected and islanded modes. The technology is patented.

21.1 Introduction

It is well known that a system with multiple power electronic converters may become unstable even if each single converter is stable individually (Du et al. 2013a; Liu et al. 2016b; Mukherjee and Strickland, 2016; Vesti et al. 2013). Hence, it is very important to guarantee the stability of a single converter while expecting that a system with multiple converters operated together is stable. The objective of this chapter is to attempt to solving this problem with a mathematical tool called passivity. A system is passive if at all times the power flowing into the system is not less than the rate of energy being stored inside. In other words, a passive system cannot generate energy or provide more energy than is stored. Any system comprising of only passive components is passive. The passivity of a system is a compositional property (Xia et al. 2017; Zhu et al. 2017) and the interconnection of passive systems

Power Electronics-Enabled Autonomous Power Systems: Next Generation Smart Grids,
First Edition. Qing-Chang Zhong.
© 2020 John Wiley & Sons Ltd. Published 2020 by John Wiley & Sons Ltd.

is passive, which meets the above expectation (to some extent). Moreover, the equilibrium of an unforced passive system, i.e. when the input is zero, is Lyapunov stable. When the input is not zero, it is possible to investigate the Lyapunov stability of its equilibrium via constructing energy-based Lyapunov functions (Maschke et al. 2000). Hence, passivity is a very promising tool for studying the stability of large-scale systems (Ghanbari et al. 2016; Yamamoto and Smith, 2016).

Guaranteeing the stability of a power electronic converter is not an easy task due to the nonlinearities of the controller, e.g. the calculation of the real power and the reactive power, and the coupling effect between frequency and voltage, etc. Although local stability can be established with small-signal analysis and linearization (Guo et al. 2014; Paquette and Divan, 2015; Pogaku, et al. 2007, Wu et al. 2016), the nonlinear dynamics of the system makes nonlinear analysis essential to obtain global stability results. A nonlinear control strategy with a power-damping property (Ashabani and Mohamed, 2014) can guarantee nonlinear system stability but it requires the knowledge of the filter parameters, which becomes impossible for large-scale power systems. Several approaches for maintaining the stability of synchronous generators are available but it is often assumed that the field-excitation current, corresponding to the voltage, is constant (Fiaz, et al. 2013, Natarajan and Weiss, 2014). In reality, this is not true because the voltage needs to be regulated as well. As shown in Chapter 13, a synchronverter improved with the bounded integral controller (Konstantopoulos et al. 2016b) is stable when the grid is stiff. It works when the grid is not stiff but the rigorous proof does require the assumption of a stiff grid. The stability of droop-controlled inverter-based microgrids with meshed topologies is analyzed in (Schiffer et al. 2014) with a particular focus on lossless microgrids. The link between power-frequency droop controllers and the Kuramoto model of phase-coupled oscillators is established in (Simpson-Porco, et al. 2013) and applied to investigate the frequency stability of droop-controlled converters. While most works study the stability under the assumption of constant voltage, the voltage stability of microgrids with converters controlled by a quadratic droop controller is investigated in (Simpson-Porco et al. 2017). In (Wang et al. 2013), sufficient conditions are derived for the voltage stability of a weak microgrid with inverter-connected sources. However, the nonlinear stability analysis of power electronic converters considering both frequency and voltage dynamics in the general case is still open.

A generic control framework based on the PH system theory (Fiaz, et al. 2013; Ortega and Romero, 2012; Ortega et al. 2008; 2001; van der Schaft and Jeltsema, 2014) and the ghost operator (Zhong, 2017b) is presented in this chapter to guarantee the passivity of the controller of a power electronic converter. The controller consists of two symmetric control loops: the torque-frequency loop to generate the frequency and the quorte-flux loop to generate the flux that corresponds to the voltage. Here, the new word *quorte* coined in (Zhong and Stefanello, 2017) represents the quantity that is dual to torque. The amplitude of the output voltage of the controller is generated by multiplying the frequency and the flux while its phase angle is generated by integrating the frequency. It is shown that these two symmetric control loops can be designed to form a passive block, which is connected through an interconnection block with the plant that represents the circuits from the output terminals of the power stages to the grid including the loads. Since most nonlinear loads such as AC/DC converters and linear loads in modern electrical systems can be described by passive PH models (Gaviria et al. 2005, Konstantopoulos and Alexandridis, 2013), in order to ensure the passivity of the overall system there is a need for the interconnection block between the controller and the plant to be lossless, according to the PH system theory (Maschke and Van Der Schaft, 1991, Ortega, van der Schaft, Castanos and Astolfi, 2008). This is addressed by applying the ghost operator introduced in Chapter 3.

21.2 Passivity and Port-Hamiltonian Systems

21.2.1 Passive Systems

Passivity can be defined for both memoryless and dynamic systems (Khalil, 2001). Assume u as the system input, y as the output with $\dim u = \dim y$, and $u^T y$ as the power flowing into the system. A memoryless system $y = h(t, u)$ is passive if

$$u^T y \geq 0 \tag{21.1}$$

for all u, and it is lossless if $u^T y = 0$. For a dynamic system, the concept of passivity involves in a non-negative function of the state vector x, called the storage function $H_s(x)$. A dynamic system is passive if the energy absorbed by the system over any period of time $[0, t]$ is greater than or equal to the increase in the energy stored in the system over the same period, i.e. if

$$\int_0^t u^T(\tau)y(\tau)d\tau \geq H_s(x(t)) - H_s(x(0)) \tag{21.2}$$

for the non-negative storage function $H_s(x)$. It is lossless if both sides of (21.2) are equal or $\frac{dH_s}{dt}(x(t)) = u^T y$.

21.2.2 Port-Hamiltonian Systems

Passivity can be investigated with different frameworks; see e.g., (Byrnes et al. 1991; Ortega, 1989; Xia et al. 2017; Zhu et al. 2017). Port-Hamiltonian system theory (van der Schaft and Jeltsema, 2014) offers a systematic mathematical framework for structural modeling, analysis and control of complex networked multi-physics systems with lumped and/or distributed parameters. It combines the historical Hamiltonian modeling approach in geometric mechanics (Holm et al. 2009) and the port-based network modeling approach in electrical engineering (Carlin, 1967; Tellegen, 1952), via geometrically associating the interconnected network with a Dirac structure (van der Schaft and Jeltsema, 2014), which is power preserving. The Hamiltonian dynamics is defined with respect to the Dirac structure and the Hamiltonian representing the total stored energy. Port-Hamiltonian systems are open dynamic systems and interact/interconnect with their environment through ports. Moreover, port-Hamiltonian systems theory can include energy-dissipating elements, which are largely absent in classical Hamiltonian systems. What is even more crucial is that compositions of Dirac structures are also Dirac structures, which means the power-preserving interconnection of PH systems (through their interconnection ports) is also a PH system. This scalable property is called interconnectivity or compositionality, and is instrumental for the modeling and control of large-scale networked systems.

At the center of a PH system, it is the Dirac structure (representing the network) having a storage port connected to energy-storing elements, a resistive port connected to energy-dissipating (resistive) elements, and an interconnection port connected to the environment (including the controller).

For a dynamic system, if (i) there are no algebraic constraints between the state variables, (ii) the interconnection port power variables can be split into input and output variables, and (iii) the resistive structure is linear and of input–output form, then the system can be described in the usual input-state–output format (Maschke et al. 2000; van der Schaft and Jeltsema, 2014) as

$$\dot{x} = (J(x) - R(x))\frac{\partial H}{\partial x}(x) + G(x)u, \tag{21.3}$$

$$y = G^T(x)\frac{\partial H}{\partial x}(x) + D(x)u, \tag{21.4}$$

where $x \in R^{n \times 1}$ is the state vector, and u and $y \in R^{m \times 1}$ are the input and the output, $H(x)$ is the Hamiltonian (function) representing the total energy of the system and $\frac{\partial H(x)}{\partial x} \in R^{n \times 1}$ is its gradient, $J(x) = -J^T(x)$ is a skew-symmetric matrix representing the network structure, $R(x)$ is a positive semi-definite symmetric matrix representing the resistive elements (damping) of the system, $G(x)$ is the input matrix, and $D(x)$ is the feed-through term. All these matrices depend smoothly on the state x but this is often omitted for simplicity in the sequel. Note that, the Hamiltonian $H(x)$ is not necessarily non-negative (nor bounded from below). The feed-through term $D(x)$ does not have to be symmetric and, in most literature, $D(x) = 0$. In this chapter, the case with $D(x) = 0$ will be followed except in Section 21.5.1, where it is non-zero and skew-symmetric for the interconnection block Σ_I.

When $D(x)$ is 0 or skew-symmetric, the time-derivative of the Hamiltonian is

$$\frac{dH(x(t))}{dt} = -\frac{\partial^T H}{\partial x}(x)R(x)\frac{\partial H}{\partial x}(x) + u^T y, \tag{21.5}$$

which characterizes the power conservation/balance property of PH systems. The product $u^T y$ is called the supply rate and has the unit of power. As a result, the Hamiltonian always satisfies

$$H(x(t)) \le H(x(0)) + \int_0^t u^T y dt \tag{21.6}$$

because of the dissipated energy associated with $R(x)$. Moreover, if the Hamiltonian $H(x)$ is bounded from below by $C > -\infty$, then the system is passive (Byrnes, et al. 1991) with the non-negative storage function $H_s(x) = H(x) - C$, the input u, and the output y. The system is lossless if $\frac{dH}{dt}(x(t)) = u^T y$. Note that $\frac{dH}{dt}(x(t))$ and $u^T y$ do not have to be 0 for a lossless system as long as they are equal. Note that when $D(x)$ is skew-symmetric, it does not appear in (21.5) or (21.6). In other words, a zero or skew-symmetric $D(x)$ does not affect the passivity or losslessness of the system.

Passivity is closely related to the Lyapunov stability. For the unforced case, i.e. when $u = 0$, if $\frac{\partial^T}{\partial x}H(x^*) = 0$ and $H(x(t)) > 0$ for every $x(t) \ne x^*$, then from (21.5) it is possible to conclude that x^* is an equilibrium of the system (21.3)–(21.4) with Lyapunov function $H(x)$, implying stability of the equilibrium state x^*. If $R(x)$ is positive definite, then the equilibrium state x^* is asymptotically stable. If the input u of a passive system is not zero, then there is no guarantee that its equilibrium is Lyapunov stable. In this case, methods, such as the one presented in (Maschke et al. 2000), can be followed to investigate the stability of its equilibrium.

Figure 21.1 Two systems with disturbances interconnected through Σ_I.

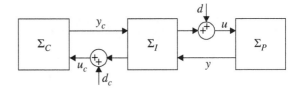

21.2.3 Passivity of Interconnected Passive Systems

Figure 21.1 shows a system consisting of two subsystems Σ_C and Σ_P interconnected through Σ_I. Σ_C has one port with port variables (u_c, y_c) and Σ_P has one port with port variables (u, y). The interconnection block Σ_I has two ports. The following result holds.

Lemma 7. (Ortega et al. 2001, Proposition 2) For two passive subsystems Σ_C and Σ_P with storage functions H_C and H_P, respectively, as shown in Figure 21.1, if the interconnection block Σ_I is lossless with the storage function H_I satisfying $\dot{H}_I = 0$, then the overall system with port variables (d, y) and (d_c, y_c) is passive.

Proof: See (Ortega, et al. 2001, Proposition 2). The proof is relatively straightforward. The supply rate of the lossless interconnection block Σ_I is

$$\dot{H}_I = \left[(u - d)^T \ (u_c - d_c)^T\right] \begin{bmatrix} y \\ y_c \end{bmatrix},$$

which means

$$u_c^T y_c + u^T y = d^T y + d_c^T y_c + \dot{H}_I. \tag{21.7}$$

Moreover, the total energy of the interconnected system

$$H = H_C + H_P + H_I$$

is bounded from below with

$$\dot{H} = \dot{H}_C + \dot{H}_P + \dot{H}_I \tag{21.8}$$

$$\leq u_c^T y_c + u^T y + \dot{H}_I \tag{21.9}$$

$$= d^T y + d_c^T y_c + 2\dot{H}_I = \left[d^T \ d_c^{\ T}\right] \begin{bmatrix} y \\ y_c \end{bmatrix}, \tag{21.10}$$

where the properties $\dot{H}_C \leq u_c^T y_c$ and $\dot{H}_P \leq u^T y$ due to the passivity of Σ_C and Σ_P and the condition $\dot{H}_I = 0$ are used. Hence, the interconnected system with the port variables (d, y) and (d_c, y_c) is passive. This concludes the proof. $\qquad\square$

Remark 8. An example of the lossless interconnection block is the well known negative feedback with $\Sigma_I = \begin{bmatrix} 0 & I \\ -I & 0 \end{bmatrix}$. For the system shown in Figure 21.1, this means $u = d + y_c$ and $u_c = d_c - y$. Indeed, the supply rate $\left[u - d \ u_c - d_c\right]^T \begin{bmatrix} y \\ y_c \end{bmatrix} = \left(\begin{bmatrix} 0 & I \\ -I & 0 \end{bmatrix} \begin{bmatrix} y \\ y_c \end{bmatrix}\right)^T \begin{bmatrix} y \\ y_c \end{bmatrix} =$

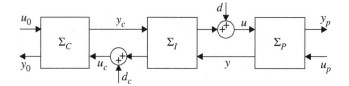

Figure 21.2 Two systems with disturbances and external ports interconnected through Σ_I.

$y_c^T y - y^T y_c$ is 0. Since there is no dynamics involved, the storage function H_I satisfies $\dot{H}_I = 0$, naturally.

This result also holds for a more general case, as shown in Figure 21.2, with both Σ_C and Σ_P having an external port.

Lemma 9. For the system shown in Figure 21.2, assume that the two subsystems Σ_C and Σ_P are passive with storage functions H_C and H_P and the interconnection block Σ_I is lossless with its storage function H_I satisfying $\dot{H}_I = 0$, then the interconnected system with port variables (d, y), (d_c, y_c), (u_0, y_0) and (u_p, y_p) is passive.

Proof: It can be proved in a similar way as Lemma 7, taking into account that $\dot{H}_C \leq u_c^T y_c + u_0^T y_0$ and $\dot{H}_P \leq u^T y + u_p^T y_p$ because of the additional external ports. □

These results are instrumental in investigating the passivity of interconnected systems because of the nice structural properties. In particular, when Σ_P itself is passive, what needs to be done is to design a passive controller Σ_C and identify a lossless interconnection block Σ_I with its energy function H_I satisfying $\dot{H}_I = 0$.

21.3 System Modeling

The system under consideration is shown in Figure 21.3. A three-phase power electronic converter is represented by ideal voltage sources $u = \begin{bmatrix} u_a & u_b & u_c \end{bmatrix}^T$ because the switching effect of the power semiconductor devices can be neglected for the purpose of control design (Zhong and Hornik, 2013). The converter may be operated in the grid-connected mode or in the islanded mode, i.e. with the main circuit breaker ON or OFF. It is worth highlighting that the voltages u_a, u_b and u_c are sinusoidal functions by design and that the grid supply voltages $v_s = \begin{bmatrix} v_{sa} & v_{sb} & v_{sc} \end{bmatrix}^T$ are also sinusoidal functions or the linear combination of sinusoidal functions when there are harmonic components in the grid supply.

The supply-side elements L_2 and R_2 account for the possibility of using an *LCL* filter and/or a coupling transformer.

Assume that the generalized dissipative PH model of the per-phase load is (Konstantopoulos and Alexandridis, 2013)

$$M_{Lw} \dot{q}_w = (J_{Lw} - R_{Lw}) q_w + G_{Lw} v_w$$

$$i_{Lw} = G_{Lw}^T q_w, \tag{21.11}$$

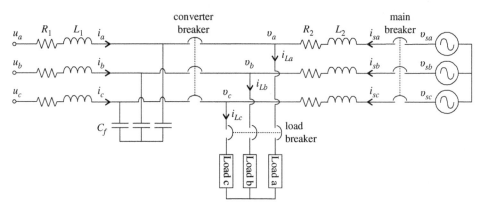

Figure 21.3 Three-phase grid-connected converter with a local load.

where $w = a$, b, c stands for the electrical phase, $q_w = \begin{bmatrix} i_{Lw} & q_{1w} & \cdots & q_{(m-1)w} \end{bmatrix}^T$ are the states with its dimension depending on the dynamics of the load, $M_{Lw} = M_{Lw}^T$ and $G_{Lw} = \begin{bmatrix} 1 & 0 & \cdots & 0 \end{bmatrix}^T$. The coupling matrix has the form $J_{Lw} = -J_{Lw}^T = \begin{bmatrix} 0 & J_{12w} \\ -J_{12w}^T & J_{22w} \end{bmatrix}$ with J_{12w} and J_{22w} $(= -J_{22w}^T)$ being $1 \times (m-1)$ and $(m-1) \times (m-1)$ matrices, respectively, and $R_{Lw} = R_{Lw}^T > 0$. From (21.11), the model of the three-phase load can be obtained straightforwardly after defining $q = \begin{bmatrix} q_a & q_b & q_c \end{bmatrix}^T$, $i_L = \begin{bmatrix} i_{La} & i_{Lb} & i_{Lc} \end{bmatrix}^T$ and $v = \begin{bmatrix} v_a & v_b & v_c \end{bmatrix}^T$ as

$$M_L \dot{q} = (J_L - R_L)q + G_L v$$

$$i_L = G_L^T q, \tag{21.12}$$

with matrices $J_L = -J_L^T$ and $R_L = R_L^T > 0$.

Defining the converter current $i = \begin{bmatrix} i_a & i_b & i_c \end{bmatrix}^T$, then the overall system shown in Figure 21.3 can be modelled as

$$L_1 \frac{di}{dt} = -R_1 i - v + u$$

$$C_f \frac{dv}{dt} = i - G_L^T q + i_s$$

$$L_2 \frac{di_s}{dt} = -v - R_2 i_s + v_s$$

$$M_L \frac{dq}{dt} = (J_L - R_L)q + G_L v. \tag{21.13}$$

Select the plant state vector as

$$x = \begin{bmatrix} L_1 i^T & C_f v^T & L_2 i_s^T & M_L q^T \end{bmatrix}^T$$

and the Hamiltonian function as

$$H_p(x) = \frac{1}{2} L_1 i^T i + \frac{1}{2} C_f v^T v + \frac{1}{2} L_2 i_s^T i_s + \frac{1}{2} q^T M_L q. \tag{21.14}$$

Then the system (21.13) can be expressed as the following PH model:

$$P_s : \begin{cases} \dot{x} = (J_p - R_p)\dfrac{\partial H_p}{\partial x} + G_p u + G_s v_s \\ i = G_p^T \dfrac{\partial H_p}{\partial x} \\ i_s = G_s^T \dfrac{\partial H_p}{\partial x} \end{cases}$$

(21.15)

with

$$\frac{\partial H_p}{\partial x} = \begin{bmatrix} i^T & v^T & i_s^T & q^T \end{bmatrix}^T,$$

$$G_p = \begin{bmatrix} I & 0 & 0 & 0 \end{bmatrix}^T, \quad G_s = \begin{bmatrix} 0 & 0 & I & 0 \end{bmatrix}^T,$$

$$J_p = \begin{bmatrix} 0 & -I & 0 & 0 \\ I & 0 & I & -G_L^T \\ 0 & -I & 0 & 0 \\ 0 & G_L & 0 & J_L \end{bmatrix}, \quad R_p = \begin{bmatrix} R_1 I & 0 & 0 & 0 \\ 0 & 0 & 0 & 0 \\ 0 & 0 & R_2 I & 0 \\ 0 & 0 & 0 & R_L I \end{bmatrix},$$

where I is the 3×3 identity matrix corresponding to three phases. Apparently, the system (21.15) is in the PH form and passive because R_p is positive semi-definite, J_p is skew symmetric and the Hamiltonian H_p in (21.14) is non-negative.

21.4 Control Framework

The control framework is shown in Figure 21.4. The controller consists of the control block Σ_C and the engendering block Σ_e. The block Σ_C has a torque-frequency loop and a quorte-flux loop including subsystems Σ_ω and Σ_φ, respectively. The subsystem Σ_ω is used to synthesize the frequency ω and the subsystem Σ_φ is used to synthesize the flux φ. The engendering block Σ_e synthesizes the generated voltage e according to the frequency ω and the flux φ, with the phase angle θ of the generated voltage e satisfying $\dot{\theta} = \omega$, and also provides the feedback for the two control loops.

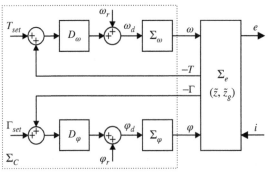

Figure 21.4 The controller for a cybersync machine with e to be supplied as u.

21.4.1 The Engendering Block Σ_e

Define

$$\tilde{z} = \sin\theta \quad \text{and} \quad \tilde{z}_g = g\tilde{z} = \cos\theta. \tag{21.16}$$

In other words, $\tilde{z}_g$ is obtained after applying the ghost operator to $\tilde{z}$, i.e. $\tilde{z}_g$ is the ghost signal of $\tilde{z}$. Note that $\tilde{z}$ and $\tilde{z}_g$ can be implemented with the dynamic system

$$\begin{bmatrix} \dot{\tilde{z}} \\ \dot{\tilde{z}}_g \end{bmatrix} = \begin{bmatrix} 0 & \omega \\ -\omega & 0 \end{bmatrix} \begin{bmatrix} \tilde{z} \\ \tilde{z}_g \end{bmatrix}, \tag{21.17}$$

with initial conditions $\tilde{z}(0) = 0$ and $\tilde{z}_g(0) = 1$. Moreover, define the following vector

$$z = \begin{bmatrix} \sin\theta & \sin\left(\theta - \frac{2\pi}{3}\right) & \sin\left(\theta + \frac{2\pi}{3}\right) \end{bmatrix}^T, \tag{21.18}$$

which is denoted as $\widetilde{\sin\theta}$ in (Zhong and Hornik, 2013; Zhong and Weiss, 2011; Zhong et al. 2014) and Chapters 4 and 8 to represent the three-phase normalized generated voltages in the *abc* reference frame. Furthermore, denote its ghost gz, i.e. after applying the ghost operator to each element of z, as z_g. Then

$$z_g = gz = \begin{bmatrix} \cos\theta & \cos\left(\theta - \frac{2\pi}{3}\right) & \cos\left(\theta + \frac{2\pi}{3}\right) \end{bmatrix}^T,$$

which is denoted as $\widetilde{\cos\theta}$ in (Zhong and Hornik, 2013; Zhong and Weiss, 2011; Zhong et al. 2014) and Chapters 4 and 8. It is easy to see that

$$\begin{bmatrix} z \\ z_g \end{bmatrix} = \begin{bmatrix} 1 & -\frac{1}{2} & -\frac{1}{2} & 0 & \frac{\sqrt{3}}{2} & -\frac{\sqrt{3}}{2} \\ 0 & -\frac{\sqrt{3}}{2} & \frac{\sqrt{3}}{2} & 1 & -\frac{1}{2} & -\frac{1}{2} \end{bmatrix}^T \begin{bmatrix} \tilde{z} \\ \tilde{z}_g \end{bmatrix}. \tag{21.19}$$

The control signal, i.e. the output voltage to be converted into pulses to drive the power electronic switches, is designed as

$$e = \begin{bmatrix} e_a & e_b & e_c \end{bmatrix}^T = Ez, \tag{21.20}$$

with

$$E = \omega\varphi$$

being the amplitude of the output voltage, which is the product of the frequency ω and the flux φ. Denote $e_g = ge$ as the ghost of voltage e. Then,

$$e_g = ge = Ez_g \tag{21.21}$$

Note that both e and e_g are in the natural reference frame.

The control of power electronic converters often involves in the calculation of instantaneous real power and reactive power. The instantaneous power theory (Akagi et al. 2007) is widely adopted for this purpose. However, there is a need to go through the $abc \rightarrow \alpha\beta0$ transformation before defining the instantaneous real power and reactive power. Moreover, it is formulated for three-phase systems and treats three phases as one unit. Hence, its application to single-phase systems is not straightforward. The ghost power theory established in (Zhong, 2017b) and discussed in Chapter 3 provides a formula to directly calculate the instantaneous reactive power without changing the definition of the instantaneous real power, offering a clear physical interpretation for reactive power. The same formula is also directly applicable to both single-phase and poly-phase systems. Actually, it can be applied to other systems involving energy or power conversion as well, including mechanical, fluid, thermal, magnetic, and chemical systems (Zhong, 2017b).

For the three-phase converter shown in Figure 21.3 with $u = e$, the real power and the reactive power at the mid-point of the conversion legs can be obtained according to the ghost power theory discussed in Chapter 3, respectively, as

$$P = e^T i, \quad Q = -e_g^T i.$$

According to (21.20) and (21.21), the real power and the reactive power are

$$P = \omega\varphi z^T i \triangleq T\omega, \quad Q = -\omega\varphi z_g^T i \triangleq \Gamma\varphi, \tag{21.22}$$

where

$$T = \varphi z^T i \tag{21.23}$$

resembles the electromagnetic torque of a synchronous machine (Zhong and Weiss, 2011) and

$$\Gamma = -\omega z_g^T i \tag{21.24}$$

is a quantity called a *quorte* as coined in (Zhong and Stefanello, 2017), which indicates its duality to the torque.

The engendering block Σ_e in Figure 21.4 can then be summarized as

$$\Sigma_e : \left\{ \begin{bmatrix} e \\ -T \\ -\Gamma \end{bmatrix} = \begin{bmatrix} 0 & \varphi z & 0 \\ -\varphi z^T & 0 & 0 \\ \omega z_g^T & 0 & 0 \end{bmatrix} \begin{bmatrix} i \\ \omega \\ \varphi \end{bmatrix} \right\}. \tag{21.25}$$

Apparently, it is not skew-symmetric or lossless so it cannot be treated as the interconnection block Σ_I in Figure 21.2, which makes it difficult to render the system consisting of Σ_C, Σ_e and P_s passive. As will be shown in Subsection 21.5.1, it can be augmented to form a lossless interconnection block Σ_I by adding a port.

21.4.2 Generation of the Desired Frequency ω_d and Flux φ_d

The purpose of the blocks Σ_ω and Σ_φ in Figure 21.4 is to generate the frequency ω and flux φ according to the desired frequency ω_d and desired flux φ_d, which are designed according to the droop control principle (Zhong, 2013c; Zhong and Zeng, 2016) as

$$\begin{aligned} \omega_d &= \omega_r + D_\omega(T_{set} - T), \\ \varphi_d &= \varphi_r + D_\varphi(\Gamma_{set} - \Gamma). \end{aligned} \tag{21.26}$$

Here, T_{set} and Γ_{set} are the set points for the torque and the quorte (corresponding to the set points for the real power and reactive power, respectively); ω_r and φ_r are the reference values for the frequency and the flux, respectively; and D_ω and D_φ are the frequency droop coefficient and the flux droop coefficient defined, respectively, as

$$D_\omega = -\frac{\Delta\omega}{\Delta T}, \quad D_\varphi = -\frac{\Delta\varphi}{\Delta\Gamma} \tag{21.27}$$

to describe the impact of the torque variation ΔT (correspondingly, the active power variation ΔP) and the quorte variation $\Delta \Gamma$ (correspondingly, the reactive power variation ΔQ) on the frequency variation $\Delta \omega$ and the flux variation $\Delta \varphi$. The reference values ω_r and φ_r can be set as

$$\omega_n = 2\pi f_n, \quad \varphi_n = \frac{\sqrt{2}V_n}{\omega_n}, \tag{21.28}$$

respectively, or obtained according to the operation modes of the system as in (Zhong and Weiss, 2011), where f_n and V_n are the nominal values of the system frequency and the phase voltage (rms). Note that D_ω and D_φ described above are static gains but they can be designed to be dynamic as well, e.g. to include an integrator.

21.4.3 Design of Σ_ω and Σ_φ to Obtain a Passive Σ_C

Lemma 10. The block Σ_C in Figure 21.4 is passive if the desired frequency and flux are designed as in (21.26) and the blocks Σ_ω and Σ_φ are designed as

$$\Sigma_\omega : \begin{cases} \dot{x}_\omega = (J_\omega - R_\omega)\dfrac{\partial H_\omega}{\partial x_\omega} + G_\omega(\dfrac{1}{D_\omega}\omega_d) \\ \omega = G_\omega^T \dfrac{\partial H_\omega}{\partial x_\omega}, \end{cases} \tag{21.29}$$

$$\Sigma_\varphi : \begin{cases} \dot{x}_\varphi = (J_\varphi - R_\varphi)\dfrac{\partial H_\varphi}{\partial x_\varphi} + G_\varphi(\dfrac{1}{D_\varphi}\varphi_d) \\ \varphi = G_\varphi^T \dfrac{\partial H_\varphi}{\partial x_\varphi}, \end{cases} \tag{21.30}$$

with non-negative Hamiltonians H_ω and H_φ, positive semi-definite damping matrices R_ω and R_φ, and skew-symmetric matrices J_ω and J_φ.

Proof: By combining the blocks Σ_ω and Σ_φ designed in (21.29)–(21.30) and considering (21.26), the block Σ_C in Figure 21.4 can be described as

$$\Sigma_C : \begin{cases} \begin{bmatrix} \dot{x}_\omega \\ \dot{x}_\varphi \end{bmatrix} = \left(\begin{bmatrix} J_\omega & 0 \\ 0 & J_\varphi \end{bmatrix} - \begin{bmatrix} R_\omega & 0 \\ 0 & R_\varphi \end{bmatrix} \right) \begin{bmatrix} \dfrac{\partial H_\omega}{\partial x_\omega} \\ \dfrac{\partial H_\varphi}{\partial x_\varphi} \end{bmatrix} \\ \qquad + \begin{bmatrix} G_\omega & 0 \\ 0 & G_\varphi \end{bmatrix} \begin{bmatrix} T_0 \\ \Gamma_0 \end{bmatrix} + \begin{bmatrix} G_\omega & 0 \\ 0 & G_\varphi \end{bmatrix} \begin{bmatrix} -T \\ -\Gamma \end{bmatrix}, \\[2ex] \begin{bmatrix} \omega \\ \varphi \end{bmatrix} = \begin{bmatrix} G_\omega^T & 0 \\ 0 & G_\varphi^T \end{bmatrix} \begin{bmatrix} \dfrac{\partial H_\omega}{\partial x_\omega} \\ \dfrac{\partial H_\varphi}{\partial x_\varphi} \end{bmatrix}, \\[2ex] \begin{bmatrix} \omega \\ \varphi \end{bmatrix} = \begin{bmatrix} G_\omega^T & 0 \\ 0 & G_\varphi^T \end{bmatrix} \begin{bmatrix} \dfrac{\partial H_\omega}{\partial x_\omega} \\ \dfrac{\partial H_\varphi}{\partial x_\varphi} \end{bmatrix}, \end{cases} \tag{21.31}$$

with

$$\begin{bmatrix} T_0 \\ \Gamma_0 \end{bmatrix} = \begin{bmatrix} \dfrac{1}{D_\omega}\omega_r + T_{set} \\ \dfrac{1}{D_\varphi}\varphi_r + \Gamma_{set} \end{bmatrix}.$$

Here, ω_r and φ_r are the reference system frequency and flux, respectively, which are often set to their rated values. T_{set} and Γ_{set} are the set points for the torque and quorte, respectively. Hence, T_0 and Γ_0 are constants, often non-zero. Note that the output equation for $\begin{bmatrix} \omega \\ \varphi \end{bmatrix}$ in (21.31) is repeated to form an additional port with the input $\begin{bmatrix} T_0 \\ \Gamma_0 \end{bmatrix}$, in addition to the input-output port with port variables ($\begin{bmatrix} -T \\ -\Gamma \end{bmatrix}$, $\begin{bmatrix} \omega \\ \varphi \end{bmatrix}$). Σ_C in (21.31) is in the standard form of the generic PH systems given in (21.3)–(21.4). The Hamiltonian H_C of the block Σ_C can be chosen as

$$H_C = H_\omega + H_\varphi.$$

Since the Hamiltonians H_ω and H_φ are non-negative, so is H_C. Moreover, as the matrices R_ω and R_φ are designed to be positive semi-definite, so is the matrix $\begin{bmatrix} R_\omega & 0 \\ 0 & R_\varphi \end{bmatrix}$. Furthermore, since J_ω and J_φ are designed to be skew-symmetric, so is the matrix $\begin{bmatrix} J_\omega & 0 \\ 0 & J_\varphi \end{bmatrix}$. Hence, Σ_C is passive with the storage function H_C, the input $\begin{bmatrix} T_0 & \Gamma_0 & -T & -\Gamma \end{bmatrix}^T$, and the output $\begin{bmatrix} \omega & \varphi & \omega & \varphi \end{bmatrix}^T$.   □

This lemma characterizes the generic form of the blocks Σ_ω and Σ_φ that is able to render the control block Σ_C passive.

21.5 Passivity of the Controller

As mentioned in Section 21.4.1, the engendering block Σ_e is not lossless. In order to solve this problem, the ghost system of the plant P_s is revealed to pair with the plant itself and form a plant pair Σ_P, as shown in Figure 21.5. Note that the input u of the plant P_s is connected to e, i.e. $u = e$, while the input of the ghost plant is connected to e_g, i.e. the ghost of the input e to the plant. Moreover, the input v_s of the ghost plant is connected to v_{sg}, i.e. the ghost of the input v_s to the plant. Since the ghost operator g satisfies $g^2 = -1$, the ghost state x_g in Figure 21.5 is the opposite of the plant state. As a result, the output of the ghost plant is the opposite of the plant output, i.e.

$$i_g = -i \quad \text{and} \quad i_{sg} = -i_s. \tag{21.32}$$

21.5.1 Losslessness of the Interconnection Block Σ_I

Adding the augmented port variables e_g and $i_g = -i$ to the engendering block Σ_e shown in Figure 21.4 results in the interconnection block Σ_I in Figure 21.5. It is lossless, as shown below.

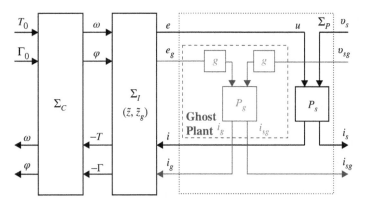

Figure 21.5 The mathematical structure of the system constructed to facilitate the passivity analysis, where the plant pair Σ_p consists of the original plant and its ghost in grey and the input to the ghost plant is the ghost of the input to the plant.

Theorem 11. *The interconnection block Σ_I in Figure 21.5 defined with the output voltage e in (21.20), the ghost voltage e_g in (21.21), the torque T in (21.23), and the quorte Γ in (21.24) is lossless with its storage function H_I satisfying $\dot{H}_I = 0$.*

Proof: The frequency ω enters the Σ_I block to form its state $x_I = \begin{bmatrix} \tilde{z} \\ \tilde{z}_g \end{bmatrix}$, according to (21.17).

Define its Hamiltonian as

$$H_I = \frac{1}{2}\tilde{z}^2 + \frac{1}{2}\tilde{z}_g^2.$$

Then, according to (21.16), there is

$$H_I = \frac{1}{2}\sin^2\theta + \frac{1}{2}\cos^2\theta = \frac{1}{2}.$$

H_I is always positive and hence is a storage function of the Σ_I block. Moreover, $\frac{\partial H_I(x_I)}{\partial x_I} = \begin{bmatrix} \tilde{z} \\ \tilde{z}_g \end{bmatrix}$ and the Σ_I block can be described in the PH form of (21.3)–(21.4) as

$$\Sigma_I : \begin{cases} \dot{x}_I = (J_I - R_I)\dfrac{\partial H_I(x_I)}{\partial x_I} + G_I u_I, \\[2mm] y_I = G_I^T \dfrac{\partial H_I(x_I)}{\partial x_I} + D_I u_I, \end{cases} \tag{21.33}$$

with $J_I = \begin{bmatrix} 0 & \omega \\ -\omega & 0 \end{bmatrix}$, $R_I = \begin{bmatrix} 0 & 0 \\ 0 & 0 \end{bmatrix}$, $G_I = \begin{bmatrix} 0 & 0 & 0 & 0 \\ 0 & 0 & 0 & 0 \end{bmatrix}$, and

$$y_I = \begin{bmatrix} e \\ e_g \\ -T \\ -\Gamma \end{bmatrix}, \quad D_I = \begin{bmatrix} 0 & 0 & \varphi z & 0 \\ 0 & 0 & 0 & \omega z_g \\ -\varphi z^T & 0 & 0 & 0 \\ 0 & -\omega z_g^T & 0 & 0 \end{bmatrix}, \quad u_I = \begin{bmatrix} i \\ i_g \\ \omega \\ \varphi \end{bmatrix},$$

after taking into account the expressions of e in (21.20), e_g in (21.21), T in (21.23), Γ in (21.24), and (21.32). Note that $G_I = 0$ and D_I is skew-symmetric. Apparently, the supply rate of the Σ_I block is

$$y_I^T u_I = u_I^T D_I^T u_I = 0.$$

Hence, Σ_I is lossless with $\dot{H}_I = 0$. This completes the proof. $\square$

Remark 12. The interconnection block Σ_I has four ports: (e, i), (e_g, i_g), $(-T, \omega)$, and $(-\Gamma, \varphi)$. The ports (e, i) and $(-T, \omega)$ represent the real power while the ports (e_g, i_g) and $(-\Gamma, \varphi)$ represent the reactive power. All the incoming real power and reactive power go out immediately without affecting the energy stored because $e^T i = T\omega$ and $e_g^T i_g = \Gamma\varphi$.

Remark 13. The interconnection block Σ_I contains dynamics $\tilde{z}$ and $\tilde{z}_g$, unlike the case of a simple negative feedback.

21.5.2 Passivity of the Cascade of Σ_C and Σ_I

Theorem 14. The controller of the power electronic converter that consists of Σ_C and Σ_I, as illustrated in Figure 21.5 with Σ_C given in (21.31) and Σ_I characterized by (21.33) and (21.19) with (21.17), is passive.

Proof: Combining the control block Σ_C (21.31) with the interconnection block Σ_I (21.33), then there is

$$
\Sigma_{C\text{-}I}: \begin{cases}
\begin{bmatrix} \dot{x}_\omega \\ \dot{x}_\varphi \end{bmatrix} = \left(\begin{bmatrix} J_\omega & 0 \\ 0 & J_\varphi \end{bmatrix} - \begin{bmatrix} R_\omega & 0 \\ 0 & R_\varphi \end{bmatrix} \right) \begin{bmatrix} \dfrac{\partial H_\omega}{\partial x_\omega} \\ \dfrac{\partial H_\varphi}{\partial x_\varphi} \end{bmatrix} \\[20pt]
\qquad + \begin{bmatrix} G_\omega & 0 \\ 0 & G_\varphi \end{bmatrix} \begin{bmatrix} T_0 \\ \Gamma_0 \end{bmatrix} \\[20pt]
\qquad - \begin{bmatrix} G_\omega & 0 \\ 0 & G_\varphi \end{bmatrix} \begin{bmatrix} \varphi z^T & 0 \\ 0 & \omega z_g^T \end{bmatrix} \begin{bmatrix} i \\ i_g \end{bmatrix}, \\[20pt]
\begin{bmatrix} \omega \\ \varphi \end{bmatrix} = \begin{bmatrix} G_\omega^T & 0 \\ 0 & G_\varphi^T \end{bmatrix} \begin{bmatrix} \dfrac{\partial H_\omega}{\partial x_\omega} \\ \dfrac{\partial H_\varphi}{\partial x_\varphi} \end{bmatrix}, \\[20pt]
\begin{bmatrix} e \\ e_g \end{bmatrix} = \begin{bmatrix} \varphi z & 0 \\ 0 & \omega z_g \end{bmatrix} \begin{bmatrix} G_\omega^T & 0 \\ 0 & G_\varphi^T \end{bmatrix} \begin{bmatrix} \dfrac{\partial H_\omega}{\partial x_\omega} \\ \dfrac{\partial H_\varphi}{\partial x_\varphi} \end{bmatrix}.
\end{cases}
\tag{21.34}
$$

Since H_φ and H_φ are non-negative and $\begin{bmatrix} R_\omega & 0 \\ 0 & R_\varphi \end{bmatrix}$ is positive semi-definite by design, $\Sigma_{\text{C-I}}$ is passive with the storage function $H_\text{C} = H_\omega + H_\varphi$, the input $\begin{bmatrix} T_0 & \Gamma_0 & -i & -i_\text{g} \end{bmatrix}^T$ and the output $\begin{bmatrix} \omega & \varphi & e & e_\text{g} \end{bmatrix}^T$.

□

Note that the negative sign before the term $\begin{bmatrix} i \\ i_\text{g} \end{bmatrix}$ is due to the fact that the current is positive when flowing out of the converter. This also guarantees the critical property of negative feedback when connected to external circuits.

21.6 Passivity of the Closed-loop System

Conjecture 15. The ghost plant of the plant P_s as shown in Figure 21.5 is passive for a passive plant P_s, with respect to the input $\begin{bmatrix} e_\text{g} & v_{\text{sg}} \end{bmatrix}^T$, the output $\begin{bmatrix} i_\text{g} & i_{\text{sg}} \end{bmatrix}^T = \begin{bmatrix} -i & -i_\text{s} \end{bmatrix}^T$, and an appropriately constructed Hamiltonian function H_g.

This seems straightforward but it has not been easy to prove. The challenge lies in constructing the Hamiltonian function H_g. Apparently, the supply rate to the ghost plant is $-e_\text{g}^T i - v_{\text{sg}}^T i_\text{s}$, which is the reactive power. It means the corresponding Hamiltonian function H_g should be related to the reactive power.

If the conjecture is proved, then the plant pair Σ_P that consists of the plant and its ghost is passive if the plant itself is passive. Moreover, the closed-loop system shown in Figure 21.5 with Σ_C given in (21.31) and Σ_I characterized by (21.33) and (21.19) with (21.17) is passive.

21.7 Sample Implementations for Blocks Σ_ω and Σ_φ

The function of the blocks Σ_ω and Σ_φ is to track the desired frequency ω_d and flux φ_d. There are many ways to achieve this. Here, two examples are illustrated. One is to adopt the standard integral controller and the other is to adopt a simple static controller.

21.7.1 Using the Standard Integral Controller (IC)

Figure 21.6 illustrates the implementation with the integrator for the blocks Σ_ω and Σ_φ, from which it is easy to find that

$$\Sigma_\omega = \frac{1}{\tau_\omega s + 1}, \qquad \Sigma_\varphi = \frac{1}{\tau_\varphi s + 1}, \tag{21.35}$$

where τ_ω and τ_φ are the time constants of the frequency and flux loops, respectively. As a rule of thumb, τ_ω and τ_φ can be chosen as $\tau_\omega = \frac{0.1}{f_\text{n}}$ s and $\tau_\varphi = \frac{1}{f_\text{n}}$ s. Select their states as $x_\omega = \frac{\tau_\omega}{D_\omega} \omega$ and $x_\varphi = \frac{\tau_\varphi}{D_\varphi} \varphi$, respectively, and the Hamiltonians as

$$H_\omega = \frac{1}{2} \frac{\tau_\omega}{D_\omega} \omega^2 \quad \text{and} \quad H_\varphi = \frac{1}{2} \frac{\tau_\varphi}{D_\varphi} \varphi^2,$$

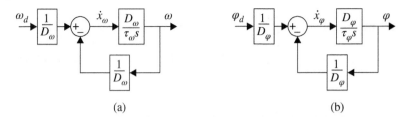

(a)   (b)

Figure 21.6 Blocks Σ_ω and Σ_φ implemented with the integral controller. (a) Σ_ω. (b) Σ_φ.

respectively. Then,

$$\frac{\partial H_\omega}{\partial x_\omega} = \omega, \quad \text{and} \quad \frac{\partial H_\varphi}{\partial x_\varphi} = \varphi.$$

As a result, Σ_ω can be written in the form of (21.29) with

$$J_\omega = 0, \quad R_\omega = \frac{1}{D_\omega}, \quad \text{and} \quad G_\omega = 1$$

while Σ_φ can be written in the form of (21.30) with

$$J_\varphi = 0, \quad R_\varphi = \frac{1}{D_\varphi}, \quad \text{and} \quad G_\varphi = 1.$$

In other words,

$$\Sigma_\omega : \quad \begin{cases} \dfrac{\tau_\omega}{D_\omega}\dot{\omega} = \left(0 - \dfrac{1}{D_\omega}\right)\omega + \dfrac{1}{D_\omega}\omega_d \\ \omega = \omega, \end{cases} \tag{21.36}$$

$$\Sigma_\varphi : \quad \begin{cases} \dfrac{\tau_\varphi}{D_\varphi}\dot{x}_\varphi = \left(0 - \dfrac{1}{D_\varphi}\right)\varphi + \dfrac{1}{D_\varphi}\varphi_d \\ \varphi = \varphi. \end{cases} \tag{21.37}$$

Apparently, their transfer functions are indeed the same as those given in (21.35).

21.7.2 Using a Static Controller

The simplest implementation is to use the static controllers

$$\Sigma_\omega = 1 \text{ and } \Sigma_\varphi = 1. \tag{21.38}$$

This is equivalent to the implementation with an integrator having zero time constant in the previous subsection.

In this case, the control signal (21.20) becomes

$$e = \varphi_d \omega_d z. \tag{21.39}$$

Remark 16. The control law (21.39) can be regarded as a modified droop controller, where the frequency and flux channels are deliberately coupled to obtain the lossless interconnection (21.33).

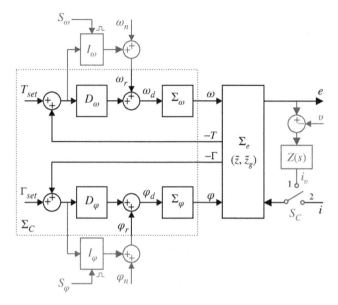

Figure 21.7 A cybersync machine equipped with regulation and self-synchronization.

21.8 Self-Synchronization and Power Regulation

The self-synchronization concept in Chapter 8 can be adapted and applied to achieve self-synchronization before being connected to the grid, as shown in Figure 21.7 with the synchronization structure easily recognizable.

A virtual impedance $Z(s)$ is added to generate a virtual current i_v before being connected to the grid, according to the voltage difference between e and v. Two integrator blocks I_ω and I_φ are introduced to force $T = T_{set}$ and $\Gamma = \Gamma_{set}$, respectively. If T_{set} and Γ_{set} are set as 0, then both T and Γ can be regulated to 0 when switch S_C is at position 1. This leads to $e = v$ and, hence, the converter is synchronized with the grid. Once the synchronization is achieved, the converter can be connected to the grid. When the converter is connected to the grid, switch S_C is thrown to position 2. Then, the converter can be operated to regulate T and Γ to the set points T_{set} and Γ_{set}, respectively, if the integrator blocks I_ω and I_φ are enabled by setting the signals S_ω and S_φ low. This operation mode is called the set mode, as introduced in (Zhong and Weiss, 2011; Zhong et al. 2014). It is equivalent to regulating the real power and reactive power around the set points $P_{set} = T_{set}\omega_n$ and $Q_{set} = \Gamma_{set}\varphi_n$. The system can also be operated in the droop mode for ω and φ if the integrator blocks I_ω and I_φ are reset by setting the signals S_ω and S_φ high, as described before. All the operation modes of the system are summarized in Table 21.1.

The integrator blocks I_ω and I_φ can be a simple integrator with a gain or a more complex block including an integrator. The gains should be small in order to make sure that the desired frequency ω_d and flux φ_d change more slowly than the tracking of the frequency and the flux. The virtual impedance $Z(s)$ can be chosen according to the guidelines provided in (Zhong et al. 2014) or Chapter 8 as a low-pass filter or other more complex impedance.

Table 21.1 Operation modes of the cybersync machine in Figure 21.7.

S_C	S_ω	S_φ	Mode
1	Low	Low	Self-synchronization
2	Low	Low	Regulation of T and Γ, i.e. P-mode and Q-mode
2	Low	High	Regulation of T, droop of φ, i.e. P-mode and Q_D-mode
2	High	Low	Droop of ω and regulation of Γ, i.e. P_D-mode and Q-mode
2	High	High	Droop of ω and φ, i.e. P_D-mode and Q_D-mode

Table 21.2 Parameters of the cybersync machine under simulation.

Parameter	Value	Parameter	Value
L_1, L_2	2.5 mH, 1 mH	P_{set}	400 W
R_1, R_2	0.5 Ω	Q_{set}	200 Var
C_f	22 μF	RL load R	54.45 Ω
V_n, f_n	110 V, 60 Hz	RL load L	48 mH
V_g, f_g	112V, 59.8 Hz	S_n	1 kVA

21.9 Simulation Results

The cybersync machine with the controller shown in Figure 21.7 is validated with computational simulations for the system shown in Figure 21.3. The system parameters are given in Table 21.2 with the control parameters determined as follows:

1. Droop of frequency: 1% of frequency variation for 100% of variation in T (hence in P) resulting in $D_\omega = \frac{1\%\omega_n}{T_n} = \frac{1\%\omega_n^2}{S_n} = 1.42$
2. Droop of flux: 5% of flux (voltage) variation for 100% of variation in Γ (hence in Q), resulting in $D_\varphi = \frac{5\%\varphi_n}{\Gamma_n} = \frac{5\%\varphi_n^2}{S_n} = 8.51 \times 10^{-6}$
3. Time constants for the controller blocks Σ_ω and Σ_φ: $\tau_\omega = 0.002$ s and $\tau_\varphi = 0.02$ s so that the frequency channel and the flux channel have different time scales
4. Gains of the integrators in the integrator blocks I_ω and I_φ: $k_\omega = 11.84$ and $k_\varphi = 4.32 \times 10^{-4}$, which are commensurate with D_ω and D_φ.

The converter has a rated power of $S_n = 1$ kVA. The set points for the torque and quorte are obtained from $T_{set} = \frac{P_{set}}{\omega_n}$ and $\Gamma_{set} = \frac{Q_{set}}{\varphi_n}$, respectively. The simulation results with both the static controller (static) and the integral controller (with IC) are shown in Figure 21.8. Both controllers perform very well. The static controller results in a faster response with larger overshoots than the integral controller, as expected, because the static controller is equivalent to the case of the integral controller with zero time constants τ_ω and τ_φ.

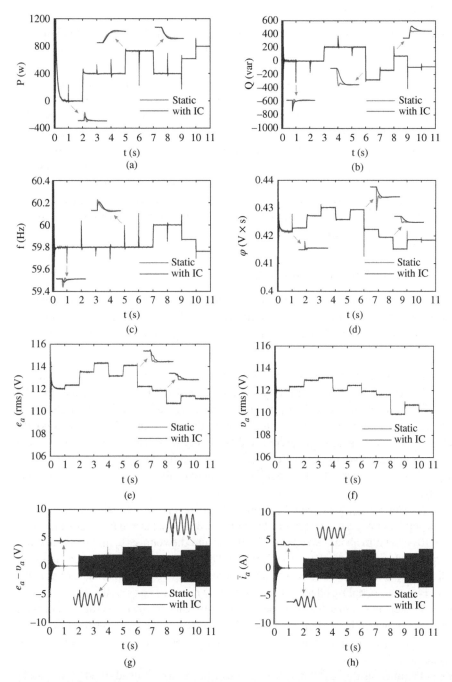

Figure 21.8 Simulation results from a cybersync machine, where the detailed transients indicated by the arrows last for five cycles with the magnitude reduced by 50%. (a) Active power P. (b) Reactive power Q. (c) Frequency f. (d) Flux φ. (e) Voltage synthesized by the converter. (f) Voltage on the load terminal. (g) Voltage difference $e_a - v_a$. (h) Current used to compute T and Γ.

21.9.1 Self-synchronization

This is mandatory before connecting the converter to the grid. As mentioned before, the self-synchronization mode is the same as the set mode for the converter to regulate T and Γ to 0 with the signals S_ω and S_φ set at low to enable the integrators but using the virtual current i_v. In order to show the robustness of the synchronization, the initial grid voltage has a phase angle of $+50°$, a frequency of 59.8 Hz, and a phase voltage of 112 V (rms value). As shown in Figure 21.8(h), the virtual current i_{va} quickly converges to 0 and so does the voltage difference $e_a - v_a$ as shown in Figure 21.8(g), with the converter voltage settling down at 112 V and 59.8 Hz. Note that for simplicity only phase a voltage and current are shown in Figure 21.8(g) and Figure 21.8(h), respectively. Also note that the current shown in Figure 21.8(h) is the combination of i_{va} until $t = 1$ s and i_a afterwards. The virtual current is large but it does not matter because it is not physical. There are also large oscillations in P, Q and f but this is not a problem either. The key is that the converter can achieve synchronization before being connected to the grid.

21.9.2 Operation after Connection to the Grid

After the converter achieves synchronization, it can be connected to the grid. In order to demonstrate the control framework, different scenarios, including operation in both the grid-connected mode with grid frequency and voltage changes and the islanded mode with load change, are tested. The results are shown in Figure 21.8 with details described below:

1. At $t = 1$s, the converter breaker is closed with T_{set} and Γ_{set} remaining at 0 to regulate T and Γ (hence P and Q) at 0 in the set mode. The physical current i is fed into the controller to calculate T and Γ. It can be seen from Figure 21.8(h) that there is a very small and short current transient, which means the grid connection is very smooth. Note that the relatively large spikes in P and Q are very short and can be ignored as long as there is no excessive current, which is the case. The filter capacitor is now supplied by the grid because i is regulated to be zero, which causes the voltages e_a and v_a to increase slightly.
2. At $t = 2$ s, step on T_{set} to achieve $P_{set} = 400$ W. The real power quickly settles down at the set point 400 W. As can be seen from Figure 21.8(h), the current gradually increases and settles down in about two cycles. The flux φ and the voltages e_a and v_a all increase in order to dispatch the required real power. The frequency f also increases in order to dispatch the increased real power but it returns to the grid frequency very quickly. There is a small coupling effect in Q but it quickly returns to 0 as well.
3. At $t = 3$ s, step on Γ_{set} to achieve $Q_{set} = 200$ var. The reactive power Q quickly increases and settles down at the set point 200 Var. The voltage further increases because of the increased reactive power. There is a small coupling effect in P but it returns to normal very quickly.
4. At $t = 4$ s, turn on the load breaker to connect the series RL load of 54.45 Ω and 48 mH (corresponding to 600 W and 200 Var). There is a small and short transient in the current i. There are some short spikes in P and Q but they quickly return to the original set points of 400 W and 200 Var because the converter is still working in the set mode. The real power and reactive power are not affected by the external condition. There is

also a small short spike in the frequency but it returns to normal very quickly as well because the grid frequency remains unchanged. The load voltage drops a bit because the load increases, which causes the flux to decrease a bit too.

5. At $t = 5$ s, the signal S_ω is changed to high to enable the frequency droop. Because the grid frequency 59.8 Hz is below the rated frequency 60 Hz, the converter increases the real power automatically by $\frac{0.2}{60} \times \frac{1000}{1\%} = 333$ W to 733 W, attempting to regulate the grid frequency. However, the grid frequency is fixed at 59.8 Hz so the frequency f quickly increases and returns back to 59.8 Hz. The current i increases accordingly. The reactive power is still operated in the set mode so the reactive power quickly returns to 200 Var after a small transient due to the coupling effect. The load voltage increases slightly because the real power dispatched increases.

6. At $t = 6$ s, the signal S_φ is changed to high to enable the flux (voltage) droop. Because the voltage e_a is about 114 V, which is above the rated voltage 110 V, the converter decreases the reactive power automatically to regulate the voltage to a value determined by the system parameters. For the system under simulation, the reactive power is reduced by 500 Var to −300 Var, with the voltage e_a settling down at 112 V. The load voltage v_a drops accordingly. The frequency is still maintained by the grid at 59.8 Hz and the real power remains more or less unchanged. Because of the change in the reactive power, the current increases slightly.

7. at $t = 7$ s, the grid frequency is changed to 60 Hz. The frequency f quickly tracks the grid frequency and settles down at 60 Hz. Because the converter is working in the frequency droop mode, the real power reduces automatically by $\frac{0.2}{60} \times \frac{1000}{1\%} = 333$ W from 733 W to 400 W, the original set point for the real power. This causes the voltage e_a to drop a bit, which then causes the reactive power to recover (increase) a bit.

8. At $t = 8$ s, the grid voltage is changed to the rated value at 110 V. The voltage v_a drops below 110 V, which causes the voltage e_a to drop too. As a consequence, the reactive power increases. The real power and the frequency remain more or less unchanged after a short transient.

9. At $t = 9$ s, the grid is changed back to 59.8 Hz and 112 V and the main breaker is opened to operate the converter in the islanded mode. The change of the grid frequency and voltage does not matter. What matters is the change of the operation from the grid-connected mode to the islanded mode. As can be seen from Figure 21.8(h), the transition is very smooth, with the current smoothly increasing to supply the full load. The real power increases by 200 W to 600 W, which causes the frequency to drop by $1\% \times \frac{200}{1000} \times 60 = 0.12$ Hz to 59.88 Hz (from 60 Hz). The reactive power of the filter capacitor is about −300 Var. Adding the load reactive power about 200 Var, the reactive power is about −100 Var. Taking into account the reactive power of the filter inductor, the total reactive power is about −90 Var, which is consistent with the value shown in Figure 21.8(b). Hence, the reactive power reduces, which causes the voltages e_a and v_a and the flux to increase slightly.

10. At $t = 10$ s, an additional 200 Ω load is connected to each phase (corresponding to 180 W). The real power increases quickly and the frequency drops by $1\% \times \frac{180}{1000} \times 60 = 0.108$ Hz to 59.772 Hz. As expected, the change in the reactive power is very small and so are the changes of flux φ and voltages e_a and v_a.

Table 21.3 Parameters of the experimental cybersync machine.

Parameter	Value	Parameter	Value
L_1, L_2	2 mH	P_{set}	100 W
C_f	22 µF	Q_{set}	100 var
V_n, f_n	120 V, 60.5 Hz	S_n	200 VA
f_{sw}	20 kHz	V_{DC}	200 V
f_s	10 kHz		

As can be seen from the zoomed-in details of the transients, the transients are very fast and often settle down within two cycles even for the IC. The frequency and the flux only change within very small ranges. The torque T and the quorte Γ look very similar to the real power P and the reactive power Q, respectively, apart from at different scales, and hence are not shown. It is worth noting that the load voltage v_a is well maintained around the rated value during the whole process although it is not directly controlled.

21.10 Experimental Results

The cybersync machine with the controller shown in Figure 21.7 is validated with a single-phase inverter connected to the grid. The system parameters are given in Table 21.3 with the control parameters determined as follows:

1. Droop of frequency: 1% of frequency variation for 100% of variation in T (hence in P) resulting in $D_\omega = \frac{1\%\omega_n}{T_n} = \frac{1\%\omega_n^2}{S_n} = 7.225$
2. Droop of flux: 10% of flux (voltage) variation for 100% of variation in Γ (hence in Q), resulting in $D_\varphi = \frac{10\%\varphi_n}{\Gamma_n} = \frac{10\%\varphi_n^2}{S_n} = 9.965 \times 10^{-5}$
3. Time constants for the controller blocks Σ_ω and Σ_φ: $\tau_\omega = 0.001$ s and $\tau_\varphi = 0.02$ s so that the frequency channel and the flux channel have different time scales
4. Gains of the integrators in the integrator blocks I_ω and I_φ: $k_\omega = 0.7225$ and $k_\varphi = 1.993 \times 10^{-4}$, which are commensurate with D_ω and D_φ.

The converter has a rated power of $S_n = 200$ VA. The set points for the torque and quorte are obtained from $T_{set} = \frac{P_{set}}{\omega_n}$ and $\Gamma_{set} = \frac{Q_{set}}{\varphi_n}$, respectively. Note that the grid frequency was about 60.7 Hz during the experiment. Moreover, because it is a single-phase converter, the ripples in T and Γ were filtered out through low-pass filters.

21.10.1 Self-synchronization

The self-synchronization mode was triggered by a manual switch on the inverter and the curves are shown in Figure 21.9(a). The voltage difference $e - v$ quickly converged to 0. Note that the synchronization started at about 120°, which is close to the worst case, but the synchronization was achieved in about 13 cycles. This is acceptable in practice.

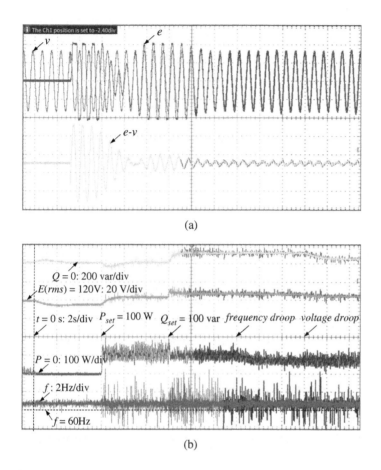

(a)

(b)

Figure 21.9 Experimental results from a cybersync machine. (a) Around synchronization. (b) After synchronization.

21.10.2 Operation after Connection to the Grid

After the converter achieved synchronization, it was connected to the grid, with the time marked as $t = 0$ s. Then different scenarios in the grid-connected mode were tested, with the results shown in Figure 21.9(b). The details are described below:

1. At $t = 0$s, the PWM signals were enabled with T_{set} and Γ_{set} remaining at 0 to regulate T and Γ (hence P and Q) at 0 in the set mode. The current was switched from the virtual current to the physical current. There was some small transient in Q due to these but P and Q were regulated to zero as expected.
2. At $t = 6$ s, step on T_{set} to achieve $P_{set} = 100$ W. The real power quickly increased and settled down at the set point 100 W. There was a small coupling effect in Q but it quickly returned to 0 as well. Note that in order to send more power to the grid, the voltage E increased as well to 123 V.
3. At $t = 12$ s, step on Γ_{set} to achieve $Q_{set} = 100$ Var. The reactive power Q quickly increased and settled down at the set point 100 Var. The voltage further increased to about 127 V

because of the increased reactive power. There was a small coupling effect in P but it returned to normal very quickly.

4. At $t = 18$ s, the signal S_ω was changed to high to enable the frequency droop. Because the grid frequency was at about 60.6 Hz, which is above the rated frequency 60.5 Hz, the converter decreased the real power automatically by $\frac{0.1}{60} \times \frac{200}{1\%} = 33$ W. The voltage dropped slightly. The reactive power was still operated in the set mode so the reactive power remained at 100 Var.

5. At $t = 24$ s, the signal S_φ is changed to high to enable the flux (voltage) droop. Because the voltage was about 124 V, which is above the rated voltage 120 V, the converter decreased the reactive power automatically by about $\frac{4/120}{10\%} \times 200 = 67$ Var.

21.11 Summary

Based on (Zhong, 2017a; Zhong and Stefanello, 2017, 2018), a new theoretical control framework to ensure the passivity of the controller of a power electronic converters is presented by using the PH theory and the ghost operator. Different passive control blocks can be adopted in the control framework and two examples, one with the integral controller and the other with a static controller, are given. Furthermore, a self-synchronization strategy has been embedded into the control framework to achieve synchronization without a dedicated synchronization unit before connecting to the grid. This also makes it possible to operate the converter in the set mode to regulate the real power and reactive power to the set points. It is worth highlighting that there is no assumption on the voltage, relaxing a common assumption needed for most control techniques.

Under the conjecture that the plant pair that consists of the plant and its ghost is passive the whole system is passive. It is expected that this will motivate further research to advance the 3G VSMs.

Part V

Case Studies

22

A Single-node System

In this chapter, a single-node system equipped with a SYNDEM smart grid research and educational kit is described. The kit is reconfigurable to obtain 10+ different topologies, covering DC/DC conversion and single-phase/three-phase DC/AC, AC/DC, and AC/DC/AC conversion, so it is ideal for carrying out research, development, and education of SYNDEM smart grids. It adopts the widely used Texas Instruments (TI) C2000 ControlCARD and is equipped with the automatic code generation tools of MATLAB®, Simulink®, and TI Code Composer Studio™ (CCS), making it possible to quickly turn computational simulations into physical experiments without writing codes.

The single-node system is equipped with 2G VSM technology and additional functions so that it can autonomously start from black, regulate voltage and frequency, detect the presence of the public grid, self-synchronize with the grid, connect to the grid, detect the loss of the grid, and island it from the grid.

22.1 SYNDEM Smart Grid Research and Educational Kit

22.1.1 Overview

In addition to the technical challenges discussed in previous chapters, the paradigm shift of power systems also imposes another unprecedented challenge: the shortage of a highly skilled workforce in the power industry worldwide. There is a strong need of speeding up the process of educating and training next-generation talented engineers and leaders, in particular, those equipped with a broad range of expertise and hands-on skills in advanced controls, power electronics, and power systems. This requires universities worldwide to revise the power engineering curriculum and increase the amount of hands-on sessions with the shortest learning curve and the highest learning efficiency. The SYNDEM smart grid research and educational kit was developed to facilitate this.

The kit is featured by MathWorks® as a reconfigurable power electronic converter for research and education in smart grids[1]. The kit can be reconfigured to obtain 10+ different power electronic converters, covering DC/DC converters and single-phase/three-phase DC/AC, AC/DC, and AC/DC/AC converters. As a result, it can be used to quickly set up research, development, and education platforms for different applications, such as solar

---

[1] https://www.mathworks.com/products/connections/product_detail/syndem-smart-grid-kit.html

Power Electronics-Enabled Autonomous Power Systems: Next Generation Smart Grids,
First Edition. Qing-Chang Zhong.

power integration, wind power integration, machine drives, energy storage systems, and flexible loads. Moreover, there is no need to spend time on coding because it adopts the widely-used Texas Instrument (TI) C2000 ControlCARD and is equipped with the automatic code generation tools of MATLAB®, Simulink®, and TI Code Composer Studio™ (CCS), making it possible to generate experimental results within hours from simulations. It comes with complete interface details and sample implementations, based on which users can easily test their own control algorithms.

The main features of the kit include

- Reconfigurable to obtain 10+ different power electronic converter topologies
- Capable of directly downloading control codes from Matlab/Simulink
- Ideal for research in smart grid, microgrid, renewable energy, EV, energy storage etc.
- Compatible with utilities around the world with 120 V or 230 V voltage, 5A current
- Versatile communication interfaces, such as RS485 and CAN, for SCADA
- Multiple DAC channels for easy debugging and monitoring of internal states
- Suitable for parallel, grid-tied, or islanded operation

22.1.2 Hardware Structure

The kit consists of one control board and up to two power boards. Figure 22.1 shows a picture of the kit with one power board. The control board is on top of the power board and auxiliary power supplies are located beneath the power board. The TI C2000 ControlCARD is inserted at the back of the control board. The control board has four switches for users to define the functions.

22.1.2.1 Power Board
The power board contains a three-leg IGBT module A1P35S12M3 and its driver circuits, relay, jumper wire connectors, current sensors, voltage sensors, inductors, capacitors, and

Figure 22.1 A photo of the SYNDEM smart grid research and educational kit.

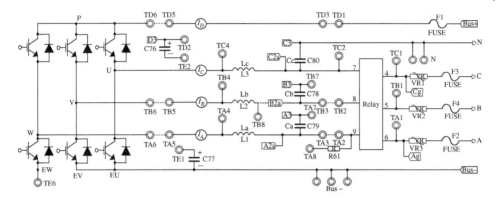

Figure 22.2   SYNDEM smart grid research and educational kit: main power circuit.

fuses. The power board accepts PWM signals from the control board through J7/J4 and transfers analog voltage/current signals to the control board through J2/J3. A diagram of the main power circuit is shown in Figure 22.2. The IGBT module A1P35S12M3 contains six 1200 V 35A devices in three legs with a common positive bus. It has a built-in NTC temperature sensor for thermal protection. There are two DC-bus 470 μF 450 V capacitors, C76 and C77, which can be configured in series or in parallel to meet the requirement of the voltage rating or the capacitance. There are three sets of inductors L1, L2, and L3 and capacitors C78, C79, C80, with two sets reconfigurable. Moreover, inductor L2 and capacitor C78 can be placed freely. A relay is included to facilitate the connection/disconnection with the grid. All the inductor currents and capacitor voltages as well as the DC-bus current and voltage are measured and sent to the control board. In addition, two voltages on the grid side are measured to facilitate grid connection. The power board is also equipped with AC fuses F2, F3, and F4, DC fuse F1, NTC thermistors VR1, VR2, and VR3 for protection and a pre-charging resistor R61. The power board is also equipped with a standalone 1600 V 35 A three-phase diode bridge with a 470 μF 450 V capacitor, making it possible to power the DC bus from an external transformer.

22.1.2.2   Control Board

The control board includes a TI C2000 TMS320F28335 ControlCARD, signal condition-ing circuits for AC signals, 22-channel 12 bit ADC, 4-channel 12 bit DAC, four switches, four LEDs, SPI, RS485 and CAN interfaces, protection circuits, and PWM circuits. Up to two power boards can be connected to one control board. This makes it possible to carry out experiments for two three-phase converters with one controller, e.g., the back-to-back converters in PMSG and DFIG WPGS.

22.1.3   Sample Conversion Topologies Attainable

The kit can be reconfigured into different converter topologies for various applications, such as solar power, wind power, energy storage systems, motor drives, electric vehicles, and flexible loads. Some sample topologies are outlined below.

22.1.3.1 DC–DC Converters

DC–DC converters are used to change the voltage level of a DC source to another level that is suitable for the load connected. A DC–DC converter can be designed to increase the voltage, decrease the voltage, or both. The ratio between the output voltage and the input voltage is called the conversion ratio. When the conversion ratio is lower than 1, the converter is called a buck (step-down) converter; when the conversion ratio is higher than 1, the converter is called a boost (step-up) converter; when the conversion ratio can be higher and lower than 1, then the converter is called a buck-boost converter. The SYNDEM kit can be adopted to realize all these three types of DC–DC converters. Figure 22.3 shows some implementations.

22.1.3.2 Uncontrolled Rectifiers

Because of the diodes in the IGBT module, it is easy to implement uncontrolled rectifiers with the SYNDEM kit. Figure 22.4 shows the implementations of a single-phase half-wave rectifier, a single-phase full-wave rectifier, and a three-phase rectifier. It is also possible to

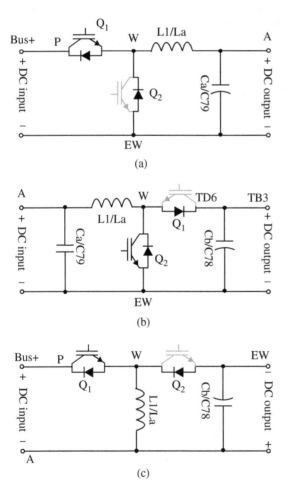

Figure 22.3 Implementation of DC–DC converters. (a) Buck (step-down) converter. (b) Boost (step-up) converter. (c) Buck-boost converter.

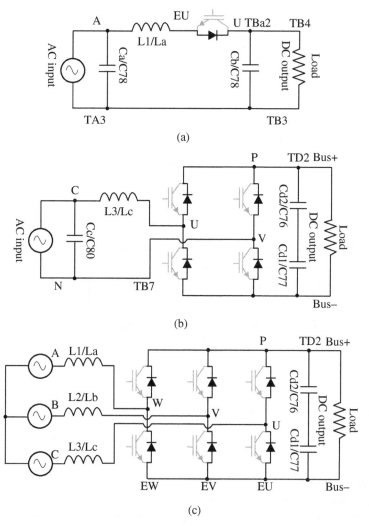

Figure 22.4 Implementation of uncontrolled rectifiers. (a) A single-phase half-wave rectifier. (b) A single-phase full-wave rectifier. (c) A three-phase rectifier.

use the standalone three-phase bridge rectifier for this purpose. When the system is configured as an uncontrolled rectifier, no PWM control command is needed from the control board. A command signal can be given to operate the relay.

22.1.3.3 PWM-controlled Rectifiers

The SYNDEM kit can be configured to form a single-phase or three-phase PWM-controlled rectifier, as shown in Figure 22.5. The circuit configuration of these PWM-controlled rectifiers remains the same as the single-phase/three-phase uncontrolled rectifiers, with the main difference being that the switches are controlled with PWM signals from the DSP.

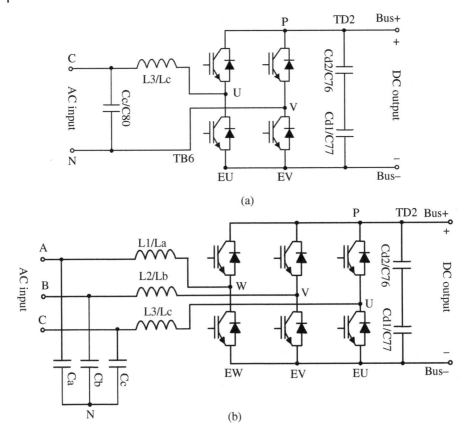

Figure 22.5 Implementation of PWM-controlled rectifiers. (a) A single-phase rectifier. (b) A three-phase rectifier.

22.1.3.4 θ-Converters

The SYNDEM kit can be used to implement other rectification topologies, for example, the θ-converter proposed in (Zhong and Ming, 2016). The configuration of the θ-converter is shown in Figure 22.6, which is a single-phase bridge converter that looks like the symbol θ. It has a common AC and DC ground, which reduces common-mode voltages and leakage currents. The DC-bus capacitor C77 provides a direct path for the double-frequency ripple current inherently existing in single-phase converters to return continuously. The output capacitor C78 only deals with switching ripples. Moreover, the DC-bus capacitor C77 is designed to store the system ripple energy with large voltage ripples. As a result, its capacitance can be reduced. The θ-converter offers more advantages than a conventional bridge converter because its two legs are controlled independently; see (Zhong and Ming, 2016) for details. The inductor L2/Lb can also be connected between P and Bus+ to form an improved θ-converter as proposed in (Ming and Zhong, 2017).

22.1.3.5 Inverters

An inverter converts an DC source into an AC output. According to the type of the DC supply, there are current-source inverters (CSI) and voltage-source inverters (VSI). Typically, an inverter is a VSI if there is a large capacitor across the DC bus and is a CSI if there is

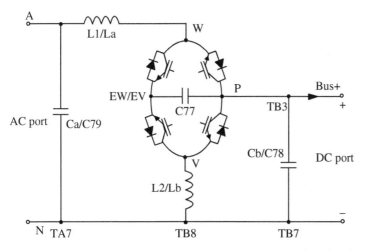

Figure 22.6   Implementation of the θ-converter.

a large inductor in series with the DC supply. The SYNDEM kit can be configured as a VSI for single-phase and three-phase applications, as shown in Figure 22.7. Depending on the control algorithm implemented, the output of the inverter can be current-controlled or voltage-controlled. Both control methods can be implemented with the SYNDEM kit because there are sensors to measure the output voltage and the output current. The inverter can be operated in the islanded mode or in the grid-connected mode. If it is operated in the grid-connected mode, the voltage on the grid side is available for synchronization. Once it is synchronized, the relay can be turned ON.

22.1.3.6   DC–DC–AC Converters
Some applications require DC–DC and DC–AC two-stage conversion. For example, a solar power system often needs a DC–DC converter on the PV side to step up the voltage before conversion into AC with a DC–AC converter. The SYNDEM kit can be configured to implement this DC–DC–AC conversion, as shown in Figure 22.8. Two separate controllers but inside the same ControlCard are needed for the DC–DC converter and the DC–AC inverter, respectively.

22.1.3.7   Single-phase AC–DC–AC Back-to-Back Converters
The SYNDEM kit can be used to implement AC–DC and DC–AC back-to-back conversion. As shown in Figure 22.9, the implementation only needs three legs, with a common neutral point provided by the θ-converter. The three legs are controlled independently.

22.1.3.8   Three-phase AC–DC–AC Back-to-Back Converters
Three-phase applications, such as wind power systems, frequency converters, and machine drives, often require three-phase back-to-back converters. One SYNDEM kit plus an additional power board can be used to form such a three-phase back-to-back converter, as shown in Figure 22.10. Both power boards can be controlled through one control board.

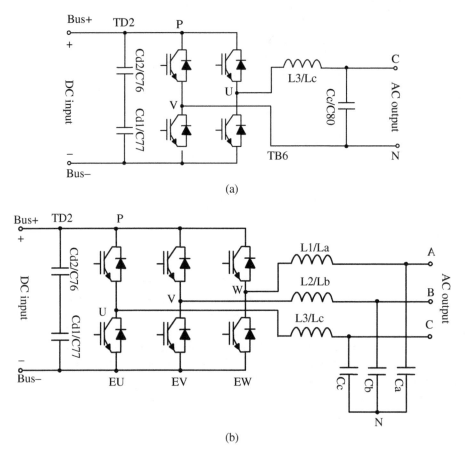

(a)

(b)

Figure 22.7 Implementation of inverters. (a) A single-phase inverter. (b) A three-phase inverter.

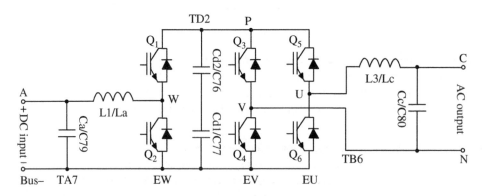

Figure 22.8 Implementation of a DC–DC–AC converter.

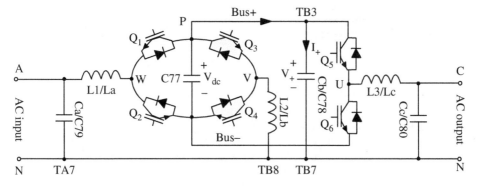

Figure 22.9 Implementation of a single-phase back-to-back converter.

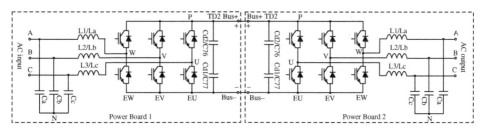

Figure 22.10 Implementation of a three-phase back-to-back converter.

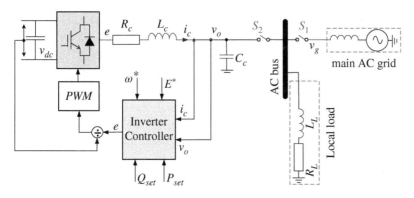

Figure 22.11 Illustrative structure of the single-node system.

22.2 Details of the Single-Node SYNDEM System

22.2.1 Description of the System

A single-node SYNDEM system is constructed with the SYNDEM kit, with the configuration shown in Figure 22.11. It is a single-phase inverter fed by a DC source. The inverter is connected through an external switch S_2 to a local AC bus, which is connected to the grid through another external switch S_1. A local load is connected to the local AC bus.

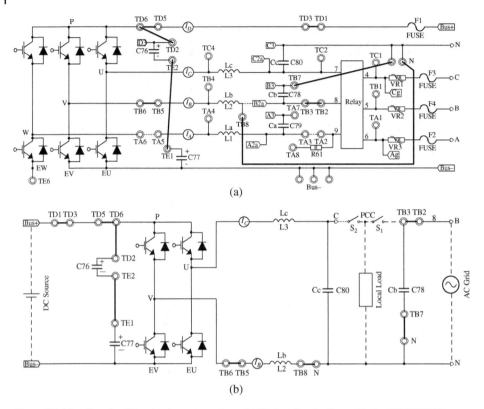

(a)

(b)

Figure 22.12 Circuit of the single-node system. (a) Wiring illustration with the SYNDEM kit power circuit. (b) Schematic diagram.

The wiring diagram for the configuration is shown in Figure 22.12, which uses four on-board voltage and current sensors: the voltage sensor across C78 to measure the AC grid voltage; the voltage sensor across the capacitor C80 to measure the output voltage v_o; a current sensor to measure the current flowing through inductor L3; and a voltage sensor to measure the voltage between Bus+ and Bus−. The wiring procedures are to:

- Connect Bus+: TD1–TD3 and TD5–TD6
- Connect the DC-bus capacitor: TD6–TD2 and TE2–TE1
- Connect the neutral line: TB6–TB5, TB8–N and TB7–N
- Connect the phase line: TB3–TB2
- Connect the DC-source: between Bus+ and Bus−
- Connect the AC grid: between B and N
- Insert switches S_1 and S_2: between C and TB3
- Connect the local load: between PCC and N.

The 2G VSM, i.e. the self-synchronized universal droop controller, discussed in Chapter 18, together with additional functions such as islanding detection (Amin et al. 2018), are implemented in MATLAB/Simulink and directly downloaded to the DSP for execution.

22.2.2 Experimental Results

Figure 22.13 shows the experimental results covering black-start, connecting local loads, grid appearance and detection, synchronizing and connecting to the grid, changing the operation condition, loss of the grid, islanding detection, etc., with details described below:

(1) The VSM started from black at about $t = 4$ s: the inverter voltage v_o was established smoothly. The reactive power Q decreased from 0 because of the filter capacitor and, consequently, the frequency f decreased slightly. Because there was no voltage present on the grid side, the relay was turned ON to pass the generated voltage to the terminal C.

(2) The switch S_2 was turned ON to connect a local load about 290 W at about $t = 8$ s: the output current i_o and the real power P increased while the inverter voltage v_o decreased slightly because of the load effect. Note that P and Q are calculated according to the filter inductor current instead of the output current i_o at the terminal.

(3) The grid voltage v_g appeared at about $t = 12$ s: the grid current i_g became non-zero because there was a capacitor C78 on the grid side. The inverter detected the presence of the grid and the islanding detection signal ID changed from 1 (no grid) to 0 (grid present). The VSM autonomously started synchronizing with the grid, without causing noticeable changes in v_o and i_o. The frequency f increased and became close to the grid frequency.

(4) The switch S_1 was turned ON to connect the system to the grid after reaching synchronization, at about $t = 16$ s: there was no noticeable change in the inverter voltage v_o or the grid voltage v_g, meaning that the grid connection was seamless. The grid took some load over, causing the output current i_o to decrease and the grid current i_g to increase. Note that P was higher than $P_{set} = 100$ W because the load drew some power from the grid and the inverter voltage was lower than the rated voltage.

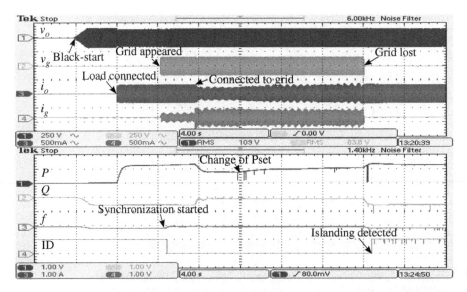

Figure 22.13 Experimental results from the single-node system equipped with a SYNDEM smart grid research and educational kit.

(5) The real power set point P_{set} of the VSM was increased to $P_{set} = 200$ W at about $t = 20$ s: there was no noticeable change in v_o. The output current i_o and real power P increased accordingly. Less power was drawn from the grid and P became close to P_{set}.

(6) The grid was lost at about $t = 32$ s: the full load was reverted back to the VSM and i_g (and v_g) became 0. The VSM detected the islanding after a short period of delay, as designed, and disconnected from the grid. There was no noticeable change in the inverter voltage v_o, meaning that the load was not affected by the loss of the grid.

The whole process was completed autonomously without communication or a PLL. The system was able to start from black and the load was not affected under random events, such as appearance of grid, change of real-power set-point, and loss of grid.

22.3 Summary

A single-node system is described after introducing the reconfigurable SYNDEM smart grid research and educational kit. It can be operated in the islanded mode and the grid-connected mode. The system is able to start from black and to regulate the voltage and the frequency. It is able to detect the appearance of the grid and start synchronization with the grid automatically. It is also able to detect the loss of the grid and then disconnect from the grid when islanding is detected. The various conversion topologies attainable from the SYNDEM kit are also briefly introduced, making it ideal to facilitate the deployment of SYNDEM smart grids through efficient research, development, and education.

Because of the versatile reconfigurability of the kit, it can be adopted to quickly build up physical smart grid testing facilities. Figure 22.14 shows such a SYNDEM microgrid facility established at Texas Tech University.

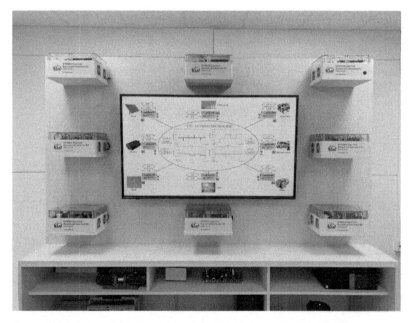

Figure 22.14 Texas Tech SYNDEM microgrid built up with eight SYNDEM smart grid research and educational kits.

23

A 100% Power Electronics Based SYNDEM Smart Grid Testbed

In this chapter, a 100% power electronics based SYNDEM smart grid testbed with eight nodes of VSMs connected to the same AC bus is described to demonstrate the operation of a SYNDEM smart grid. Experimental results show that the SYNDEM smart grid framework is very effective and all the VSM nodes, including wind power, solar power, DC loads, and AC loads, can work together to regulate the SYNDEM grid frequency and voltage, without relying on an ICT system.

23.1 Description of the Testbed

23.1.1 Overall Structure

Figure 23.1 illustrates the world's first SYNDEM smart grid testbed. It is a single-phase 120 V 60 Hz grid with eight nodes of VSMs connected to its AC-bus, i.e. the oval line in the middle of Figure 23.1(a). There is nothing else directly connected to its AC bus. Hence, this SYN-DEM grid is 100% power electronics-based. The eight nodes of VSMs include two nodes of wind power, two nodes of solar power, two nodes of energy bridges for grid integration, one node of AC loads, and one node of DC loads. Additionally, there are four more power electronic converters: two to simulate two wind turbine generator systems and two to simulate two strings of solar panels, respectively. Thus, there are in total 12 power electronic converters on the panel. The function of the two energy bridges is to exchange power between the SYNDEM grid and the utility grid or disconnect the SYNDEM grid from the utility grid. As a result, the SYNDEM grid can operate in the grid-tied mode or in the islanded mode. The SYNDEM grid can autonomously regulate its frequency and voltage without the need of any communication network. In order to monitor the operation of the SYNDEM grid, an RS485 network is set up. This corresponds to the case when the dashed arrow in the overall SYNDEM architecture shown in Figure 2.5 is not used, which enhances security.

23.1.2 VSM Topologies Adopted

In general, each node, whether supplying or consuming, needs a DC/AC converter on the AC-bus side of the SYNDEM grid, which is operated as a 2G VSM. Some nodes also need another stage of conversion cascaded to the DC-bus, which can be a DC/DC converter or an AC/DC converter depending on the type of the node. Because the testbed is a single-phase

Power Electronics-Enabled Autonomous Power Systems: Next Generation Smart Grids,
First Edition. Qing-Chang Zhong.
© 2020 John Wiley & Sons Ltd. Published 2020 by John Wiley & Sons Ltd.

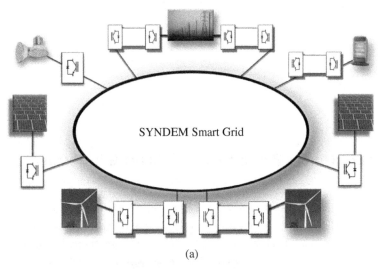

(a)

(b)

Figure 23.1 Illinois Tech SYNDEM smart grid testbed. (a) System structure. (b) Photo of the testbed.

system, each VSM needs to have a neutral line. Both the θ-converter proposed in (Zhong and Ming, 2016) and the Beijing converter proposed in (Zhong et al. 2017) can provide an independently controlled neutral line and are adopted in the testbed to implement the VSMs.

23.1.2.1 θ-Converter

As shown in Figure 23.2 or Figure 22.5, a θ-converter (Zhong and Ming, 2016) consists of two legs: one conversion leg and one neutral leg with only four switches. The conversion leg consists of switches Q_1 and Q_2, inductor L_g, and DC-bus capacitor C. The neutral leg

Figure 23.2 Topology of a
θ-converter.

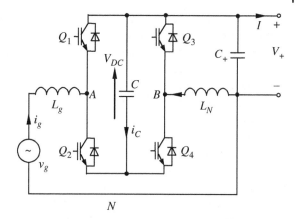

consists of switches Q_3 and Q_4, inductor L_N, and capacitor C_+. The DC-bus capacitor C is to store the ripple energy in the converter while the capacitor C_+ only needs to filter out the switching ripples. Hence, both capacitors can have significantly reduced capacitance.

The θ-converter offers more advantages than a conventional bridge converter because its two legs are controlled independently. The conversion leg is responsible for regulating the power exchanged with the grid (or the AC-bus) and the neutral leg is mainly responsible for regulating the DC voltage V_+ while providing a neutral point N. The θ-converter can be operated in the rectification or inversion mode, without any restriction on the power factor. More details about θ-converters can be found in (Zhong and Ming, 2016).

23.1.2.2 Beijing Converter
Figure 23.3 shows the topology of a Beijing converter (Zhong et al. 2017). It is formed by adding a small auxiliary capacitor C_- into a conventional full-bridge converter between the neutral line N and the negative pole of the DC bus. Similar to the θ-converter, the four switches are arranged to form one conversion leg and one neutral leg. The conversion leg

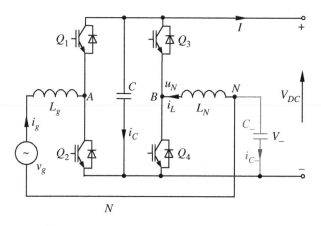

Figure 23.3 Topology of a Beijing converter.

consists of switches Q_1 and Q_2, inductor L_g, and DC-bus capacitor C. The neutral leg consists of switches Q_3 and Q_4, inductor L_N, and capacitor C_-. The two legs are also independently controlled, with the ripple energy diverted to and stored in the auxiliary capacitor C_-. The voltage V_- is allowed to vary in a wide range so the capacitor C_- can be chosen small. Moreover, the DC-bus capacitor is only needed to deal with switching ripples so the capacitance can be small too. More details about Beijing converters can be found in (Zhong et al. 2017).

Note that it does not really matter whether the capacitor C_- is connected to the negative pole of the DC bus or the positive pole of the DC bus as in a θ-converter. The major difference of the Beijing converter from the θ-converter lies in how the control strategy diverts the ripple energy. The Beijing converter maintains the whole DC-bus voltage with small ripples with the ripple energy stored in C_- so it is more suitable for applications that require a higher DC voltage. The θ-converter maintains V_+, which is only a part of the DC-bus voltage, with small ripples with the ripple energy stored in C so it is more suitable for applications that require a lower DC voltage. Both converters are able to provide an independently controlled neutral point N and are bidirectional. Obviously, more conversion legs can be added if needed. For example, two more conversion legs can be added to form a three-phase converter with an independently controlled neutral line.

23.1.3 Individual Nodes

23.1.3.1 Energy Bridges

The two energy bridge nodes connect/disconnect the utility grid and the SYNDEM grid. Each energy bridge here is a single-phase back-to-back converter, which consists of a Beijing converter connected to the SYNDEM grid and an additional conversion leg connected to the utility grid, as shown in Figure 23.4. The conversion leg of the Beijing converter and the additional conversion leg share a common neutral line provided by the Beijing converter.

The conversion leg of the Beijing converter is controlled as a 2G VSM while the conversion leg for the utility grid is controlled with a PI controller to regulate the DC-bus voltage $V_{DC} = V_+ - V_-$, where V_+ and V_- are the voltages of the positive and negative poles of the DC bus with respect to the neutral line N. Any power imbalance due to the regulation of the SYNDEM grid voltage and frequency is met by the utility grid.

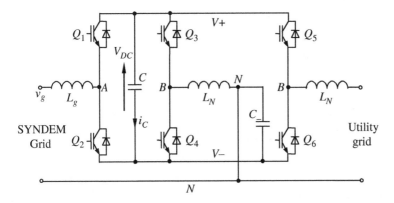

Figure 23.4 Back-to-back converter formed by a Beijing converter and a conversion leg.

23.1.3.2 Solar Power Nodes

In this system, no physical solar panels are used. Instead, each solar power node is connected to a solar simulator, which is a DC/DC converter that generates a DC output voltage of around 200 V from a 48 V DC power supply, according to the $V-I$ characteristics of some solar panels. Both solar simulators are mounted on the panel.

Each solar power node contains a DC/DC converter and a θ-converter cascaded through the DC-bus. The θ-converter is connected to the SYNDEM grid and is controlled as a 2G VSM. Because of the limitation in the hardware, no MPPT algorithm is implemented but this does not affect the system operation.

23.1.3.3 Wind Power Nodes

Similarly, no physical wind turbines are used. Instead, each wind power node is connected to a wind simulator, which is a single-phase AC/DC/AC back-to-back converter that generates an AC output voltage according to the $V-I$ characteristics of some wind power generation systems. Both wind simulators are mounted on the panel.

Each wind power node is a single-phase AC/DC/AC back-to-back converter. It has a topology similar to the one used in energy bridges, as shown in Figure 23.4, which consists of a Beijing converter connected to the SYNDEM grid and an additional conversion leg connected to a wind simulator instead of the utility grid. The Beijing converter is controlled as a 2G VSM and the additional conversion leg is controlled by a PI controller to maintain the DC-bus voltage. An MPPT control algorithm is adopted to generate the real power reference for the 2G VSM. If the wind profile in the wind simulator changes, the real power generated will change accordingly.

23.1.3.4 DC-Load Node

The DC-load node adopts the θ-converter shown in Figure 23.2, with a purely resistive DC load connected to the DC output voltage V_+. The DC-bus capacitor C serves the purpose of buffering the power imbalance between the VSM and the load, with the control strategy discussed in Chapter 19.

23.1.3.5 AC-Load Node

The AC-load node adopts the topology shown in Figure 23.5, which is a θ-converter cascaded with an additional conversion leg to generate an AC voltage for AC loads. Similar

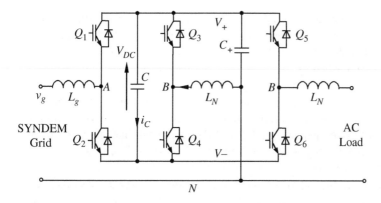

Figure 23.5 Back-to-back converter formed by a θ-converter and a conversion leg.

to the DC-load node, the DC-bus capacitor C serves the purpose of buffering the power imbalance between the VSM and the load, adopting the control strategy discussed in Chapter 19. In this way, the operation of the load is not affected by short-term disturbances in the SYNDEM grid. The additional conversion leg can be controlled in other ways to accommodate longer-term disturbances if needed.

23.2 Experimental Results

Some experimental results from the testbed are shown below to demonstrate the autonomous operation of a SYNDEM smart grid without relying on a communication network for frequency and voltage regulation. The droop coefficients are set as 10% voltage change corresponding to 100% change of real power and 0.1% frequency change corresponding to 100% change of the reactive power. Each VSM is rated at 300 VA.

23.2.1 Operation of Energy Bridges

Each of the energy bridges can start from black to establish the SYNDEM grid voltage according to the droop function. The energy bridges are designed to minimize the power exchange between the SYNDEM grid and the utility grid, i.e. with a set point of 0 W.

The experimental results are shown in Figure 23.6. The system is started at around $t = 20$ s. The SYNDEM grid voltage V_g (RMS) is established smoothly, reaching the steady-state value at around $t = 22$ s as designed. Since there is no load or supplies connected to the SYNDEM grid, the power exchanged with the utility grid is about 0 W, as shown in Figure 23.6(a). The SYNDEM grid voltage V_g is very close to 120 V according to the droop coefficient. The frequency f is about 59.98 Hz because the reactive power is about -100 Var, due to the filter of the VSM. As can be seen from Figure 23.6(b), the black start of the energy bridge is very smooth, with the SYNDEM grid voltage established without much overshoot. There is no excessive current either. Figure 23.6(c) shows the zoomed-in view of the steady-state SYNDEM grid voltage and the energy bridge current. The voltage quality is very good.

23.2.2 Operation of Solar Power Nodes

Figure 23.7(a) shows the responses of the solar VSM and Figure 23.7(b) shows the responses of the the energy bridge when a solar power node is integrated to the SYNDEM grid. The PWM signals of the solar VSM are enabled at around $t = 7$ s after it synchronizes with the SYNDEM grid voltage established by the energy bridge. After a short transition, the solar VSM regulates the real power and reactive power in the set mode to the power references $P_{set} = 0$ and $Q_{set} = 0$ well. The energy bridge draws about 25 W to cover the losses while receiving the reactive power 200 Var from the filler capacitors, making the SYNDEM grid voltage about 119 V and the frequency f about 59.96 Hz, as shown in Figure 23.7(b). At around $t = 16$ s, the solar VSM P_{set} is changed to 80 W while it is still operated in the set mode, and P increases and reaches 80 W. Since more power from solar is injected into the SYNDEM grid, the energy bridge stops drawing power from the utility grid and starts injecting power (50 W) into the utility grid. As a result, the SYNDEM grid voltage V_g increases.

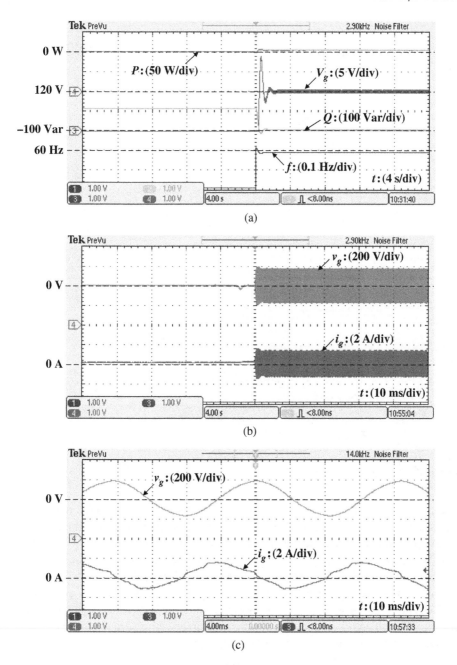

Figure 23.6 Operation of the energy bridge to black start the SYNDEM grid. (a) P, Q, V_g and f. (b) Instantaneous SYNDEM grid voltage v_g and current i_g. (c) Zoomed-in v_g and current i_g.

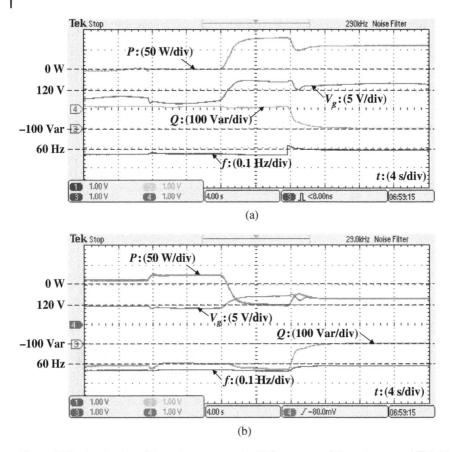

Figure 23.7 Integration of the solar power node. (a) Responses of the solar power VSM. (b) Responses of the energy bridge VSM.

This leads to a small change in the reactive power of the energy bridge, which makes the frequency change slightly. At around $t = 24$ s, the solar VSM enters into the droop mode. The reactive powers of both the solar VSM and the energy bridge change to about -100 Var, which pushes up the frequency from 59.96 Hz to 59.98 Hz, close to the rated frequency. The real powers of the solar VSM and the energy bridge also reduce and reach a new balance, pushing the voltage down and closer to the rated voltage 120 V because the real power reference of the energy bridge is set as $P_{set} = 0$. Note that the solar VSM and the energy bridge have slightly different voltages and real power because of the voltage drop on the cable and also measurement errors, but the difference is acceptable.

23.2.3 Operation of Wind Power Nodes

Figure 23.8(a) shows the responses of the wind VSM and Figure 23.8(b) shows the responses of the energy bridge when a wind power node is integrated to the SYNDEM grid. The PWM signals of the wind VSM are enabled at around $t = 8$ s after it synchronizes with the SYNDEM grid voltage established by the energy bridge. After a short transition, the wind VSM

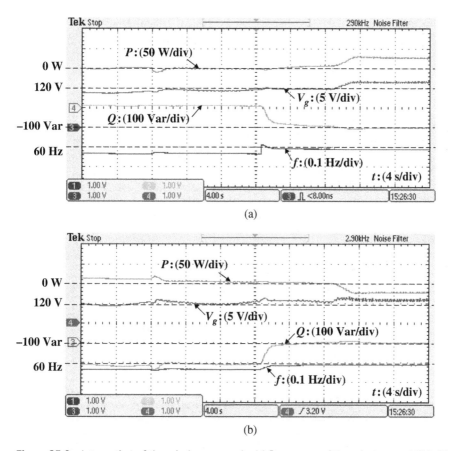

Figure 23.8 Integration of the wind power node. (a) Responses of the wind power VSM. (b) Responses of the energy bridge VSM.

continues working in the set mode with power references $P_{set} = 0$ and $Q_{set} = 0$. Indeed, as shown in Figure 23.8(a), the real power and the reactive power are controlled to be around 0, which means the energy bridge has to cover the losses of the system and the reactive power of the filters, etc. Indeed, the reactive power of the energy bridge is about -200 Var, making the frequency to be about 59.96 Hz. At around $t = 21$ s, the VSM is changed to operate in the droop mode. There is not much change in the $P \sim V_g$ loop because V_g is at around the rated value. However, the reactive powers of the wind VSM and the energy bridge all change to -100 Var, which leads to a change in the frequency f from 59.96 Hz to 59.98 Hz. At around $t = 29$ s, the MPPT function is enabled and the real power P increases to about 30 W, which pushes up the voltage to about 121.5 V and forces the energy bridge to inject real power to the utility grid, as shown in Figure 23.8(b).

In order to better understand the operation of the wind power node, Figure 23.9(a) shows the voltage and frequency of the wind simulator output when the wind speed S_w changes from 11.5 m s^{-1} to 10.5 m s^{-1} in a sinusoid manner and Figure 23.9(b) shows the performance of the wind VSM. The real power P generated by the wind VSM follows the wind speed S_w well. The SYNDEM grid voltage V_g also follows the trend in P but stays

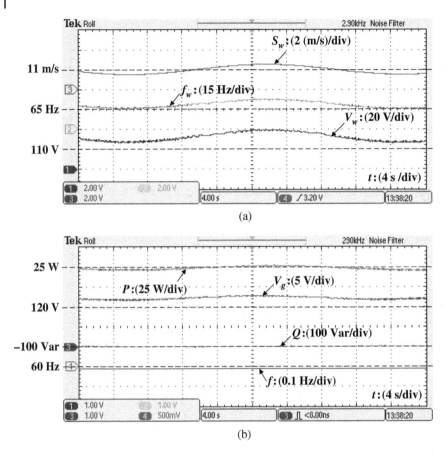

Figure 23.9 Performance of the wind power node when the wind speed S_w changes. (a) Wind simulator performance. (b) Wind VSM performance.

within the designed range. There is not much change in the frequency and the reactive power.

23.2.4 Operation of the DC-Load Node

Figure 23.10(a) shows the responses of the DC-load VSM and Figure 23.10(b) shows the responses of the energy bridge when the DC-load node is integrated to the SYNDEM grid. The DC-load node is connected to the SYNDEM grid at about $t = 7$ s in the droop mode, drawing real power and reactive power with visible transients to establish the DC-bus voltage. Indeed, the real power drawn by the energy bridge from the utility grid has a spike and the grid voltage drops and then recovers, as normally seen when connecting a load. After it settles down, the DC-load VSM consumes about 20 W with the terminal voltage being about 116 V. The real power of the energy bridge is about 40 W, which covers the losses in the SYNDEM grid, and the corresponding SYNDEM grid voltage is around 119 V. Note that the voltage difference between the energy bridge and the DC-load VSM is due to the voltage drop on the cable (and measurement errors). The DC-load node and the energy bridge have

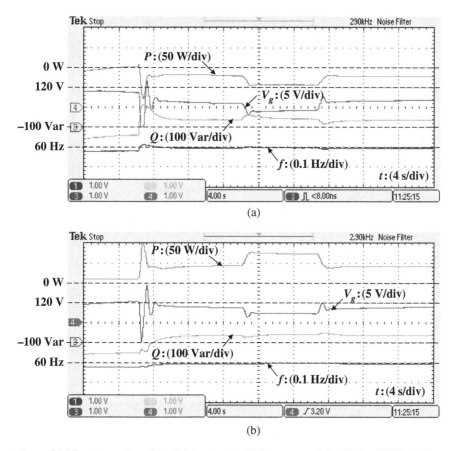

Figure 23.10 Integration of the DC-load node. (a) Responses of the DC-load VSM. (b) Responses of the energy bridge VSM.

similar reactive power of −60 Var, with the frequency settling down at around 59.99 Hz. At about $t = 18$ s, a 1200 Ω resistor is connected to the DC bus of the DC-load node. The DC output voltage of the node is regulated at 200 V, meaning that the load power is $\frac{200^2}{1200} = 33$ W. Adding the losses, the real power of the DC-load VSM is about 50 W. The real power of the energy bridge is about 75 W, which is consistent. The SYNDEM grid voltage is expected to drop by $\frac{75}{300} \times \frac{10\%}{100\%} \times 120 = 3$ V from 120 V. Indeed, as shown in Figure 23.10(b), the voltage is about 117 V. At about $t = 27$ s, the 1200 Ω resistor is removed. The system settles down after some transients, with the SYNDEM frequency being around 59.99 Hz. Note that the rated voltage of the DC-load VSM is set at 115V instead of 120V.

23.2.5 Operation of the AC-Load Node

Figure 23.11(a) shows the responses of the AC-load VSM and Figure 23.11(b) shows the responses of the energy bridge when the AC-load node is connected to the SYNDEM grid. The AC-load node is connected to the SYNDEM grid at $t = 1$ s in the droop mode with the PWM signal for the rectifier leg enabled, drawing real power and reactive power with

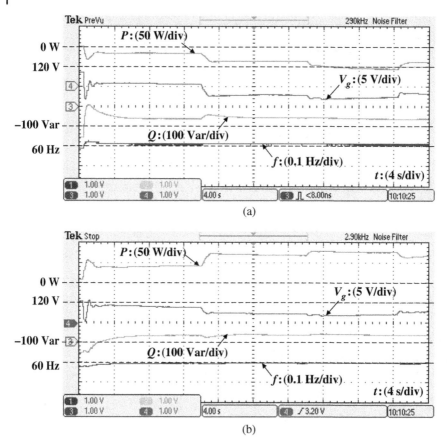

Figure 23.11 Integration of the AC-load node. (a) Responses of the AC-load VSM. (b) Responses of the energy bridge VSM.

visible transients to establish the DC-bus voltage, as expected. The node consumes about 20 W while the energy bridge provides about 40 W to the SYNDEM grid. The difference of 20 W is due to losses in the line as well as some measurement errors. The output voltage of the energy bridge is about 119 V while the voltage of the AC-load node is 116 V. The AC-load node and the energy bridge have similar reactive power of −60 Var, with the frequency settling down at around 59.99 Hz. At about $t = 14$ s, the PWM signal for the inversion leg is enabled to build up the output AC voltage for the AC load. As shown in Figure 23.11(a), the AC-load VSM consumes about 40 W, with the additional power coming from the losses of the inversion leg and the LC filter on the AC output side. At about $t = 27$ s, a 500 Ω resistor is connected to the output of the inversion leg. Indeed, the real power drawn by the AC-load VSM increased to around 60 W, which corresponds to the power consumed by the AC load, which is about $\frac{110^2}{500} = 24$ W because the load voltage is around 110V. The real power of the energy bridge increases to about 80 W. At about $t = 36$ s, the AC load is disconnected and the system returns to the original operation. Note that the frequency is maintained very well because the reactive power does not change much while the voltage changes according to the real power, as designed. The more real power the AC-load VSM draws, the lower the voltage (but still within the designed range).

23.2.6 Operation of the Whole Testbed

Figure 23.12(a) illustrates the behavior of the first energy bridge when the whole testbed is in operation. This energy bridge is set up at about $t = 70$ s to black start the SYNDEM grid and establish the voltage in the droop mode with its power references set at zero. It is clearly shown that the SYNDEM grid voltage is maintained well at 120 V. The reactive power is about -100 Var, making the frequency about 59.98 Hz.

At about $t = 142$ s, the second energy bridge is connected to the SYNDEM grid. Because there is a soft-start resistor, the first energy bridge provides about 55 W power to the SYN-DEM grid. The reactive power is about -150 Var, making the frequency about 59.97 Hz. When the soft-start resistor is bypassed at about $t = 168$ s, the real power of the first energy

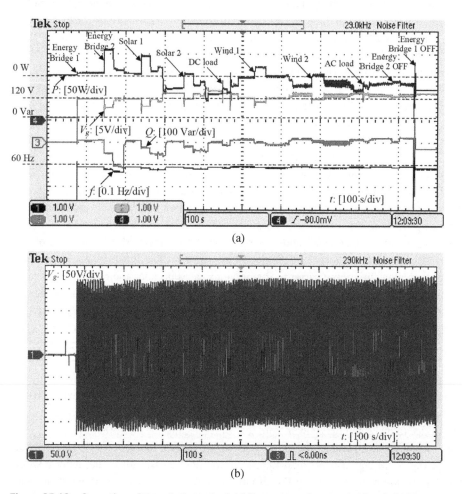

(a)

(b)

Figure 23.12 Operation of the whole testbed. (a) Responses of energy bridge 1. (b) The instantaneous voltage of the SYNDEM grid.

bridge reduces to around the original value. The reactive power further reduces to slightly over −200 Var as expected because the filter capacitor is now fully connected. This brings the frequency to 59.96 Hz, which follows the droop coefficient designed. At about $t = 200$ s, the PWM signal of the second energy bridge is enabled. Both energy bridges work in parallel in the droop mode, without any problems in maintaining voltage and frequency. The reactive power changes to −100 Var because the the second energy bridge also shares the reactive power needed by the filters, which brings the frequency to 59.98 Hz.

At about $t = 240$ s, the first solar node is connected to the SYNDEM grid. There is also a soft-start resistor at the output, which causes the real power of the energy bridge to increase. Because there are two energy bridges in parallel operation, the real power increase seen from the first energy bridge is less than that when the second energy bridge is connected. At about $t = 260$ s, the soft-start resistor is bypassed and the real power of the energy bridge drops. At about $t = 280$ s, the PWM signal of the first solar node is enabled in the set mode with its power reference set at 0. At about $t = 292$ s, its power reference is set at 80 W. Since there is no load connected to the SYNDEM grid, this real power is shared by the two energy bridges and injected to the utility. Indeed, there is a drop of 40 W real power in the first energy bridge, raising the voltage to about 122V. After 8 s, the solar node is changed to operate in the droop mode. The real power delivered by the solar node is reduced slightly.

At about $t = 350$ s, the second solar node is connected to the SYNDEM grid. It has the same parameters as the first solar node. Note that during the soft-start process, the real power increase of the first energy bridge is smaller because now there are two energy bridges and one solar node to share the real power. The same is true for the reactive power. At about $t = 400$ s, its real power reference is set to 80 W. Accordingly, the real power of the first energy bridge drops further by about 40 W and the SYNDEM grid voltage increases further. When the droop mode is enabled for the second solar node after 8 s, the real power injected by the first energy bridge decreases more than the case when the first solar node changes the operation mode to the droop mode because the SYNDEM grid voltage is now higher.

At about $t = 450$ s, the DC-load node is connected to the SYNDEM grid. Because of the soft-start resistor, the first energy bridge injects less power to the utility grid but the amount changed is smaller with comparison to the previous cases because the real power is now shared by two energy bridges and two solar nodes. Moreover, the filter capacitor for the DC load is 10 μF instead of 20 μF, as in the energy bridges, solar nodes, and wind nodes. At about $t = 470$ s, its PWM signal is enabled. The first energy bridge injects less power to the grid because the loss of the DC-load node is increased. At about $t = 490$ s, a load is connected to its DC port, which pushes the real power exchanged with the utility grid to be close to zero and the SYNDEM grid voltage close to 120 V. Note that because of the smaller filter capacitor the reactive power of the first energy bridge (and other nodes) changes slightly.

At about $t = 530$ s, the first wind node is connected to the SYNDEM grid. The first energy bridge draws some real power from the utility because of the soft-start resistor and reduces its real power to about 0 when the resistor is bypassed at about $t = 560$ s. At about $t = 620$ s, the PWM signal is enabled to operate the node in the set mode. After a short period, its operation is changed to the droop mode with its power reference generated from the MPPT algorithm. The SYNDEM voltage increases due to the increased power generation.

At about $t = 680$ s, the second wind node is connected in a similar way. More real power is injected into the utility grid and the SYNDEM grid voltage is increased accordingly.

At about $t = 808$ s, the AC-load node is connected to the SYNDEM grid. The first energy bridge sends about 20 W to the utility grid again.

At about $t = 880$ s, the second energy bridge is turned OFF so only the first energy bridge is connected the utility grid. The real power sent to the utility grid by the first energy bridge reduces slightly.

At about $t = 940$ s, the first energy bridge is turned OFF to island the SYNDEM grid from the utility grid. Since Figure 23.12(a) illustrates the signals from the first energy bridge, the signals are lost. As can be seen from Figure 23.12(a), there are no visible changes in the SYNDEM grid voltage when the SYNDEM grid is islanded from the utility grid.

It is clear that the SYNDEM grid voltage and frequency are maintained stable within the ranges, respectively, during the whole process without relying on an ICT system when different generation facilities and loads are connected and even when the operation mode is changed from grid-tied to islanded. Note that no special efforts are made to calibrate the sensors to high accuracy but the experimental results are still very consistent.

23.3 Summary

Based on (Zhong and Lyu, 2019a), a 100% power electronics based SYNDEM smart grid testbed is described. Consisting of eight VSMs connected to the AC bus, it is able to start from black, autonomously regulate the frequency and the voltage without relying on an ICT system, and seamlessly connect and disconnect from the utility grid.

24

A Home Grid

In this chapter, a home grid based on the SYNDEM framework built up at the Llano River Field Station, Texas Tech University Center at Junction, Texas, is described. The home grid consists of five DER units including four 3 kW solar units and one 3 kW wind unit, with battery storage built in each unit, and one energy bridge for grid connection. The energy bridge has two VSMs connected back-to-back with a common DC bus. All the units are equipped with self-synchronized universal droop controller with many advanced functions, such as black start, grid forming, self-synchronization without a PLL, voltage and frequency regulation, power sharing, and power quality control. The home grid can work with or without the public grid, with seamless mode change between grid-tied operation and islanded operation. All the functions are achieved without using communication network and, hence, potential cyber-attacks through the home grid are completely avoided. Field operation results of the home grid are presented to demonstrate the autonomous operation of the system.

24.1 Description of the Home Grid

An aerial photo of the home at the Llano River Field Station, Texas Tech University Center at Junction, Texas, is shown in Figure 24.1. The home grid is designed according to the SYNDEM grid architecture, with its one-line diagram shown in Figure 24.2(a) and its backbone shown in Figure 24.2(b). It consists of four solar units, one wind unit, and one energy bridge. Each solar unit consists of two strings of 1.5 kW solar panels, a built-in battery pack, and a Syndem PV inverter. The rated voltage of each PV string is 162.5 V. Each battery pack contains 18 rechargeable lead-acid batteries (12 V @ 20 Ah each) in series. The battery packs store energy when the load demand is less than the generation and release energy when the load demand is more than the generation. The PV inverters extract the maximum electricity from the solar panels and collectively maintain the stability of the home grid, with the battery pack as a buffer. The wind unit has a similar design as the solar units but connecting to a 3 kW wind turbine. The energy bridge consists of two interconnected back-to-back inverters with a common DC bus and its main function is to behave as a system backup when the renewable resources and the battery storage are insufficient to power the home. The home grid can be connected to or disconnected from the public grid via the energy bridge. All the six units are coupled together on the AC side to power household loads.

Power Electronics-Enabled Autonomous Power Systems: Next Generation Smart Grids,
First Edition. Qing-Chang Zhong.
© 2020 John Wiley & Sons Ltd. Published 2020 by John Wiley & Sons Ltd.

Figure 24.1   The home field at the Texas Tech University Center at Junction, Texas.

Because the inverters have a single phase, an isolation transformer is adopted to generate a split-phase AC grid. The home grid can operate independently without the public grid, enhancing resilience.

The core control algorithm adopted is the self-synchronized universal droop controller described in Chapter 18. All the units collectively establish the voltage and the frequency of the home grid and maintain its stability, without relying on a communication network.

24.2   Results from Field Operations

The home grid reliably operates to power the home under different scenarios. Four typical cases are selected below to demonstrate the autonomous operation of the system.

24.2.1   Black start and Grid forming

The black start and grid forming capabilities are demonstrated by the results shown in Figure 24.3. The isolation transformer and some household loads, e.g., lights, laptops, etc. are initially connected to the home grid. At $t = 2$ s, PV inverter I starts to form the grid to supply electricity to household loads, as shown in Figure 24.3(a). The peak-to-peak output voltage of PV inverter I is around 340 V, which corresponds to 120 V RMS. The corresponding current of the PV inverter I is shown in Figure 24.3(a) as well. The current has some transients initially, mainly caused by the initialization of the isolation transformer, but it settles down quickly. At around $t = 6$ s, PV inverter II synchronizes with the home grid and then connects to it. After the connection of PV inverter II, the current of PV inverter I decreases, and the current of PV inverter II increases to a value similar to that of PV inverter

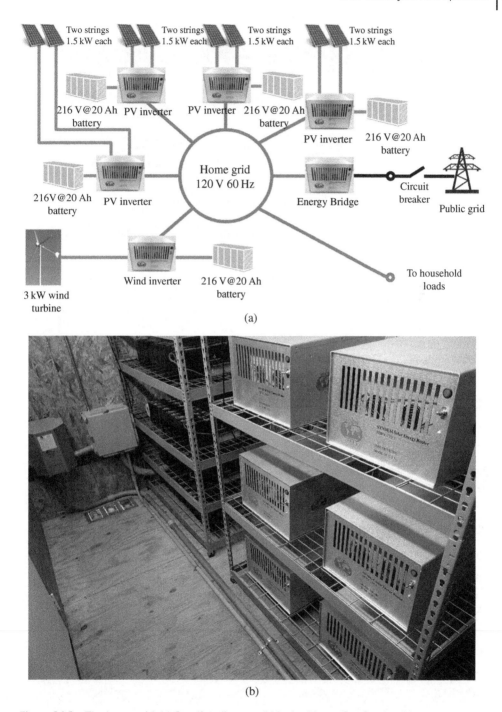

(a)

(b)

Figure 24.2 The home grid. (a) One-line diagram. (b) Its backbone: five Syndem inverters, one Syndem energy bridge, and battery packs.

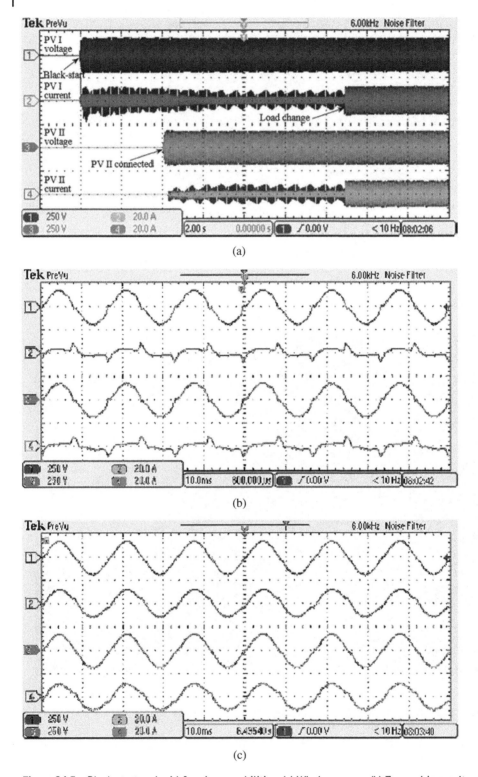

Figure 24.3 Black-start and grid-forming capabilities. (a) Whole process. (b) Zoomed-in results at around $t = 10$ s. (c) Zoomed-in results at around $t = 18$ s.

I because of the power sharing mechanism. Figure 24.3(b) shows the zoomed-in results at around $t = 10$ s. The two inverters have very consistent voltages and currents and work together to stabilize the grid. There are some harmonics in the output currents of both inverters, which is caused by the nonlinearity of the isolation transformer and the household loads. Note that both inverters share the harmonic currents well. At $t = 15$ s, a large resistive load is added to the system, the currents of both inverters increase, as shown in Figure 24.3(a). The zoomed-in results at around $t = 18$ s are shown in Figure 24.3(c). With the connection of the resistive load, the current quality is improved. The voltage quality is improved as well. Both inverters continue sharing the current well after the load change.

24.2.2 From Islanded to Grid-tied Operation

The connection of the energy bridge to the public grid is demonstrated with the results shown in Figure 24.4. The public grid is always ON. At $t = 10$ s, the energy bridge starts to

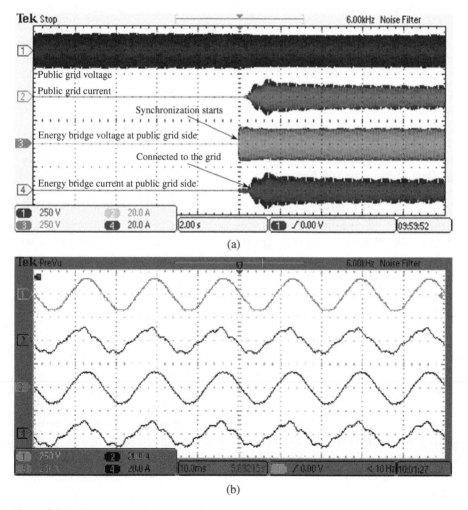

Figure 24.4 From islanded to grid-tied operation. (a) Whole process. (b) Zoomed-in results at around $t = 18$ s.

synchronize with the public grid. There is some initial current in the energy bridge current (i.e. the inductor current) during the synchronization process, as shown in Figure 24.4(a), which is caused by the filter capacitor. The energy bridge is connected to the public grid at about $t = 10.4$ s. Simultaneously, the energy bridge also connects and delivers power to the home grid, which causes the increase of the energy bridge current because some power is sent to the home grid. The corresponding public grid current increases as well. The zoomed-in results at around $t = 18$ s are shown in Figure 24.4(b). The energy bridge voltage at the public grid side is synchronized with the public grid voltage with almost the same phase angle, because of the synchronization mechanism of the VSM. The public-grid current is similar to the energy-bridge current because the filter capacitor is small.

24.2.3 Seamless Mode Change when the Public Grid is Lost and Recovered

The results are shown in Figure 24.5. Initially, the home grid is in continuous operation in the islanded mode with the public grid ON. At $t = 2$ s, the energy bridge starts to connect the home grid to the public grid and the home grid enters into the grid-tied operation mode. It absorbs power from the public grid and delivers power to the home grid, with the energy bridge output current at the home-grid side determined by the load on the home grid according to the droop function. The public grid current increases accordingly to keep the power balance of the common DC bus of the back-to-back converters. At about $t = 8$ s, the public grid is lost, and the energy bridge shuts down immediately. Both the public-grid current and the energy-bridge current become zero. The home grid enters into the islanded mode and is not affected by the loss of the public grid. When the public grid is recovered at about $t = 14$ s, the energy bridge is re-synchronized with the public grid and the home grid before connecting the home grid to the public grid autonomously. Figure 24.5(b) shows the zoomed-in results at around $t = 18$ s. The home-grid voltage is different from the public-grid voltage, which demonstrates the independent operation of the home grid from the public grid. The energy bridge output current has some harmonics, which are caused by the split-phase isolation transformer and the nonlinear loads in the home grid. The home grid can seamlessly change the operation mode between grid-tied and islanded modes and maintain independent operation regardless of the presence of the public grid.

24.2.4 Voltage/Frequency Regulation and Power Sharing

In this case, all renewable units work together to power the home grid, with the results shown in Figure 24.6. All data are sent out by the Syndem inverters and recorded by a personal computer through an RS485 channel. Initially, the home grid is connected to the transformer and some household loads. With the capabilities of the black start and the synchronization demonstrated in Section 24.2.1, PV inverter I starts to form the grid and other units are then connected one by one with PV inverter II at about $t = 6$ min, PV inverter III at about $t = 11$ min, PV inverter IV at about $t = 17$ min, and the wind inverter at about $t = 22$ min, as shown in Figure 24.6(a). All connected inverters can share the real power

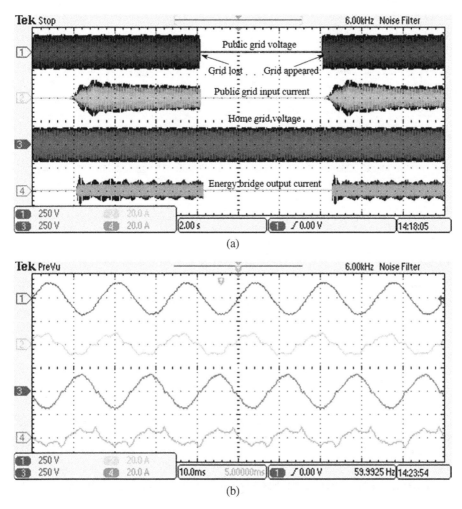

Figure 24.5 Seamless mode change when the public grid is lost and then recovered. (a) Whole process. (b) Zoomed-in results at around $t = 18$ s.

and reactive power well, as shown in Figure 24.6(a) and Figure 24.6(c). At about $t = 25$ min, an electric fan is turned ON, which causes the increase of the reactive power and the real power. At about $t = 38$ min, a large load is added to the system, which results in an increase in the real power and a decrease in the reactive power. It can be noticed that all connected inverters can still share the loads during the load change. The home-grid voltage and frequency are well regulated during the whole normal operation, as shown in Figure 24.6(b) and Figure 24.6(d), with voltage variations less than 1.8 V and frequency variations less than 0.005 Hz because all inverters work together to regulate both voltage and frequency collectively without relying on a communication network.

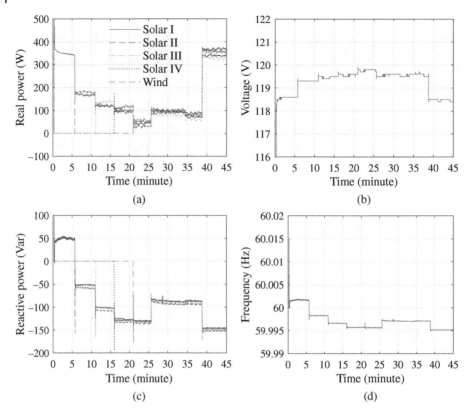

Figure 24.6 Power sharing and regulation of the voltage and frequency of the home grid. (a) Real power. (b) Voltage. (c) Reactive power. (d) Frequency.

24.3 Unexpected Problems Emerged During the Field Trial

The field trial also revealed some unexpected problems. There were three major problems: the extremely high nonlinearity of the transformer, the high nonlinearity of the household loads, and the large inrush current of the air-conditioning unit.

Figure 24.7(a) shows the inverter voltage and the current when there is only one inverter in operation and the transformer has no load connected. The current of the transformer presents very large harmonics, which normally results in high THD in the voltage. Even with such excessive high current harmonics, the Syndem inverters can maintain good voltage quality. With more inverters added into the system, the voltage quality is improved further. Figure 24.7(b) shows the case when two inverters are connected together. The voltage quality is improved and the two inverters share the current harmonics well.

Figure 24.8 shows the home grid voltage and the split-phase currents of the household loads. The current of the phase with the isolation transformer contains significant harmonics due to the transformer; the current of the phase without the transformer also contains some visible harmonics. Although the current quality is very low, the SYNDEM inverters are able to maintain good voltage quality with low THD.

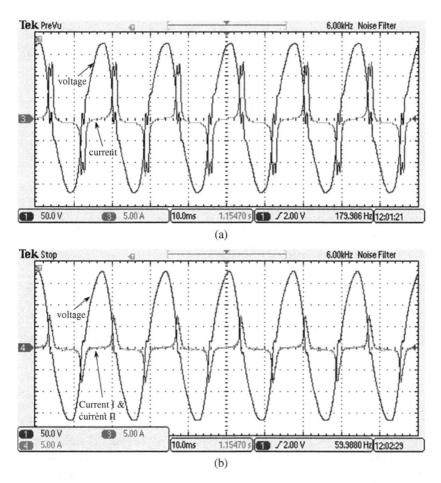

Figure 24.7 The nonlinearity of the transformer. (a) With one inverter. (b) With two inverters.

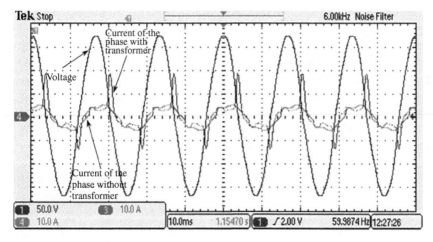

Figure 24.8 The nonlinearity of household loads.

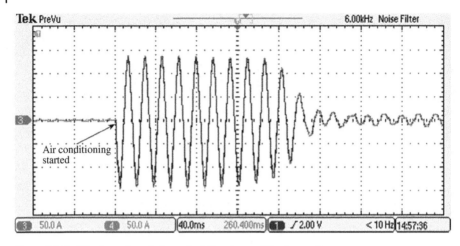

Figure 24.9 The large inrush current of the air-conditioning unit.

Figure 24.9 shows the large inrush current when turning ON the air conditioning unit, which is measured when connected to the public grid. This is far beyond the designed capacity of the home grid and current protection is triggered to avoid any damage when the unit is turned ON. An inrush current limiter is added to the air conditioning unit to suppress the inrush current.

24.4 Summary

Based on (Zhong et al. 2019), this chapter describes a home grid developed for a smart home at the Llano River Field Station, Texas Tech University Center at Junction, Texas, based on the SYNDEM grid architecture. The home grid contains four solar units, one wind unit, and one energy bridge. The autonomous operation of the home grid to achieve tight voltage and frequency regulation is demonstrated without relying on a communication network. Many advanced functions, such as black start, grid forming, self-synchronization without a PLL, voltage and frequency regulation, power sharing, power quality control, and power balance, have been demonstrated through field results. The home grid can operate with or without the public grid, with seamless mode change between grid-tied operation and islanded operation. Some unexpected problems, such as the high nonlinearity of the isolation transformer, the high nonlinearity of household loads, and the large inrush current of air-conditioning units, are described as well. The field trial demonstrated friendly integration of DERs without relying on communication networks, which improves grid stability, reliability, security, resiliency, and sustainability.

25

Texas Panhandle Wind Power System

The Texas Panhandle area, the northernmost region in the US State of Texas, has an abundance of wind power. A lot of wind farms are there. However, it suffers from a severe problem of exporting the wind power generated to load centers far from Panhandle. In this chapter, an attempt is made to illustrate how the SYNDEM smart grid architecture and its underpinning technologies could remove the export limit imposed on the wind farms in Panhandle so that they can export the wind power generated at the full capacity without causing problems to the grid. Note that, while every effort is made to make the study as close to the real situation as possible, many assumptions and simplifications have to be made. Further studies with high-fidelity models and details are needed to fully understand the problem.

25.1 Geographical Description

The Texas Panhandle, situated in the US State of Texas, is a rectangular area consisting of the northernmost 26 counties in the state. It is bordered by New Mexico to the west and Oklahoma to the north and east. The southern boundary of Panhandle ends with the line forming the southern boundary of Swisher County. The land area of Panhandle is 25,610 sq mi, which is nearly 10% of the state's total. There is an additional 62.75 sq mi of water area. Literally, it is the handle of the pan-shaped state of Texas.

The landscape of Panhandle consists of a flat surface, canyon, river, and lake. The high plains cover almost all areas of the panhandle except the undulating southeastern part from where the rolling plains begin. These two landscapes are separated by the high plains escarpment commonly called the Caprock. The upper tributaries of the Red River and the Canadian River cross the region. The Canadian goes across the high plains to isolate the southern part, the Llano Estacado. The surface of the Llano Estacado is flat, which is one of the world's flattest areas of such size. Palo Duro Canyon is situated in the south of the City of Amarillo. This canyon carved by the Prairie Dog Town Fork Red River is the second largest canyon in the United States. The Lake Meredith situated in the North of Amarillo is a reservoir created by Sanford Dam on the Canadian River. The lake, along with the Ogallala Aquifer, is the region's most valuable resource with an enormous store of relict water. It provides drinking water and irrigation for this moderately dry area of this high plains of Panhandle (Rathjen, 2018).

Power Electronics-Enabled Autonomous Power Systems: Next Generation Smart Grids,
First Edition. Qing-Chang Zhong.
© 2020 John Wiley & Sons Ltd. Published 2020 by John Wiley & Sons Ltd.

The unique landscape of the area creates a strong and steady wind all through the year. The Panhandle area has become one of the fastest-growing wind power-producing regions in the United States.

25.2 System Structure

The Texas State Legislature introduced the concept of a Competitive Renewable Energy Zone (CREZ) in 2005 to construct necessary transmission capacity to connect areas with abundant wind resources to more highly populated parts of the State. The Panhandle Renewable Energy Zone (PREZ), shown in Figure 25.1(a), is one of the five zones and has the highest wind capacity factor. Before the PREZ project, there was no transmission

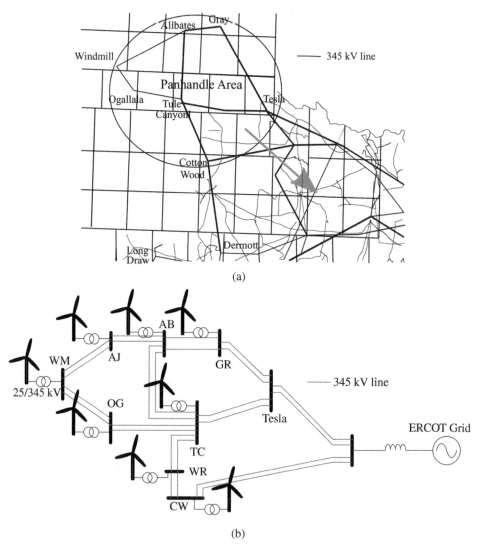

Figure 25.1 Panhandle wind power system. (a) Geographical illustration. (b) Wind farms in the Panhandle Renewable Energy Zone.

facility in the Panhandle area and no load or generation in the area was connected to the Texas main grid.

After the initiation of the CREZ project, the Electric Reliability Council of Texas (ERCOT) received many requests to build wind farms in the area. The number of wind projects with a signed interconnection agreement had reached 6 GW by October 2014, which exceeded the "Initial Build" capacity of 2.4 GW (ABB, 2010). The installed capacity of wind power in the panhandle reached 3997 MW with another 4022 MW expected to come online by 2020 (ERCOT, 2017). Figure 25.1(b) illustrates some major wind farms in Panhandle.

25.3 Main Challenges

The Panhandle grid requires a long-distance transmission line to the load center. There is no direct connection of the synchronous generator (SG) to the Panhandle grid either. Hence, the panhandle grid requires a large amount of reactive power support in order to export the wind power generated. Initially, the reactive power support was only installed for exporting 2400 MW wind power. Several studies conducted by ERCOT, LCG Consulting, and Sharyland have recommended employing reactive power compensation devices, such as synchronous condensers (SCs) and static var compensators (SVCs), and to construct new transmission lines and/or double the existing transmission lines in order to increase the power export capability (ERCOT, 2014).

In order to maintain the stability of the Texas grid, ERCOT introduced the Panhandle Export Limit to limit wind generation in the Panhandle. It had the second highest congestion rent on the ERCOT system in 2017 and is expected to become the constraint with the highest congestion. There is a pressing demand to adopt advanced technologies to fully release the export capacity of the transmission lines.

Some challenges of the panhandle grid as identified in (Schmall et al. 2015) include:

- Due to the long geographical distance from major SGs and load centers, the Panhandle grid is extremely weak for integrating a large amount of wind farms. It has very low short circuit ratios and high voltage sensitivity of dV/dQ, which requires special coordination of various complex control systems. Improper control could lead to undamped oscillations, over-voltage cascading or voltage collapse.
- The above mentioned compensation devices can improve voltage stability. However, in a highly compensated weak grid, voltage collapse can occur within the normal operating voltage range, e.g., 0.95 to 1.05 pu, resulting in voltage stability risks in real-time operations. Compensation devices such as static capacitors or SVCs contribute to this effect and may have limited effectiveness to further increase export capability.
- The Panhandle power system is dominated by power electronic converters, which are effectively connected to a common point of interconnection (POI). There are strong interactions among the wind farms in the area, which make the situation even worse.

25.4 Overview of Control Strategies Compared

The wind turbines are assumed to be full-scale type IV systems, as shown in Figure 25.2. Each unit employs a permanent-magnet synchronous generator (PMSG) to convert the wind power into AC electricity, a rotor-side converter (RSC) to convert the AC

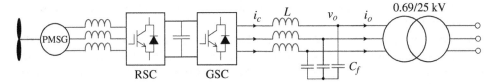

Figure 25.2 Connection of a wind power generation system to the grid.

Table 25.1 Parameters of the back-to-back converters.

Parameter	Value
Rated power	1.5 MVA
Rated AC voltage	690 V
Rated DC-link voltage	1300 V
Inverter side inductance	0.08 pu
Parasitic resistance	0.0025 pu
Filter capacitance	0.0185 pu

electricity into DC electricity, and a grid-side converter (GSC) to convert the DC electricity into AC electricity that is compatible with the grid. The RSC and the GSC are connected back-to-back to share a common DC-bus. Standard two-level VSCs are assumed for RSC and GSC. The parameters of the back-to-back converters are assumed to be as shown in Table 25.1.

25.4.1 VSM Control

The VSM control strategy adopted in this study is the self-synchronized robust droop control for inverters with inductive impedance, without using a PLL. Figure 25.3(a) shows the controller for the RSC and Figure 25.3(b) shows the controller for the GSC.

The main task of the RSC controller is to operate the RSC as a PWM rectifier to regulate the DC-bus voltage while the main task of the GSC controller is to export the power around the maximum power available to the grid while taking part in the regulation of the grid frequency and voltage. The real power reference P_{set} of the GSC controller can be set according to the maximum power from an MPPT algorithm.

The MPPT algorithm adopted in this study is outlined below.

When the speed of wind changes, a wind turbine should operate at its maximum power point. This can be achieved by maintaining the tip speed ratio to the value that maximizes its aerodynamic efficiency. The power obtained from the wind is

$$P_{\mathrm{w}} = \frac{1}{2}\rho\pi R^2 v_{\mathrm{w}}^3 C_{\mathrm{p}}(\lambda, \beta), \tag{25.1}$$

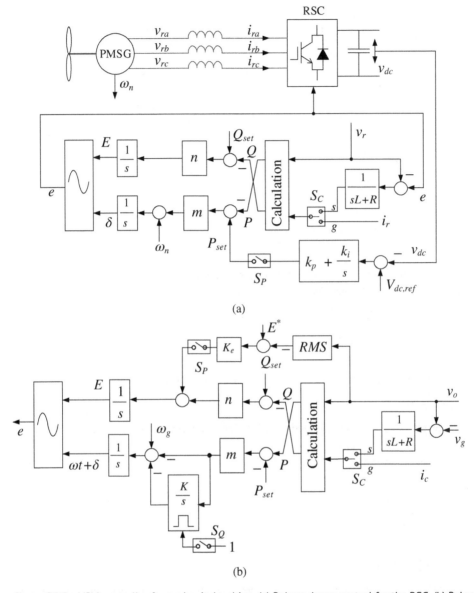

Figure 25.3 VSM controller for each wind turbine. (a) Robust droop control for the RSC. (b) Robust droop control for the GSC.

where ρ is the air density, R is the turbine radius, v_w is the wind speed and $C_p(\lambda, \beta)$ is the power coefficient depending on the blade angle β and the tip speed ratio λ defined as

$$\lambda = \frac{\omega_n R}{v_w} \tag{25.2}$$

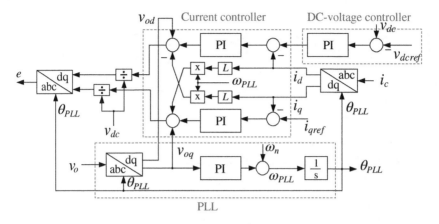

Figure 25.4 Standard DQ controller for the GSC.

with ω_n being the turbine rotor speed. For different wind speeds, the optimal values of the tip speed and the pitch angle need to be followed in order to achieve the maximum C_p and the maximum power output. Thus, the maximum power can be calculated as

$$P_w = \frac{1}{2}\rho\pi R^2 \left(\frac{R\omega_n}{\lambda_{opt}}\right)^3 C_{pm}. \tag{25.3}$$

Taking losses into account, in this study, the power that can be extracted is assumed to be

$$P_{set} = 0.95 P_w. \tag{25.4}$$

The power coefficient is calculated by using the following formula (Sun et al. 2003):

$$C_p(\lambda, \beta) = 0.22 \left(\frac{116}{\beta} - 0.4 - 5\right) e^{-\frac{12.5}{\beta}} \tag{25.5}$$

with

$$\beta = \frac{1}{\frac{1}{\lambda+0.08\lambda} - \frac{0.035}{\lambda^3+1}}.$$

25.4.2 DQ Control

The traditional control strategy for a wind turbine adopts a current controller in the DQ reference frame. In this study, the DQ controller shown in Figure 25.4 is adopted for the GSC. It includes a DC-bus voltage controller to maintain the DC-bus voltage through injecting the right amount of power to the grid. Its synchronization with the grid is achieved through a PLL.

The RSC controller is an MPPT controller to control the RSC current so that the maximum power available is extracted. No reactive power injection is considered here.

25.5 Simulation Results

Figure 25.5 shows the simulated Panhandle system. The total installed wind capacity is 4 GW and is connected to substations with different capacities, shown in Table 25.2. To simplify the system model, multiple wind units are lumped into one unit with the same generation capacity. For example, the White River wind farm has a capacity of 702 MW and

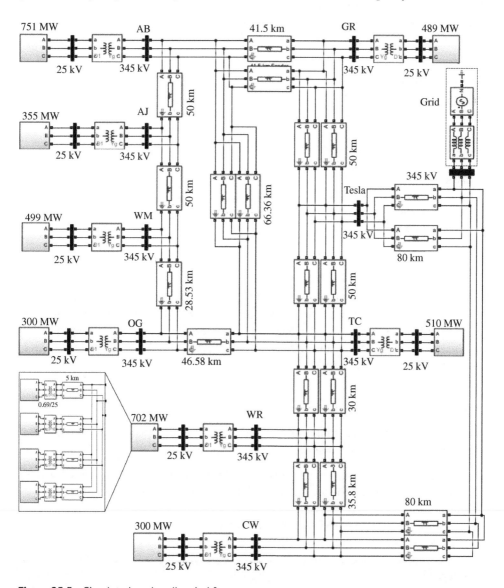

Figure 25.5 Simulated panhandle wind farms.

Table 25.2 Installed wind capacity in Panhandle.

Substation	Wind power capacity (MW)
Alibates (AB)	751
White River (WR)	702
Tule Canyon (TC)	510
Windmill (WM)	499
Gray (GR)	489
AJ Swope (AJ)	355
Cotton Wood (CW)	300
Ogallala (OG)	300

Table 25.3 Parameters of 345 kV transmission lines.

Parameter	Value
Resistance per km	0.02 Ω
Inductance per km	0.8532 mH
Capacitance per km	0.0135 μF

it is represented by 4 units with the capacity of 175.5 MW each. The wind turbine generators are connected to 25 kV AC collection buses through 0.69/25 kV, 60 Hz transformers and 5 km transmission lines and further onto the 345 kV grid with 25/345 kV transformers. The parameters of the 345 kV transmission lines are given in Table 25.3. No other reactive power support devices are included in the study.

25.5.1 VSM Control

Figure 25.6 shows the simulation results of one of the units under the VSM control in the droop mode, according to the following sequence of actions:

1. Start the system but keep all the IGBTs off with the initial wind speed of 12 m s^{-1} so that the RSC operates as a diode rectifier to establish the DC-bus voltage with a chopper DC load to precharge the DC-bus capacitor
2. The GSC is operated in the self-synchronization mode with $P_{set} = 0$ and $Q_{set} = 0$ to synchronize with the grid voltage with the circuit breaker OFF.
3. Start operating the IGBTs of the RSC at $t = 0.5$ s with $Q_{set} = 0$ to regulate the DC-bus voltage to 1300 V.
4. Start operating the IGBTs of the GSC at $t = 1$ s with $P_{set} = 0$ and $Q_{set} = 0$ and turn the circuit breaker ON to connect the GSC to the grid.
5. At $t = 2$ s, the power reference of the GSC is set to $P_{set} = 0.95P_w$ with $Q_{set} = 0$.
6. At $t = 3.5$ s, a 50 % grid voltage drop is applied, which is recovered after 5 cycles.

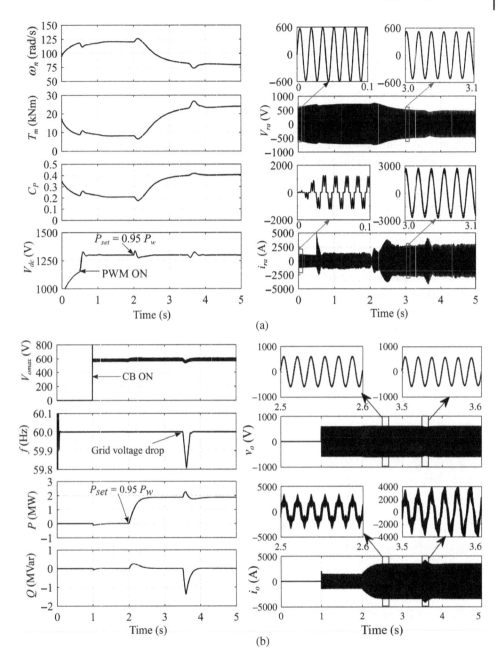

Figure 25.6 Simulation results from a single unit. (a) Dynamic response of the RSC. (b) Dynamic response of the GSC.

As shown in Figure 25.6, the unit works properly. When the PWM signals of the RSC are turned ON at $t = 0.5$ s, the DC-bus voltage is regulated well at 1300 V. When the circuit breaker CB is turned ON at $t = 1$ s to connect the GSC to the grid, the dynamics in P and Q is very small, which means the GSC is synchronized with the grid. Note that the grid-side current i_o becomes non-zero because of the filter capacitor. When the MPPT algorithm is started at $t = 2$ s, the wind turbine slows down but its torque increases while the power coefficient gradually reaches the maximum value and the real power sent to the grid gradually increases to the maximum power. The voltage increases slightly in order to export the power. There is a slight drop in the DC-bus voltage but its recovery is fast. When the voltage drops 50% at $t = 3.5$ s, there is a 0.2Hz frequency drop and a sudden increase in reactive power, as expected.

The same VSM controller is deployed in other wind turbines in the panhandle system shown in Figure 25.5 and the system is operated in the droop mode. The circuit breakers are turned ON at 1 s to connect all the GSCs to the grid with $P_{set} = 0$ and $Q_{set} = 0$. Assuming the maximum power from the wind turbines is available, the power P_{set} to be exported is gradually increased from 0% to 50%, 60%, ..., 100%. The resulting voltage, frequency, real power and reactive power at the 345 kV buses of the wind farms are shown in the left column of Figure 25.7. All the voltages at these buses are maintained close to 1 pu while the frequency is assumed to be maintained by the Texas grid at 60 Hz.

The results of a single unit in the Wind Mill wind farm are shown as an example in the left column of Figure 25.8. The dynamic response of reactive power has a noticeable feature. When the wind power unit does not export much real power, the transmission line

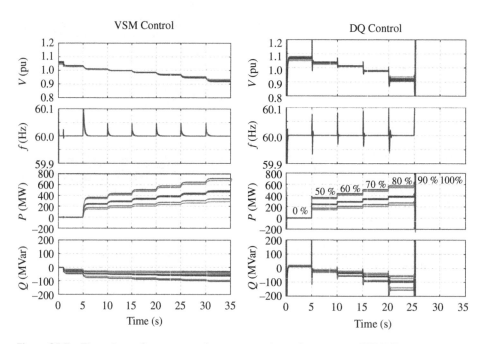

Figure 25.7 The voltage, frequency, active power, and reactive power at 345 kV buses.

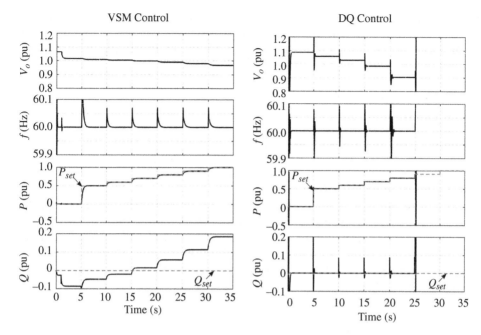

Figure 25.8 Panhandle wind power system: the voltage, frequency, active power, and reactive power at the PCC of a single unit at the Wind Mill wind farm.

generates reactive power due to the Ferranti effect and there is a relative voltage rise that is proportional to the square of the line length and the square of frequency. This pushes the voltage up to over 1.0 pu. Hence, the wind unit has to absorb reactive power, which is indeed the case, as shown in the left column of Figure 25.8. With the increase of the power exported, the voltage magnitude decreases. When the exported power reaches 70% of the capacity, the voltage drops below 1.0 pu and the wind power unit starts generating reactive power in order to boost the voltage up. The more the real power exported, the more the reactive power generated.

The frequency of the grid is assumed to stay at 60 Hz in the simulation; the wind farms delivered the amount of real power set by P_{set}, although in the droop mode.

The simulation demonstrates that VSM technology can reach 100% export capacity without causing voltage collapse.

25.5.2 DQ Control

The traditional DQ current controller discussed in Section 25.4.2 is deployed in all wind units in the panhandle power system shown in Figure 25.5. The simulation results following the same simulation procedure as done for the case with VSM control are shown in the right column of Figure 25.7 for different buses and in the right column of Figure 25.8 for a single unit in the Wind Mill wind farm. The system can export up to 80% power of the capacity and collapses when the power exported reaches 90%.

Table 25.4 Comparison of export capabilities under different control strategies.

Control strategy	Power exported without causing collapse
VSM control	100%
DQ control (PLL bandwidth: 9 Hz)	80%
DQ control (PLL bandwidth: 30 Hz)	60%

The system stability strongly depends on the PLL bandwidth. In the case discussed above, the PLL bandwidth is 9 Hz. If the PLL bandwidth is increased to improve the frequency tracking performance, the maximum power that can be exported actually goes down to 60% of the capacity. The VSM control does not use a PLL so this problem does not exist.

Figures 25.7 and 25.8 show that the VSM controller has much better capability of voltage and frequency regulation than DQ control. The export capabilities of the VSM control and the traditional DQ current control are summarized in Table 25.4, which shows clearly the advantages of the VSM technologies.

25.6 Summary and Conclusions

Based on (Amin and Zhong, 2019), the Texas Panhandle wind power system is described. The Panhandle area has an abundance of wind power but has suffered from the challenges of exporting limited wind power to the Texas main grid because of the long distance to major power plants and load centers, interactions among wind turbines in the same area, etc. Simulations have shown that adopting the SYNDEM theoretical framework and the VSM control technology could remove the export limit of power to reach 100% of export capacity while the traditional DQ current control technology can only reach 80% of export capacity when the PLL bandwidth is 9 Hz or 60% of export capacity when the PLL bandwidth is 30 Hz, demonstrating huge potential for the SYNDEM theoretical framework and its underpinning VSM technologies.

Bibliography

ABB 2010 *ERCOT CREZ Reactive Power Compensation Study*.

AEMO (2017). Black system South Australia 28 September 2016. Technical report, Australian Energy Market Operator. Available at https://www.aemo.com.au/-/media/Files/Electricity/ NEM/Market_Notices_and_Events/Power_System_Incident_Reports/2017/Integrated-Final-Report-SA-Black-System-28-September-2016.pdf.

Ahrens CD (2017). Transition to very high share of renewables in Germany. *CSEE Journal of Power and Energy Systems*. **3**(1), 17–25.

Akagi H, Watanabe E and Aredes M (2007). *Instantaneous Power Theory and Applications to Power Conditioning*. Wiley-IEEE Press.

Akhmatov V and Eriksen P (2007). A large wind power system in almost island operation: A Danish case study. *IEEE Trans. Power Syst.* **22**(3), 937–943.

Al-Nabi E, Wu B, Zargari NR and Sood V (2012). Input power factor compensation for high-power CSC fed PMSM drive using d-axis stator current control. *IEEE Trans. Ind. Electron.* **59**(2), 752–761.

Alamri BR and Alamri AR (2009). Technical review of energy storage technologies when integrated with intermittent renewable energy. *Proc. Int. Conf. Sustainable Power Generation and Supply*. pp. 1–5.

Ambrose J (2019). National Grid 'had three blackout near-misses in three months'. https:// www.theguardian.com/business/2019/aug/16/national-grid-blackout-report-avoidable-faults-blamed.

Amin M (2008). Challenges in reliability, security, efficiency, and resilience of energy infra-structure: Toward smart self-healing electric power grid. *Proc. IEEE Power and Energy Society General Meeting - Conversion and Delivery of Electrical Energy in the 21st Century*. pp. 1–5.

Amin M and Zhong QC (2019). Application of virtual synchronous machines to the Texas Panhandle wind power system. Technical report.

Amin M, Zhong QC, Lyu Z, Zhang L, Li Z and Shahidehpour M (2018). An anti-islanding protection for inverters in distributed generation. *Proc. IECON 2018*. pp. 1669–1674.

Amin SM and Wollenberg BF (2005). Toward a smart grid: power delivery for the 21st century. *IEEE Power and Energy Magazine*. **3**(5), 34–41.

Anaya-Lara O, Hughes FM, Jenkins N and Strbac G (2006). Contribution of DFIG-based wind farms to power system short-term frequency regulation. *Transmission and Distribution IEE Proceedings-Generation*. **153**(2), 164–170.

Power Electronics-Enabled Autonomous Power Systems: Next Generation Smart Grids, First Edition. Qing-Chang Zhong. © 2020 John Wiley & Sons Ltd. Published 2020 by John Wiley & Sons Ltd.

Ancuti R, Boldea I and Andreescu GD (2010). Sensorless v/f control of high-speed surface permanent magnet synchronous motor drives with two novel stabilising loops for fast dynamics and robustness. *IET Proc. Electric Power Appl.* **4**(3), 149–157.

Anderson C and Brown CE (2010). The functions and dysfunctions of hierarchy. *Research in Organizational Behavior*. **30**, 55 –89.

Aouini R, Marinescu B, Kilani KB and Elleuch M (2016). Synchronverter-based emulation and control of HVDC transmission. *IEEE Trans. Power Syst.* **31**(1), 278–286.

Arrillaga J (2008). *High voltage direct current transmission.* London, UK: Institution of Electrical Engineers.

Arrow KJ (1951). *Social Choice and Individual Values.* New York: Wiley.

Arrow KJ (2012). *Social Choice and Individual Values.* Yale University Press.

Ashabani M and Mohamed YARI (2014). Integrating VSCs to Weak Grids by Nonlinear Power Damping Controller With Self-Synchronization Capability. *IEEE Trans. Power Syst.* **29**(2), 805–814.

Ashabani S and Mohamed YR (2012). A flexible control strategy for grid-connected and islanded microgrids with enhanced stability using nonlinear microgrid stabilizer. *IEEE Trans. Smart Grid*. **3**(3), 1291–1301.

Bae Y and Kim RY (2014). Suppression of common-mode voltage using a multicentral photovoltaic inverter topology with synchronized PWM. *IEEE Trans. Ind. Electron.* **61**(9), 4722–4733.

Barater D, Buticchi G, Lorenzani E and Concari C (2014).Active common-mode filter for ground leakage current reduction in grid-connected PV converters operating with arbitrary power factor. *IEEE Trans. Ind. Electron.* **61**(8), 3940–3950.

Baroudi J, Dinavahi V and Knight A (2005). A review of power converter topologies for wind generators. *Proc. of IEEE International Conference on Electric Machines and Drives.* pp. 458–465.

Bascetta L, Magnani G, Rocco P and Zanchettin AM (2010). Performance limitations in field-oriented control for asynchronous machines with low resolution position sensing. *IEEE Trans. Control Syst. Technol.* **18**(3), 559–573.

Beck HP and Hesse R (2007). Virtual synchronous machine *Proc. of the 9th International Conference on Electrical Power Quality and Utilisation (EPQU).* pp. 1–6.

Bevrani H and Shokoohi S (2013). An intelligent droop control for simultaneous voltage and frequency regulation in islanded microgrids. *IEEE Trans. Smart Grid*. **4**(3), 1505–1513.

Bina M and Bhat A (2008). Averaging technique for the modeling of STATCOM and active filters. *IEEE Trans. Power Electron.* **23**(2), 723–734.

Blaabjerg F and Ma K (2013). Future on power electronics for wind turbine systems. *IEEE Journal of Emerging and Selected Topics in Power Electronics* **1**(3), 139–152.

Blaabjerg F, Teodorescu R, Liserre M and Timbus A (2006). Overview of control and grid synchronization for distributed power generation systems. *IEEE Trans. Ind. Electron.* **53**(5), 1398–1409.

Bohn C and Atherton DP (1995) An analysis package comparing PID anti-windup strategies. *IEEE Control Systems.* **15**(2), 34–40.

Boldea I (2008). Control issues in adjustable speed drives. *IEEE Ind. Electron. Mag.* **2**(3), 32–50.

Bollen MH (2000). *Understanding Power Quality Problems: Voltage Sags and Interruptions.* Wiley-IEEE Press.

Bose B (1993) Variable frequency drives-technology and applications. *Proc. of IEEE International Symposium on International Industrial Electronics (ISIE)*, pp. 1–18.

Bose B (2001). *Modern Power Electronics and AC Drives*. Englewood Cliffs, NJ: Prentice-Hall.

Bose B (2009). Power electronics and motor drives recent progress and perspective. *IEEE Trans. Ind. Electron.* **56**(2), 581–588.

Bottrell N and Green TC (2014). Comparison of current-limiting strategies during fault ride-through of inverters to prevent latch-up and wind-up. *IEEE Trans. Power Electron.* **29**(7), 3786–3797.

Brabandere KD, Bolsens B, den Keybus JV, Woyte A, Driesen J and Belmans R (2007). A voltage and frequency droop control method for parallel inverters. *IEEE Trans. Power Electron.* **22**(4), 1107–1115.

Bracewell RN (1999) *The Fourier Transform and Its Applications*. McGraw-Hill Higher Education.

Breazeale L and Ayyanar R (2015). A photovoltaic array transformer-less inverter with film capacitors and silicon carbide transistors. *IEEE Trans. Power Electron.* **30**(3), 1297–1305.

Brennan J (2016). *Against Democracy*. Princeton University Press.

Brewer R (2014). How renewable energy could leave you mired in blackouts. *The Motley Fool*. https://www.fool.com/investing/general/2014/10/19/renewable-energy-could-leave-you-mired-in-blackout.aspx.

Bueno E, Cobreces S, Rodriguez F, Hernandez A and Espinosa F (2008). Design of a back-to-back NPC converter interface for wind turbines with squirrel-cage induction generator. *IEEE Trans. Energy Convers.* **23**(3), 932–945.

Buticchi G, Barater D, Lorenzani E, Concari C and Franceschini G (2014). A nine-level grid-connected converter topology for single-phase transformerless PV systems. *IEEE Trans. Ind. Electron.* **61**(8), 3951–3960.

Byrnes CI, Isidori A and Willems JC (1991) Passivity, feedback equivalence, and the global stabilization of minimum phase nonlinear systems. *IEEE Trans. Automat. Contr.* **36**(11), 1228–1240.

Cai W, Jiang L, Liu B, Duan S and Zou C (2015). A power decoupling method based on four-switch three-port DC/DC/AC converter in DC microgrid. *IEEE Transactions on Industry Applications*. **51**(1), 336–343.

Cardenas R and Pena R (2004). Sensorless vector control of induction machines for variable-speed wind energy applications. *IEEE Trans. Energy Convers.* **19**(1), 196–205.

Carlin HJ (1967) Network theory without circuit elements. *Proceedings of the IEEE*. **55**(4), 482–497.

Carlin HJ (1969) Correction to "network theory without circuit elements". *Proceedings of the IEEE*. **57**(6), 1171.

Carroll O (2019). US cyber attack: Did America really try to override the Russian power grid? https://www.independent.co.uk/news/world/europe/us-cyber-attack-russia-power-grid-war-kremlin-a8964506.html.

Cavalcanti M, de Oliveira K, de Farias A, Neves F, Azevedo G and Camboim F (2010). Modulation techniques to eliminate leakage currents in transformerless three-phase photovoltaic systems. *IEEE Trans. Ind. Electron.* **57**(4), 1360–1368.

Cecati C, Dell'Aquila A, Lecci A and Liserre M (2005). Implementation issues of a fuzzy-logic-based three-phase active rectifier employing only voltage sensors. *IEEE Trans. Ind. Electron.* **52**(2), 378 –385.

Chatterjee D (2011). A novel magnetizing-curve identification and computer storage technique for induction machines suitable for online application. *IEEE Trans. Ind. Electron.* **58**(12), 5336–5343.

Chaudhary SK, Teodorescu R, Rodriguez P and Kjar PC (2009). Chopper controlled resistors in VSC-HVDC transmission for WPP with full-scale converters. *2009 IEEE PES/IAS Conference on Sustainable Alternative Energy (SAE)*, pp. 1–8, Valencia, Spain.

Chen B and Joos G (2008). Direct power control of active filters with averaged switching frequency regulation. *IEEE Trans. Power Del.* **23**(6), 2729–2737.

Chen D, Xu Y and Huang AQ (2017). Integration of DC microgrids as virtual synchronous machines into the AC grid. *IEEE Transactions on Industrial Electronics.* **64**(9), 7455–7466.

Chen WH, Yang J, Guo L and Li S (2016). Disturbance observer-based control and related methods: An overview. *IEEE Trans. Ind. Electron.* **63**(2), 1083–1095.

Chen WL, Liang WG and Gau HS (2010). Design of a mode decoupling STATCOM for voltage control of wind-driven induction generator systems. *IEEE Trans. Power Del.* **25**(3), 1758–1767.

Chen Y, Hesse R, Turschner D and Beck HP (2011). Improving the grid power quality using virtual synchronous machines. *Proc. of International Conference on Power Engineering, Energy and Electrical Drives (POWERENG)*, pp. 1–6.

Chen Z, Guerrero J and Blaabjerg F (2009). A review of the state of the art of power electronics for wind turbines. *IEEE Trans. Power Electron.* **24**(8), 1859–1875.

Chua L (1971) Memristor-the missing circuit element. *IEEE Trans. Circuit Theory.* **18**(5), 507–519.

Chuang Tzu (2016). *The Writings of Chuang Tzu.* Translated by James Legge. CreateSpace Independent Publishing Platform.

Cichowlas M, Malinowski M, Kazmierkowski M, Sobczuk D, Rodriguez P and Pou J (2005). Active filtering function of three-phase PWM boost rectifier under different line voltage conditions. *IEEE Trans. Ind. Electron.* **52**(2), 410–419.

Ciobotaru M, Teodorescu R and Blaabjerg F (2006). A new single-phase PLL structure based on second order generalized integrator. *Proc. of the 37th IEEE Power Electronics Specialists Conference (PESC)*, pp. 1–6.

Coelho EAA, Cortizo PC and Garcia PFD (1999) Small signal stability for single phase inverter connected to stiff AC system. *Proc. of the IEEE Industry Applications Conference (IAS)*, vol. **4**, pp. 2180–2187.

Costanzo GT, Zhu G, Anjos MF and Savard G (2012). A system architecture for autonomous demand side load management in smart buildings. *IEEE Transactions on Smart Grid* **3**(4), 2157–2165.

da Silva C, Pereira R, da Silva L, Lambert-Torres G, Bose B and Ahn S (2010). A digital PLL scheme for three-phase system using modified synchronous reference frame. *IEEE Trans. Ind. Electron.* **57**(11), 3814–3821.

Dag O and Mirafzal B (2016). On stability of islanded low-inertia microgrids *Power Systems Conference (PSC), 2016 Clemson University*, pp. 1–7 IEEE.

Dahono P (2003). A method to damp oscillations on the input LC filter of current-type AC-DC PWM converters by using a virtual resistor. *Proc. of the 25th International Telecommunications Energy Conference (INTELEC'03)*, pp. 757–761.

Dahono P (2004). A control method for DC-DC converter that has an LCL output filter based on new virtual capacitor and resistor concepts. *Proc. of the 35th Annual Power Electronics Specialists Conference*, pp. 36–42.

Dahono P, Bahar Y, Sato Y and Kataoka T (2001). Damping of transient oscillations on the output LC filter of PWM inverters by using a virtual resistor *Proc. of the 4th IEEE International Conference on Power Electronics and Drive Systems*, pp. 403–407.

de Leon F, Qaseer L and Cohen J (2012). AC power theory from Poynting theorem: Identification of the power components of magnetic saturating and hysteretic circuits. *IEEE Transactions on Power Delivery* **27**(3), 1548–1556.

Delille G, Francois B and Malarange G (2012). Dynamic frequency control support by energy storage to reduce the impact of wind and solar generation on isolated power system's inertia. *IEEE Transactions on Sustainable Energy* **3**(4), 931–939.

Deo S, Jain C and Singh B (2015). A PLL-less scheme for single-phase grid interfaced load compensating solar PV generation system.*IEEE Trans. Ind. Informat.* **11**(3), 692–699.

Depenbrock M (1988). Direct self-control (DSC) of inverter-fed induction machine. *IEEE Trans. Power Electron.* **3**(4), 420–429.

Dixon J and Ooi BT (1988). Indirect current control of a unity power factor sinusoidal current boost type three-phase rectifier. *IEEE Trans. Ind. Electron.* **35**(4), 508–515.

DOE (2009). Smart grid system report. Technical report, The U.S. Department of Energy. http://energy.gov/sites/prod/files/oeprod/DocumentsandMedia/SGSR_Annex_A-B_090707_lowres.pdf.

DOE (2015). Wind vision: A new era for wind power in the United States. Technical report. http://www.energy.gov/sites/prod/files/WindVision_Report_final.pdf.

Dong D, Boroyevich D, Mattavelli P and Cvetkovic I (2011). A high-performance single-phase phase-locked-loop with fast line-voltage amplitude tracking. *Proc. of the (26th Annual IEEE Applied Power Electronics Conference and Exposition (APEC)*, pp. 1622–1628.

Dong D, Luo F, Boroyevich D and Mattavelli P (2012). Leakage current reduction in a single-phase bidirectional AC–DC full-bridge inverter. *IEEE Trans. Power Electron.* **27**(10), 4281–4291.

Dong D, Wen B, Boroyevich D, Mattavelli P and Xue Y (2015). Analysis of phase-locked loop low-frequency stability in three-phase grid-connected power converters considering impedance interactions. *IEEE Trans. Ind. Electron.* **62**(1), 310–321.

Dong D, Wen B, Mattavelli P, Boroyevich D and Xue Y (2013). Grid-synchronization modeling and its stability analysis for multi-paralleled three-phase inverter systems *Applied Power Electronics Conference and Exposition (APEC), (2013 Twenty-Eighth Annual IEEE*, pp. 439–446.

Dong S, Chi Y and Li Y (2016). Active voltage feedback control for hybrid multiterminal HVDC system adopting improved synchronverters. *IEEE Trans. Power Del.* **31**(2), 445–455.

Driesen J and Visscher K (2008). Virtual synchronous generators. *Proc. of IEEE Power and Energy Society General Meeting*, pp. 1–3.

Du W, Zhang J, Zhang Y and Qian Z (2013a). Stability criterion for cascaded system with constant power load. *IEEE Trans. Power Electron.* **28**(4), 1843–1851.

Du Y, Guerrero JM, Chang L, Su J and Mao M (2013b). Modeling, analysis, and design of a frequency-droop-based virtual synchronous generator for microgrid applications. *Proc. IEEE ECCE Asia Downunder*, pp. 643–649.

Duarte J, Van Zwam A, Wijnands C and Vandenput A (1999). Reference frames fit for controlling PWM rectifiers. *IEEE Trans. Ind. Electron.* **46**(3), 628–630.

Duckwitz D and Fischer B (2017). Modeling and design of df/dt -based inertia control for power converters. *IEEE J. Emerg. Sel. Topics Power Electron.* **5**(4), 1553–1564.

Eder-Neuhauser P, Zseby T and Fabini J (2016). Resilience and security: A qualitative survey of urban smart grid architectures. *IEEE Access* **4**, 839–848.

Ekanayake J and Jenkins N (2004 Comparison of the response of doubly fed and fixed-speed induction generator wind turbines to changes in network frequency. *IEEE Transactions on Energy conversion.* **19**(4), 800–802.

Ekanayake J, Jenkins N, Liyanage K, Wu J and Yokoyama A (2012). *Smart Grid: Technology and Applications.* John Wiley & Sons.

El-Moursi M, Bak-Jensen B and Abdel-Rahman M (2010). Novel STATCOM controller for mitigating SSR and damping power system oscillations in a series compensated wind park. *IEEE Trans. Power Electron.* **25**(2), 429–441.

El Moursi MS, Xiao W and Kirtley JL (2013). Fault ride through capability for grid interfacing large scale PV power plants. *IET Generation, Transmission & Distribution.* **7**(9), 1027–1036.

Elizondo J and Kirtley JL (2014). Effect of dg and induction motor load power rating on microgrid transient behavior. *Innovative Smart Grid Technologies Conference (ISGT), (2014 IEEE PES*, pp. 1–5 IEEE.

Ellison A (1965). *Electromechanical Energy Conversion.* London: George G. Harrap Co. Ltd.

England C (2016). France to shut down all coal-fired power plants by (2023. https://www .independent.co.uk/news/world/europe/france-close-coal-plants-shut-down-2023-global-warming-climate-change-a7422966.html.

ERCOT (2014). *Panhandle Renewable Energy Zone (PREZ) Study report.*

ERCOT (2017). *ERCOT Capacity, Demand and Reserve Report.*

Ericsen T, Hingorani N and Khersonsky Y (2006). Power electronics and future marine electrical systems. *IEEE Trans. Ind. Appl.* **42**(1), 155 –163.

Erlicki MS and Emanuel-Eigeles A (1968). New aspects of power factor improvement Part I—theoretical basis. *IEEE Transactions on Industry and General Applications* **IGA-4**(4), 441–446.

Escobar G, Martinez-Montejano M, Valdez A, Martinez P and Hernandez-Gomez M (2011ERCOT (2017). Fixed-reference-frame phase-locked loop for grid synchronization under unbalanced operation. *IEEE Trans. Ind. Electron.* **58**(5), 1943–1951.

Escobar G, Stankovic A and Mattavelli P (2004). An adaptive controller in stationary reference frame for D-STATCOM in unbalanced operation. *IEEE Trans. Ind. Electron.* **51**(2), 401–409.

Escobar G, Stankovic A, Carrasco J, Galvan E and Ortega R (2003). Analysis and design of direct power control (DPC) for a three-phase synchronous rectifier via output regulation subspaces. *IEEE Trans. Power Del.* **18**(3), 823–830.

Evangelou S, Limebeer DJN, Sharp RS and Smith MC (2006). Control of motorcycle steering instabilities. *IEEE Control Systems* **26**(5), 78–88.

Fairley P (2015). Hawaii's solar push strains the grid. *MIT Technology Review.* https://www .technologyreview.com/s/534266/hawaiis-solar-push-strains-the-grid/.

Fairley P (2016). Why Southern China broke up its power grid [news]. *IEEE Spectrum* **53**(12), 13–14.

Fang X, Misra S, Xue G and Yang D (2012). Smart grid - the new and improved power grid: A survey. *IEEE Commun. Surveys Tuts.* **14**(4), 944–980.

Farhangi H (2010). The path of the smart grid. *IEEE Power Energy Mag.* **8**(1), 18–28.

Farrell J (2011). Democratizing the electricity system: A vision for the (21st century grid. Technical report, Institute for Local Self-Reliance. https://ilsr.org/democratizing-electricity-system-vision-21st-century-grid/.

Femia N, Lisi G, Petrone G, Spagnuolo G and Vitelli M (2008). Distributed maximum power point tracking of photovoltaic arrays: Novel approach and system analysis. *IEEE Trans. Ind. Electron.* **55**(7), 2610–2621.

Femia N, Petrone G, Spagnuolo G and Vitelli M (2005). Optimization of perturb and observe maximum power point tracking method. *IEEE Trans. Power Electron.* **20**(4), 963–973.

FERC (2018 Essential reliability services and the evolving bulk-power system — primary frequency response. Technical report. FERC Order No. 842, available at https://www.ferc.gov/media/news-releases/2018/2018-1/02-15-18-E-2.asp#.WohkZ6OZmAV.

Fiaz S, Zonetti D, Ortega R, Scherpen JMA and Van der Schaft AJ (2013). A port-Hamiltonian approach to power network modeling and analysis. *European Journal of Control.* **19**(6), 477–485.

Firestone FA (1933) A new analogy between mechanical and electrical systems. *The Journal of the Acoustical Society of America.* **4**(3), 249–267.

Fitzgerald A, Kingsley C and Umans S (2003). *Electric Machinery.* New York, NY: McGraw-Hill.

Freddy Tan K, Abd Rahim N, Hew W and Che H (2015). Modulation techniques to reduce leakage current in three-phase transformerless H7 photovoltaic inverter. *IEEE Trans. Ind. Electron.* **62**(1), 322–331.

Friedman TL (2005). *The World Is Flat: A Brief History of the Twenty-first Century.* Farrar, Straus and Giroux.

Gao F and Iravani MR (2008). A control strategy for a distributed generation unit in grid-connected and autonomous modes of operation. *IEEE Trans. Power Del.* **23**(2), 850–859.

Gatlan S (2019). U.S. airlines cancel, delay flights because of Aerodata outage. https://www.bleepingcomputer.com/news/technology/us-airlines-cancel-delay-flights-because-of-aerodata-outage/.

Gaviria C, Fossas E and Grino R (2005). Robust controller for a full-bridge rectifier using the IDA approach and GSSA modeling. *IEEE Trans. Circuits Syst. I.* **52**(3), 609–616.

Ghaffari A, Krstic M and Seshagiri S (2014). Power optimization for photovoltaic microconverters using multivariable Newton-based extremum seeking. *IEEE Trans. Control Syst. Technol.* **22**(6), 2141–2149.

Ghanbari V, Wu P and Antsaklis PJ (2016). Large-scale dissipative and passive control systems and the role of star and cyclic symmetries. *IEEE Trans. Autom. Control.* **61**(11), 3676–3680.

Giordano AA and Hsu FM (1985) *Least square estimation with applications to digital signal processing.* John Wiley & Sons, New York.

Grainger J and Stevenson W (1994) *Power System Analysis.* New York, NY: McGraw-Hill.

Grayling A (2017). *Democracy and Its Crisis.* Oneworld Publications.

Gu Y, Li W, Zhao Y, Yang B, Li C and He X (2013). Transformerless inverter with virtual DC bus concept for cost-effective grid-connected PV power systems. *IEEE Trans. Power Electron.* **28**(2), 793–805.

Guan M, Pan W, Zhang J, Hao Q, Cheng J and Zheng X (2015). Synchronous generator emulation control strategy for voltage source converter (VSC) stations. *IEEE Trans. Power Syst.* **30**(6), 3093–3101.

Guerrero J, Hang L and Uceda J (2008). Control of distributed uninterruptible power supply systems. *IEEE Trans. Ind. Electron.* **55**(8), 2845–2859.

Guerrero J, Matas J, de Vicuna L, Castilla M and Miret J (2006). Wireless-control strategy for parallel operation of distributed-generation inverters. *IEEE Trans. Ind. Electron.* **53**(5), 1461–1470.

Guerrero JM, de Vicuna LG, Matas J, Castilla M and Miret J (2004). A wireless controller to enhance dynamic performance of parallel inverters in distributed generation systems. *IEEE Transactions on Power Electronics.* **19**(5), 1205–1213.

Guerrero JM, G. de Vicuna L, Matas J, Castilla M and Miret J (2005). Output impedance design of parallel-connected UPS inverters with wireless load-sharing control. *IEEE Trans. Ind. Electron.* **52**(4), 1126–1135.

Gulez K, Adam AA and Pastaci H (2008). Torque ripple and EMI noise minimization in PMSM using active filter topology and field-oriented control. *IEEE Trans. Ind. Electron.* **55**(1), 251–257.

Guo W, Xiao L and Dai S (2012). Enhancing low-voltage ride-through capability and smoothing output power of DFIG with a superconducting fault-current limiter–Magnetic energy storage system. *IEEE Trans. Energy Convers.* **27**(2), 277–295.

Guo X, Cavalcanti MC, Farias AM and Guerrero JM (2013). Single-carrier modulation for neutral-point-clamped inverters in three-phase transformerless photovoltaic systems. *IEEE Trans. Power Electron.* **28**(6), 2635–2637.

Guo X, Liu W, Zhang X, Sun X, Lu Z and Guerrero JM (2015). Flexible control strategy for grid-connected inverter under unbalanced grid faults without PLL. *IEEE Trans. Power Electron.* **30**(4), 1773–1778.

Guo X, Lu Z, Wang B, Sun X, Wang L and Guerrero JM (2014). Dynamic phasors-based modeling and stability analysis of droop-controlled inverters for microgrid applications. *IEEE Trans. Smart Grid* **5**(6), 2980–2987.

Guo X, Wu W and Chen Z (2011). Multiple-complex coefficient-filter-based phase-locked loop and synchronization technique for three-phase grid-interfaced converters in distributed utility networks. *IEEE Trans. Ind. Electron.* **58**(4), 1194–1204.

Gyugyi L (1988) Power electronics in electric utilities: Static VAR compensators *Proc. of IEEE*, pp. 483–494.

Haj-ahmed MA and Illindala MS (2014). The influence of inverter-based DGs and their controllers on distribution network protection. *IEEE Trans. Ind. Appl.* **50**(4), 2928–2937.

Hansen A and Michalke G (2009). Multi-pole permanent magnet synchronous generator wind turbines' grid support capability in uninterrupted operation during grid faults. *IET Proc. Renewable Power Generation.* **3**(3), 333–348.

Hansen S, Malinowski M, Blaabjerg F and Kazmierkowski M (2000). Sensorless control strategies for PWM rectifier *Proc. of Applied Power Electronics Conference and Exposition (APEC)*, pp. 832–838.

Haque ME, Saw YC and Chowdhury MM (2014). Advanced control scheme for an IPM synchronous generator-based gearless variable speed wind turbine. *IEEE Trans. Sust. Energy.* **5**(2), 354–362.

Harnefors L, Bongiorno M and Lundberg S (2007). Input-admittance calculation and shaping for controlled voltage-source converters. *IEEE Trans. Ind. Electron.* **54**(6), 3323–3334.

Hatua K, Jain A, Banerjee D and Ranganathan V (2012). Active damping of output *LC* filter resonance for vector-controlled VSI-Fed AC motor drives. *IEEE Trans. Ind. Electron.* **59**(1), 334–342.

Heldwein M, Nussbaumer T and Kolar J (2010). Common mode modelling and filter design for a three-phase buck-type pulse width modulated rectifier system. *IET Power Electronics.* **3**(2), 209–218.

Hingorani N and Gyugyi L (1999) *Understanding FACTS: Concepts and Technology of Flexible AC Transmission Systems.* Wiley-IEEE Press.

Hingorani NG (1988) Power electronics in electric utilities: role of power electronics in future power systems. *Proc. IEEE.* **76**(4), 481–482.

Hochgraf C and Lasseter R (1998) STATCOM controls for operation with unbalanced voltages. *IEEE Trans. Power Del.* **13**(2), 538–544.

Holm DD (2011a). *Geometric Mechanics Part I: Dynamics and Symmetry.* Imperial College Press.

Holm DD (2011b). *Geometric Mechanics Part II: Rotating, Translating and Rolling.* Imperial College Press.

Holm DD, Schmah T and Stoica C (2009). *Geometric Mechanics and Symmetry: From Finite to Infinite Dimensions.* Oxford University Press.

Hong CM, Chen CH and Tu CS (2013). Maximum power point tracking-based control algorithm for PMSG wind generation system without mechanical sensors. *Energy Conversion and Management* **69**(0), 58–67.

Horita Y, Morishima N, Kai M, Onishi M, Masui T and Noguchi M (2010). Single-phase STATCOM for feeding system of Tokaido Shinkansen. *Proc. of International Power Electronics Conference(IPEC)*, pp. 2165–2170.

Hornik T and Zhong QC (2011a). A current control strategy for voltage-source inverters in microgrids based on H^∞ and repetitive control. *IEEE Trans. Power Electron.* **26**(3), 943–952.

Hornik T and Zhong QC (2011b). KTA synchronverter project. Technical report. University of Liverpool.

Hutt R (2016). What are the 10 biggest global challenges? https://www.weforum.org/agenda/2016/01/what-are-the-10-biggest-global-challenges.

IEC (1993) *IEC60364. Assessment of General Characteristics.* International Electrotechnical Commission.

Illian H. et al. (2017). Measurement, monitoring, and reliability issues related to primary governing frequency response. Technical report. IEEE Power & Energy Society Technical Report TR-24. Available at goo.gl/FcYcSa.

Im WS, Wang C, Liu W, Liu L and Kim JM (2016). Distributed virtual inertia based control of multiple photovoltaic systems in autonomous microgrid. *IEEE/CAA Journal of Automatica Sinica.* **4**(3), 512–519.

Jain AK and Ranganathan VT (2011). Modeling and field oriented control of salient pole wound field synchronous machine in stator flux coordinates. *IEEE Trans. Ind. Electron.* **58**(3), 960–970.

Jeltsema D and Doria-Cerezo A (2012). Port-Hamiltonian formulation of systems with memory. *Proceedings of the IEEE* **100**(6), 1928–1937.

Jiang JZ and Smith MC (2011). Regular positive-real functions and five-element network synthesis for electrical and mechanical networks. *IEEE Trans. Autom. Control* **56**(6), 1275–1290.

Johansson M (1999) *The Hilbert transform* Master's thesis VÄaxjÄo University.

Johnson K, Fingersh L, Balas M and Pao L (2004). Methods for increasing region 2 power capture on a variable speed wind turbine. *Journal of Solar Energy Engineering.* **126**(4), 1092–1100.

Jovcic D, Lamont L and Xu L (2003). VSC transmission model for analytical studies. *Proc. of IEEE Power Engineering Society General Meeting*, vol. **3**, pp. 1737–1742.

Karady G and Holbert K (2004). *Electrical Energy Conversion and Transport: An Interactive Computer-Based Approach.* IEEE Press Series on Power Engineering. Wiley–IEEE Press.

Karimi-Ghartemani M (2015). Universal integrated synchronization and control for single-phase DC/AC converters. *IEEE Trans. Power Electron.* **30**(3), 1544–1557.

Karimi-Ghartemani M and Iravani M (2001). A new phase-locked loop (PLL) system. *Proc. of the 44th IEEE 2001 Midwest Symposium on Circuits and Systems (MWSCAS)*, pp. 421–424.

Karimi-Ghartemani M and Iravani M (2002). A nonlinear adaptive filter for online signal analysis in power systems: Applications. *IEEE Trans. Power Del.* **17**(2), 617–622.

Karimi-Ghartemani M and Ziarani A (2003). Periodic orbit analysis of two dynamical systems for electrical engineering applications. *Journal of Engineering Mathematics.* **45**, 135–154.

Karimi-Ghartemani M and Ziarani A (2004). Performance characterization of a non-linear system as both an adaptive notch filter and a phase-locked loop. *Int. J. Adapt. Control Signal Process.* **18**, 23–53.

Karimi-Ghartemani M, Khajehoddin A, Jain P and Bakhshai A (2011). Problems of startup and phase jumps in PLL systems. *IEEE Trans. Power Electron.* **27**(4), 1830–1838.

Karschny D (1998) Wechselrichter. German patent, DE19 642 522 C1.

Katiraei F and Iravani MR (2006). Power management strategies for a microgrid with multiple distributed generation units. *IEEE transactions on power systems.* **21**(4), 1821–1831.

Kazmierkowski M and Malesani L (1998). Current control techniques for three-phase voltage-source PWM converters: a survey. *IEEE Trans. Ind. Electron.* **45**(5), 691–703.

Kazmierkowski M, Franquelo L, Rodriguez J, Perez M and Leon J (2011). High-performance motor drives. *IEEE Ind. Electron. Mag.* **5**(3), 6–26.

Khalil HK (2001). *Nonlinear Systems.* Prentice Hall.

Khan H, Dasouki S, Sreeram V, Iu H and Mishra Y (2013). Universal active and reactive power control of electronically interfaced distributed generation sources in virtual power plants operating in gridconnected and islanding modes. *IET Generation, Transmission & Distribution.* **7**(8), 885–897.

Khoucha F, Lagoun S, Marouani K, Kheloui A and El Hachemi Benbouzid M (2010). Hybrid cascaded H-bridge multilevel-inverter induction-motor-drive direct torque control for automotive applications. *IEEE Trans. Ind. Electron.* **57**(3), 892–899.

Killi M and Samanta S (2015). An adaptive voltage-sensor-based MPPT for photovoltaic systems with SEPIC converter including steady-state and drift analysis. *IEEE Trans. Ind. Electron.* **62**(12), 7609–7619.

Kim KH, Jeung YC, Lee DC and Kim HG (2010). Robust control of PMSG wind turbine systems with back-to-back PWM converters. *Proc. of the (2nd Power Electronics for Distributed Generation Systems (PEDG)*, pp. 433–437.

King F (2009). *Hilbert Transforms* vol. 1,2 of *Encyclopedia of Mathematics and its Applications*. Cambridge University Press.

Kirby N, Luckett M, Xu L and Siepmann W (2001). HVDC transmission for large offshore windfarms. *Proc. of the 7th Conf. on AC-DC Power Transmission*, pp. 162–168.

Konstantopoulos G, Zhong QC, Ren B and Krstic M (2015). Boundedness of synchronverters *Proc. of the European Control Conference (ECC (2015)*, pp. 1050–1055, Linz, Austria.

Konstantopoulos GC and Alexandridis AT (2013). Generalized nonlinear stabilizing controllers for Hamiltonian-passive systems with switching devices. *IEEE Trans. Control Syst. Technol.* **21**(4), 1479–1488.

Konstantopoulos GC and Alexandridis AT (2014). Full-scale Modeling, Control and Analysis of Grid-Connected Wind Turbine Induction Generators with Back-to-Back AC/DC/AC Converters. *IEEE Journal of Emerging and Selected Topics in Power Electronics* **2**(4), 739–748.

Konstantopoulos GC, Zhong QC and Ming WL (2016a). PLL-less nonlinear current-limiting controller for single-phase grid-tied inverters: Design, stability analysis and operation under grid faults. *IEEE Trans. Ind. Electron.* **63**(9), 5582–5591.

Konstantopoulos GC, Zhong QC, Ren B and Krstic M (2016b). Bounded integral control of input-to-state practically stable nonlinear systems to guarantee closed-loop stability. *IEEE Trans. Autom. Control.* **61**(12), 4196–4202.

Koutroulis E and Kalaitzakis K (2006). Design of a maximum power tracking system for wind-energy-conversion applications. *IEEE Trans. Ind. Electron.* **53**(2), 486–494.

Kundur PS (1994). *Power System Stability and Control*. New York: Mc-Graw-Hills.

Laaksonen HJ (2010). Protection principles for future microgrids. *IEEE Trans. Power Electron.* **25**(12), 2910–2918.

Lacroix B and Calvas R (1995) Earthing systems worldwide and evolutions. Technical report, Cahier Technique, no.173, Schneider Electric. Available at http://electrical-engineering-portal.com/res/System-earthings-worldwide-and-evolutions.pdf.

Lalor G, Mullane A and O'Malley M (2005). Frequency control and wind turbine technologies. *IEEE Trans. Power Syst.* **20**(4), 1905–1913.

Lao Tzu (2016a). *Tao Te Ching*. Translated by James Legge, CreateSpace Independent Publishing Platform.

Lao Tzu (2016b).*Tao Te Ching*. Translated by Charles Johnston. CreateSpace Independent Publishing Platform.

Lee CK, Chaudhuri NR, Chaudhuri B and Hui SYR (2013). Droop control of distributed electric springs for stabilizing future power grid. *IEEE Trans. Smart Grid.* **4**(3), 1558–1566.

Lee DC and Lim DS (2002). AC voltage and current sensorless control of three-phase PWM rectifiers. *IEEE Power Electron. Lett.* **17**(6), 883 –890.

Lee RM, Assante MJ and Conway T (2016). Analysis of the Cyber Attack on the Ukrainian Power Grid. https://www.nerc.com/pa/CI/ESISAC/Documents/E-ISAC_SANS_Ukraine_DUC_18Mar2016.pdf.

Leidhold R (2011). Position sensorless control of PM synchronous motors based on zero-sequence carrier injection. *IEEE Trans. Ind. Electron.* **58**(12), 5371–5379.

Leith DJ and Leithead WE (1997) Implementation of wind turbine controllers. *International Journal of Control.* **66**(3), 349–380.

Li K, Liu J, Zhao G and Wang Z (2006). Control and optimization of VCVS static var generators for voltage unbalance mitigation *Proc. of the (21st IEEE Annual Applied Power Electronics Conference and Exposition (APEC)*, p. 6

Li S, Haskew TA, Swatloski RP and Gathings W (2012). Optimal and direct-current vector control of direct-driven PMSG wind turbines. *IEEE Trans. Power Electron.* **27**(5), 2325–2337.

Li S, Yang J, Chen WH and Chen X (2014). *Disturbance Observer-Based Control: Methods and Applications.* CRC Press.

Li Y (2009) Control and resonance damping of voltage-source and current-source converters with LC filters. *IEEE Trans. Ind. Electron.* **56**(5), 1511–1521.

Li Y and Kao CN (2009). An accurate power control strategy for power-electronics-interfaced distributed generation units operating in a low-voltage multibus microgrid. *IEEE Trans. Power Electron.* **24**(12), 2977–2988.

Linus RM and Damodharan P (2015). Maximum power point tracking method using a modified perturb and observe algorithm for grid connected wind energy conversion systems. *IET Renewable Power Generation.* **9**(6), 682–689.

Liu H and Li S (2012). Speed control for PMSM servo system using predictive functional control and extended state observer. *IEEE Trans. Ind. Electron.* **59**(2), 1171–1183.

Liu J, Miura Y and Ise T (2016a). Comparison of dynamic characteristics between virtual synchronous generator and droop control in inverter-based distributed generators. *IEEE Trans. Power Electron.* **31**(5), 3600–3611.

Liu S, Liu PX and Wang X (2016b).Stability analysis of grid-interfacing inverter control in distribution systems with multiple photovoltaic-based distributed generators. *IEEE Trans. Ind. Electron.* **63**(12), 7339–7348.

Liu W, Wang K, Chung H and Chuang S (2015). Modeling and design of series voltage compensator for reduction of DC-link capacitance in grid-tie solar inverter. *IEEE Trans. Power Electron.* **30**(5), 2534–2548.

Lu Z, Ye X, Qiao Y and Min Y (2015). Initial exploration of wind farm cluster hierarchical coordinated dispatch based on virtual power generator concept. *CSEE Journal of Power and Energy Systems.* **1**(2), 62–67.

Luiz da Silva E, Hedgecock J, Mello J and Ferreira da Luz J (2001). Practical cost-based approach for the voltage ancillary service. *IEEE Trans. Power Syst.* **16**(4), 806–812.

Ma Z, Zhong QC and Yan J (2012). Synchronverter-based control strategies for three-phase PWM rectifiers *Proc. of the 7th IEEE Conference on Industrial Electronics and Applications (ICIEA)*, Singapore.

Malinowski M, Jasinski M and Kazmierkowski M (2004). Simple direct power control of three-phase PWM rectifier using space-vector modulation (DPC-SVM). *IEEE Trans. Ind. Electron.* **51**(2), 447–454.

Maschke B and Van Der Schaft A (1991) Port controlled Hamiltonian systems: modeling origins and system theoretic properties *IFAC Symp. Nonlinear Control Systems Design*, pp. 359–365.

Maschke B, Ortega R and Schaft AJVD (2000). Energy-based Lyapunov functions for forced Hamiltonian systems with dissipation. *IEEE Transactions on Automatic Control.* **45**(8), 1498–1502.

Mcelroy MB and Chen X (2017). Wind and solar power in the united states: status and prospects. *CSEE Journal of Power and Energy Systems.* **3**(1), 1–6.

McGranaghan M, Mueller D and Samotyj M (1993) Voltage sags in industrial systems. *IEEE Trans. Ind. Appl.* **29**(2), 397–403.

Mengoni M, Zarri L, Tani A, Serra G and Casadei D (2008). Stator flux vector control of induction motor drive in the field weakening region. *IEEE Trans. Power Electron.* **23**(2), 941–949.

Ming WL and Zhong QC (2014). Synchronverter-based transformerless PV inverters *The 40th Annual Conference of IEEE Industrial Electronics Society*, pp. 4396–4401.

Ming WL and Zhong QC (2017). Current-stress reduction for the neutral inductor of θ-converters. *IEEE Trans. Power Electron.* **32**(4), 2794–2807.

Mizoguchi T, Nozaki T and Ohnishi K (2013). The power factor in mechanical system. *2013 IEEE International Conference on Mechatronics (ICM)*, pp. 576–581.

Mohamed YARI and Lee TK (2006). Adaptive self-tuning MTPA vector controller for IPMSM drive system. *IEEE Trans. Energy Convers.* **21**(3), 636–644.

Mohod S and Aware M (2010). A STATCOM-control scheme for grid connected wind energy system for power quality improvement. *IEEE Syst. J.* **4**(3), 346–352.

Mohseni M, Islam S and Masoum M (2011). Enhanced hysteresis-based current regulators in vector control of DFIG wind turbines. *IEEE Trans. Power Electron.* **26**(1), 223–234.

Mohsenian-Rad AH, Wong VW, Jatskevich J, Schober R and Leon-Garcia A (2010). Autonomous demand-side management based on game-theoretic energy consumption scheduling for the future smart grid. *IEEE transactions on Smart Grid* **1**(3), 320–331.

Momoh J (2012). *Smart Grid: Fundamentals of Design and Analysis*. John Wiley & Sons.

Montanari AA and Gole AM (2017). Enhanced instantaneous power theory for control of grid connected voltage sourced converters under unbalanced conditions. *IEEE Trans. Power Electron.* **32**(8), 6652–6660.

Montoya DG, Ramos-Paja CA and Giral R (2016). Improved design of sliding-mode controllers based on the requirements of MPPT techniques. *IEEE Trans. Power Electron.* **31**(1), 235–247.

Moore K (2011). The decline but not fall of hierarchy – what young people really want. *Forbes*. http://www.forbes.com/sites/karlmoore/2011/06/14/the-decline-but-not-fall-of-hierarchy-what-young-people-really-want/.

Mori S, Matsuno K, Hasegawa T, Ohnishi S, Takeda M, Seto M, Murakami S and Ishiguro F (1993). Development of a large static VAR generator using self-commutated inverters for improving power system stability. *IEEE Trans. Power Syst.* **8**(1), 371–377.

Morimoto S, Sanada M and Takeda Y (1994) Effects and compensation of magnetic saturation in flux-weakening controlled permanent magnet synchronous motor drives. *IEEE Trans. Ind. Appl.* **30**(6), 1632–1637.

Morren J, de Haan SW and Ferreira J (2006a). Contribution of DG units to primary frequency control. *International Transactions on Electrical Energy Systems.* **16**(5), 507–521.

Morren J, de Haan SWH, Kling WL and Ferreira JA (2006b). Wind turbines emulating inertia and supporting primary frequency control. *IEEE Trans. Power Syst.* **21**(1), 433–434.

Moule AC (1924) The Chinese south-pointing carriage. *T'oung Pao* **23**(2/3), 83–98.

Mukherjee N and Strickland D (2016). Control of cascaded DC-DC converter-based hybrid battery energy storage systems; Part I: Stability issue. *IEEE Trans. Ind. Electron.* **63**(4), 2340–2349.

NAE (2000). Electrification. Technical report. Available at http://www.greatachievements.org/?id=2949.

Natarajan V and Weiss G (2014). Almost global asymptotic stability of a constant field current synchronous machine connected to an infinite bus *2014 IEEE 53rd Annual Conference on Decision and Control (CDC2014)*, pp. 3272–3279, Los Angeles, CA, USA.

National Grid (2016). *The Grid Code*. National Grid Electricity Transmission plc.

Nayak O, Gole A, Chapman D and Davies J (1994). Dynamic performance of static and synchronous compensators at an HVDC inverter bus in a very weak AC system. *IEEE Trans. Power Syst.* **9**(3), 1350–1358.

Needham J (1986). *Science and Civilization in China*. Caves Books, Ltd., Taipei. Volume 4, Part (2): page 298.

Newton I (2014). *The Principia:Mathematical Principles of Natural Philosophy*. University of California Press.

Nguyen PL, Zhong QC, Blaabjerg F and Guerrero J (2012). Synchronverter-based operation of STATCOM to mimic synchronous condensers. *Proc. of the 7th IEEE Conference on Industrial Electronics and Applications (ICIEA)*, pp. 942–947, Singapore.

Nobelprize.org (2017). The (2017 Nobel Prize in Physiology or Medicine - Advanced information: Discoveries of molecular mechanisms controlling the circadian rhythm. Nobel Media AB 2014.

Noguchi T, Tomiki H, Kondo S and Takahashi I (1998) Direct power control of PWM converter without power-source voltage sensors. *IEEE Trans. Ind. Appl.* **34**(3), 473–479.

Norouzi A and Sharaf A (2005). Two control schemes to enhance the dynamic performance of the STATCOM and SSSC. *IEEE Trans. Power Del.* **20**(1), 435–442.

Ogata K and Yang Y (1970). *Modern Control Engineering*. Englewood Cliffs, NJ: Prentice-Hall.

Olivares DE, Mehrizi-Sani A, Etemadi AH, Canizares CA, Iravani R, Kazerani M, Hajimiragha AH, Gomis-Bellmunt O, Saeedifard M, Palma-Behnke R, Jimenez-Estevez GA and Hatziargyriou ND (2014). Trends in microgrid control. *IEEE Trans. Smart Grid.* **5**(4), 1905–1919.

Omron (2013). PID-preventive transformerless PV inverter. Technical report, Omron Corporation. http://industrial.omron.co.za/en/solutions/green-automation/pid-causes-70-power-reduction.

Ortega R (1989). Passivity properties for stabilization of cascaded nonlinear systems. *Automatica.* **27**(2), 423–424.

Ortega R and Romero JG (2012). Robust integral control of port-Hamiltonian systems: The case of non-passive outputs with unmatched disturbances. *Systems & Control Letters.* **61**(1), 11–17.

Ortega R, van der Schaft A, Castanos F and Astolfi A (2008). Control by interconnection and standard passivity-based control of port-Hamiltonian systems. *IEEE Transactions on Automatic Control.* **53**(11), 2527–2542.

Ortega R, van der Schaft AJ, Mareels I and Maschke B (2001). Putting energy back in control. *IEEE Control Systems* **21**(2), 18–33.

Overman TM, Sackman RW, Davis TL and Cohen BS (2011). High-assurance smart grid: A three-part model for smart grid control systems. *Proc. IEEE.* **99**(6), 1046–1062.

Pacas M (2011). Sensorless drives in industrial applications. *IEEE Ind. Electron. Mag.* **5**(2), 16–23.

Palensky P and Dietrich D (2011). Demand side management: Demand response, intelligent energy systems, and smart loads. *IEEE Transactions on Industrial Informatics.* **7**(3), 381–388.

Papaefthymiou SV, Karamanou EG, Papathanassiou SA and Papadopoulos MP (2010). A wind-hydro-pumped storage station leading to high RES penetration in the autonomous island system of Ikaria. *IEEE Transactions on Sustainable Energy.* **1**(3), 163–172.

Paquette AD and Divan DM (2015). Virtual impedance current limiting for inverters in microgrids with synchronous generators. *IEEE Trans. Ind. Appl.* **51**(2), 1630–1638.

Patel C, Ramchand R, Sivakumar K, Das A and Gopakumar K (2011). A rotor flux estimation during zero and active vector periods using current error space vector from a hysteresis controller for a sensorless vector control of IM drive. *IEEE Trans. Ind. Electron.* **58**(6), 2334–2344.

Pellegrino G, Armando E and Guglielmi P (2009). Direct flux field-oriented control of IPM drives with variable DC link in the field-weakening region. *IEEE Trans. Ind. Appl.* **45**(5), 1619–1627.

Pena R, Clare J and Asher G (1996a). Doubly fed induction generator using back-to-back PWM converters and its application to variable-speed wind-energy generation. *IEE Proc. Electric Power Appli.* **143**(3), 231–241.

Pena R, Clare J and Asher G (1996b). A doubly fed induction generator using back-to-back PWM converters supplying an isolated load from a variable speed wind turbine. *IEE Proc. Electric Power Appli.* **143**(5), 380–387.

Peng FZ and Lai JS (1996) Generalized instantaneous reactive power theory for three-phase power systems. *IEEE Transactions on Instrumentation and Measurement.* **45**(1), 293–297.

Pilawa-Podgurski RCN and Perreault DJ (2013). Submodule integrated distributed maximum power point tracking for solar photovoltaic applications. *IEEE Trans. Power Electron.* **28**(6), 2957–2967.

Pinson P, Mitridati L, Ordoudis C and Ostergaard J (2017). Towards fully renewable energy systems: Experience and trends in denmark. *CSEE Journal of Power and Energy Systems.* **3**(1), 26–35.

Pogaku N, Prodanović M and Green TC (2007). Modeling, analysis and testing of autonomous operation of an inverter-based microgrid. *IEEE Trans. Power Electron.* **22**(2), 613–625.

Poole D (2011). *Linear algebra: A modern introduction* 3rd edn. Boston, MA: Brooks/Cole-Cengage, Learning.

Portillo R, Prats M, Leon J, Sanchez J, Carrasco J, Galvan E and Franquelo L (2006). Modeling strategy for back-to-back three-level converters applied to high-power wind turbines. *IEEE Trans. Ind. Electron.* **53**(5), 1483–1491.

Qiao W, Venayagamoorthy G and Harley R (2009). Real-time implementation of a STATCOM on a wind farm equipped with doubly fed induction generators. *IEEE Trans. Ind. Appl.* **45**(1), 98–107.

Rathjen FW (2018). *The Handbook of Texas-Panhandle*. Texas State Historial Association. https://tshaonline.org/handbook/online/articles/ryp01

Rebeiro RS and Uddin MN (2012). Performance analysis of an FLC-based online adaptation of both hysteresis and PI controllers for IPMSM drive. *IEEE Trans. Ind. Appl.* **48**(1), 12–19.

Regan H and McLaughlin EC (2019). A blackout left tens of millions in South America without power. Officials still don't know what caused it. https://www.cnn.com/2019/06/17/world/power-restored-argentina-uruguay-paraguay-intl-hnk/index.html.

Rengifo C, Kaddar B, Aoustin Y and Chevallereau C (2012). Reactive power compensation in mechanical systems *The (2nd Joint International Conference on Multibody System Dynamics, Stuttgart, Germany*, pp. 576–581.

Rezaei E, Ebrahimi M and Tabesh A (2016). Control of DFIG wind power generators in unbalanced microgrids based on instantaneous power theory. *IEEE Transactions on Smart Grid* **8**(5), 2278–2286.

Rifkin J (2011). *The Third Industrial Revolution: How Lateral Power Is Transforming Energy, the Economy, and the World*. Palgrave Macmillan.

Rodriguez J, Dixon J, Espinoza J, Pontt J and Lezana P (2005). PWM regenerative rectifiers: State of the art. *IEEE Trans. Ind. Electron.* **52**(1), 5–22.

Rodriguez J, Kennel R, Espinoza J, Trincado M, Silva C and Rojas C (2012). High-performance control strategies for electrical drives: An experimental assessment. *IEEE Trans. Ind. Electron.* **59**(2), 812–820.

Rodriguez P, Pou J, Bergas J, Candela J, Burgos R and Boroyevich D (2007a). Decoupled double synchronous reference frame PLL for power converters control. *IEEE Trans. Power Electron.* **22**(2), 584–592.

Rodriguez P, Timbus AV, Teodorescu R, Liserre M and Blaabjerg F (2007b). Flexible active power control of distributed power generation systems during grid faults. *IEEE Transactions on Industrial Electronics.* **54**(5), 2583–2592.

Rush P and Smith I (1978). Run-up and synchronisation of a large synchronous compensator. *IEE Proc. Electric Power Appli.* **1**(3), 91–99.

Saad-Saoud Z, Lisboa M, Ekanayake J, Jenkins N and Strbac G (1998) Application of STATCOMs to wind farms. *IEE Proc. Generation, Transmission and Distribution.* **145**(5), 511–516.

Sallam A and Malik O (2011). *Electric Distribution Systems*. John Wiley & Sons.

Sanger DE and Perlroth N (2019). U.S. escalates online attacks on Russia's power grid. https://www.nytimes.com/2019/06/15/us/politics/trump-cyber-russia-grid.html.

Santos Filho R, Seixas P, Cortizo P, Torres L and Souza A (2008). Comparison of three single-phase PLL algorithms for UPS applications. *IEEE Trans. Ind. Electron.* **55**(8), 2923–2932.

Sao C and Lehn P (2005). Autonomous load sharing of voltage source converters. *IEEE Trans. Power Del.* **20**(2), 1009–1016.

Schäfer B, Witthaut D, Timme M and Latora V (2018).Dynamically induced cascading failures in power grids. *Nature Communications.* **9**, 1975.

Schiffer J, Ortega R, Astolfi A, Raisch J and Sezi T (2014). Conditions for stability of droop-controlled inverter-based microgrids. *Automatica* **50**(10), 2457–2469.

Schmall J, Huang SH, Li Y, Billo J, Conto J and Zhang Y (2015). Voltage stability of large-scale wind plants integrated in weak networks: An ercot case study *2015 IEEE Power Energy Society General Meeting*, pp. 1–5.

Seixas M, Melicio R and Mendes V (2014). Fifth harmonic and sag impact on PMSG wind turbines with a balancing new strategy for capacitor voltages. *Energy Conversion and Management.* **79**(0), 721–730.

Shan DL, Song SZ, Ma JW and Wang XB (2010). Direct power control of PWM rectifiers based on virtual flux*Proc. of the International Conference on Computer Application and System Modeling (ICCASM)*, pp. 613–616.

Shao SY, Abdi E, Barati F and McMahon R (2009). Stator-flux-oriented vector control for brushless doubly fed induction generator. *IEEE Trans. Ind. Electron.* **56**(10), 4220–4228.

Shariatpanah H, Fadaeinedjad R and Rashidinejad M (2013). A new model for PMSG-based wind turbine with yaw control. *IEEE Trans. Energy Convers.* **28**(4), 929–937.

Shetty PP (2013). How differential gearbox works. https://www.youtube.com/watch?v=lN_xGRt_vVY.

Shinnaka S (2008). A robust single-phase PLL system with stable and fast tracking. *IEEE Trans. Ind. Appl.* **44**(2), 624–633.

Shinnaka S (2011 A novel fast-tracking D-Estimation method for single-phase signals. *IEEE Trans. Power Electron.* **26**(4), 1081–1088.

Shyu KK, Lin JK, Pham VT, Yang MJ and Wang TW (2010). Global minimum torque ripple design for direct torque control of induction motor drives. *IEEE Trans. Ind. Electron.* **57**(9), 3148–3156.

Simpson-Porco JW, Dörfler F and Bullo F (2013). Synchronization and power sharing for droop-controlled inverters in islanded microgrids. *Automatica.* **49**(9), 2603–2611.

Simpson-Porco JW, Dörfler F and Bullo F (2017). Voltage stabilization in microgrids via quadratic droop control. *IEEE Transactions on Automatic Control.* **62**(3), 1239–1253.

Singh B, Chandra A and Al-Haddad K (2000a). DSP-based indirect-current-controlled STATCOM. I. Evaluation of current control techniques. *IEE Proc. Electric Power Appli.* **147**(2), 107–112.

Singh B, Chandra A and Al-Haddad K (2000b). DSP-based indirect-current-controlled STATCOM. II. Multifunctional capabilities. *IEE Proc. Electric Power Appli.* **147**(2), 113–118.

Singh B, Saha R, Chandra A and Al-Haddad K (2009). Static synchronous compensators (STATCOM): A review. *IET Proc. Power Electron.* **2**(4), 297–324.

Singh M, Khadkikar V and Chandra A (2011). Grid synchronisation with harmonics and reactive power compensation capability of a permanent magnet synchronous generator-based variable speed wind energy conversion system. *IET Proc. Power Electron.***4**(1), 122–130.

Smith M (2002). Synthesis of mechanical networks: The inerter. *IEEE Trans. Autom. Control.* **47**(10), 1648–1662.

Sneyers B, Novotny DW and Lipo TA (1985) Field weakening in buried permanent magnet AC motor drives. *IEEE Trans. Ind. Appl.* **21**(2), 398–407.

Song Q and Liu W (2009). Control of a cascade STATCOM with star configuration under unbalanced conditions. *IEEE Trans. Power Electron.* **24**(1), 45–58.

Soni N, Doolla S and Chandorkar MC (2013). Improvement of transient response in microgrids using virtual inertia. *IEEE Transactions on Power Delivery.* **28**(3), 1830–1838.

Srinivasan R and Oruganti R (1998) A unity power factor converter using half-bridge boost topology. *IEEE Trans. Power Electron.* **13**(3), 487–500.

Strogatz SH (2004). *Sync: How Order Emerges From Chaos In the Universe, Nature, and Daily Life.*Hachette Books.

Suetake M, da Silva I and Goedtel A (2011). Embedded DSP-Based compact fuzzy system and its application for induction-motor v/f speed control. *IEEE Trans. Ind. Electron.* **58**(3), 750–760.

Sun J (2011). Impedance-based stability criterion for grid-connected inverters. *IEEE Trans. Power Electron.* **26**(11), 3075–3078.

Sun T, Chen Z and Blaabjerg F (2003). Voltage recovery of grid-connected wind turbines after a short-circuit fault. *Proc. of Industrial Electronics Society (IECON)*, pp. 2723–2728.

Sun X, Tian Y and Chen Z (2014). Adaptive decoupled power control method for inverter connected DG. *IET Proc. Renewable Power Generation* **8**(2), 171–182.

Sun XD, Koh KH, Yu BG and Matsui M (2009). Fuzzy-logic-based v/f control of an induction motor for a DC grid power-leveling system using flywheel energy storage equipment. *IEEE Trans. Ind. Electron.* **56**(8), 3161–3168.

Svensson J (2001). Synchronisation methods for grid-connected voltage source converters. *IEE Proc. Generation, Transmission and Distribution* **148**(3), 229–235.

Takahashi I and Noguchi T (1986). A new quick-response and high-efficiency control strategy of an induction motor. *IEEE Trans. Ind. Appl.* **22**(5), 820–827.

Taylor E, Burkes K and Cheung K (2017). U.S. DoE solid state power substation roadmapping workshop: Summary report. Technical report. https://www.energy.gov/sites/prod/files /2017/09/f36/SSPS Roadmapping Workshop - Proceedings - Final.pdf.

Tellegen B (1952) A general network theorem, with applications. *Philips Research Reports* **7**, 259–269.

Teng JH, Huang WH, Hsu TA and Wang CY (2016). Novel and fast maximum power point tracking for photovoltaic generation. *IEEE Trans. Ind. Electron.* **63**(8), 4955–4966.

Teodorescu R and Blaabjerg F (2004). Flexible control of small wind turbines with grid failure detection operating in stand-alone and grid-connected mode. *IEEE Trans. Power Electron.* **19**(5), 1323–1332.

Thacker T, Boroyevich D, Burgos R and Wang F (2011). Phase-locked loop noise reduction via phase detector implementation for single-phase systems. *IEEE Trans. Ind. Electron.* **58**(6), 2482–2490.

Tielens P and Hertem DV (2017). Receding horizon control of wind power to provide frequency regulation. *IEEE Trans. Power Syst.* **32**(4), 2663–2672.

Tilli A and Tonielli A (1998). Sequential design of hysteresis current controller for three-phase inverter. *IEEE Trans. Ind. Electron.* **45**(5), 771–781.

Trentin A, Zanchetta P, Gerada C, Clare J and Wheeler PW (2009). Optimized commissioning method for enhanced vector control of high-power induction motor drives. *IEEE Trans. Ind. Electron.* **56**(5), 1708–1717.

Tuladhar A, Jin H, Unger T and Mauch K (1997). Parallel operation of single phase inverter modules with no control interconnections. *Proc. of the 12th IEEE Applied Power Electronics Conference and Exposition*, pp. 94–100.

Tymerski R, Vorperian V, Lee F and Baumann W (1989). Nonlinear modeling of the PWM switch. *IEEE Trans. Power Electron.* **4**(2), 225–233.

UN (2018 The Paris Agreement. https://unfccc.int/process-and-meetings/the-paris- agreement/the-paris-agreement.

van der Schaft A and Jeltsema D (2014). Port-Hamiltonian systems theory: An introductory overview. *Foundations and Trends in Systems and Control* **1**(2-3), 173–378.

Vandoorn TL, De Kooning JDM, Meersman B, Guerrero JM and Vandevelde L (2012). Automatic power-sharing modification of p/v droop controllers in low-voltage resistive microgrids. *IEEE Trans. Power Del.* **27**(4), 2318–2325.

Vaughan A (2018). UK government spells out plan to shut down coal plants. https://www .theguardian.com/business/2018/jan/05/uk-coal-fired-power-plants-close-2025.

Vesti S, Suntio T, Oliver J, Prieto R and Cobos J (2013). Impedance-based stability and transient-performance assessment applying maximum peak criteria. *IEEE Trans. Power Electron.* **28**(5), 2099–2104.

Walker J (1981). *Large Synchronous Machines: Design, Manufacture and Operation*. Oxford University Press.

Wang S, Hu J and Yuan X (2015a). DFIG-based wind turbines with virtual synchronous control: Inertia support in weak grid. *Proc. IEEE Power Energy Society General Meeting*, pp. 1–14.

Wang S, Hu J, Yuan X and Sun L (2015b). On inertial dynamics of virtual-synchronous-controlled DFIG-based wind turbines. *IEEE Transactions on Energy Conversion.* **30**(4), 1691–1702.

Wang Y and Li Y (2011). Grid synchronization PLL based on cascaded delayed signal cancellation. *IEEE Trans. Power Electron.* **26**(7), 1987–1997.

Wang Z, Xia M and Lemmon M (2013). Voltage stability of weak power distribution networks with inverter connected sources *2013 American Control Conference*, pp. 6577–6582.

Weisstein EW (1999). Hilbert transform. From MathWorld–A Wolfram Web Resource. http://mathworld.wolfram.com/HilbertTransform.html.

Wen B, Boroyevich D, Burgos R, Mattavelli P and Shen Z (2015). Small-signal stability analysis of three-phase AC systems in the presence of constant power loads based on measured d-q frame impedances. *IEEE Trans. Power Electron.* **30**(10), 5952–5963.

Wiggins S (2003). *Introduction to Applied Nonlinear Dynamical Systems and Chaos.* 2nd edn. New York: Springer.

WIKIpedia (2018). Differential (mechanical device). https://en.wikipedia.org/wiki/Differential_(mechanical_device).

Wildi T (2005). *Electrical Machines, Drives and Power Systems.* 6th edn. Prentice-Hall.

Wiles JC (2012). Photovoltaic system grounding. Technical report, New Mexico State University. Available at http://www.solarabcs.org/about/publications/reports/systemgrounding/pdfs/SystemGrounding_studyreport.pdf.

Winter W, Elkington K, Bareux G and Kostevc J (2015). Pushing the limits: Europe's new grid: innovative tools to combat transmission bottlenecks and reduced inertia. *IEEE Power and Energy Magazine* **13**(1), 60–74.

Wu FF, Moslehi K and Bose A (2005). Power system control centers: Past, present, and future. *Proc. IEEE.* **93**(11), 1890–1908.

Wu H, Ruan X, Yang D, Chen X, Zhao W, Lv Z and Zhong QC (2016). Small-signal modeling and parameters design for virtual synchronous generators. *IEEE Trans. Ind. Electron.* **63**(7), 4292–4303.

Xia M, Antsaklis PJ, Gupta V and Zhu F (2017). Passivity and dissipativity analysis of a system and its approximation. *IEEE Trans. Autom. Control.* **62**(2), 620–635.

Xiao H and Xie S (2012). Transformerless split-inductor neutral point clamped three-level PV grid-connected inverter. *IEEE Trans. Power Electron.* **27**(4), 1799–1808.

Xiao H, Liu X and Lan K (2014). Zero-voltage-transition full-bridge topologies for transformerless photovoltaic grid-connected inverter. *IEEE Trans. Ind. Electron.* **61**(10), 5393–5401.

Xie H, Zheng S and Ni M (2017). Microgrid development in China: A method for renewable energy and energy storage capacity configuration in a megawatt-level isolated microgrid. *IEEE Electrification Magazine* **5**(2), 28–35.

Xin H, Huang L, Zhang L, Wang Z and Hu J (2016). Synchronous instability mechanism of P-f droop-controlled voltage source converter caused by current saturation. *IEEE Trans. Power Syst.* **31**(6), 5206–5207.

Xu Y, Tolbert L, Kueck J and Rizy D (2010). Voltage and current unbalance compensation using a static var compensator. *IET Proc. Power Electron.* **3**(6), 977–988.

Yamamoto K and Smith MC (2016). Bounded disturbance amplification for mass chains with passive interconnection. *IEEE Trans. Autom. Control* **61**(6), 1565–1574.

Yang D, Ruan X and Wu H (2014). Impedance shaping of the grid-connected inverter with LCl filter to improve its adaptability to the weak grid condition. *IEEE Trans. Power Electron.* **29**(11), 5795–5805.

Yang Y, Nishikawa T and Motter AE (2017). Small vulnerable sets determine large network cascades in power grids. *Science.*

Yao W, Chen M, Matas J, Guerrero J and Qian ZM (2011). Design and analysis of the droop control method for parallel inverters considering the impact of the complex impedance on the power sharing. *IEEE Trans. Ind. Electron.* **58**(2), 576–588.

Yaramasu V, Wu B, Sen PC, Kouro S and Narimani M (2015). High-power wind energy conversion systems: State-of-the-art and emerging technologies. *Proc. IEEE.* **103**(5), 740–788.

Yazdanpanahi H, Li YW and Xu W (2012). A new control strategy to mitigate the impact of inverter-based DGs on protection system. *IEEE Trans. on Smart Grid.* **3**(3), 1427–1436.

Ye F, Qian Y and Hu RQ (2017). *Smart Grid Communication Infrastructures: Big Data, Cloud Computing, and Security.* IEEE.

Yuan X, Merk W, Stemmler H and Allmeling J (2002). Stationary-frame generalized integrators for current control of active power filters with zero steady-state error for current harmonics of concern under unbalanced and distorted operating conditions. *IEEE Trans. Ind. Appl.* **38**(2), 523–532.

Yuan X, Wang F, Boroyevich D, Li Y and Burgos R (2009). DC-link voltage control of a full power converter for wind generator operating in weak-grid systems. *IEEE Trans. Power Electron.* **24**(9), 2178–2192.

Zaccarian L and Teel AR (2004). Nonlinear scheduled anti-windup design for linear systems. *IEEE Trans. Autom. Control* **49**(11), 2055–2061.

Zaid SA, Mahgoub OA and El-Metwally KA (2010). Implementation of a new fast direct torque control algorithm for induction motor drives. *IET Proc. Electric Power Appl.* **4**(5), 305–313.

Zeineldin HH, El-Saadany EF and Salama MMA (2006). Protective relay coordination for micro-grid operation using particle swarm optimization. *2006 Large Engineering Systems Conference on Power Engineering*, pp. 152–157, Halifax, NS.

Zhang L, Harnefors L and Nee HP (2010). Power-synchronization control of grid-connected voltage-source converters. *IEEE Trans. Power Syst.* **25**(2), 809–820.

Zhang LD, Nee HP and Harnefors L (2011a). Analysis of stability limitations of a VSC-HVDC link using power-synchronization control. *IEEE Trans. Power Syst.* **26**(3), 1326–1337.

Zhang Q, Sun XD, Zhong YR, Matsui M and Ren BY (2011b). Analysis and design of a digital phase-locked loop for single-phase grid-connected power conversion systems. *IEEE Trans. Ind. Electron.* **58**(8), 3581–3592.

Zhang Z, Zhao Y, Qiao W and Qu L (2014). A space-vector-modulated sensorless direct-torque control for direct-drive PMSG wind turbines. *IEEE Trans. Ind. Appl.* **50**(4), 2331–2341.

Zhao H, Hong M, Lin W and Loparo KA (2019). Voltage and frequency regulation of microgrid with battery energy storage systems. *IEEE Trans. Smart Grid.* **10**(1), 414–424.

Zhao J, Lyu X, Fu Y, Hu X and Li F (2016). Coordinated microgrid frequency regulation based on DFIG variable coefficient using virtual inertia and primary frequency control. *IEEE Transactions on Energy Conversion.* **31**(3), 833–845.

Zhao Y, Chai J and Sun X (2015). Virtual synchronous control of grid-connected DFIG-based wind turbines. *Proc. IEEE Applied Power Electronics Conf. and Exposition (APEC)*, pp. 2980–2983.

Zheng LB, Fletcher JE, Williams BW and He XN (2011).A novel direct torque control scheme for a sensorless five-phase induction motor drive. *IEEE Trans. Ind. Electron.***58**(2), 503–513.

Zhong L, Rahman MF, Hu WY and Lim KW (1997). Analysis of direct torque control in permanent magnet synchronous motor drives. *IEEE Trans. Power Electron.* **12**(3), 528–536.

Zhong QC (2010a). Four-quadrant operation of AC machines powered by inverters that mimic synchronous generators. *Proc. of the 5th IET International Conference on Power Electronics, Machines and Drives.*

Zhong QC (2010b). Speed-sensorless AC Ward Leonard drive systems. *Proc. of International Symposium on Power Electronics Electrical Drives Automation and Motion (SPEEDAM)*, pp. 1512–1517.

Zhong QC (2013a).AC Ward Leonard drive systems: Revisiting the four-quadrant operation of AC machines. *European Journal of Control.* **19**(5), 426–435.

Zhong QC (2013b). How to achieve completely autonomous power in the next generation of smart grids. Technical report. http://smartgrid.ieee.org/september-2013/973-how-to-achieve-completely-autonomous-power-in-the-next-generation-of-smart-grids.

Zhong QC (2013c). Robust droop controller for accurate proportional load sharing among inverters operated in parallel. *IEEE Trans. Ind. Electron.* **60**(4), 1281–1290.

Zhong QC (2016a). Operating doubly-fed induction generators as virtual synchronous generators. UK patent GB2554954.

Zhong QC (2016b). Virtual synchronous machines: A unified interface for smart grid integration. *IEEE Power Electronics Magazine.* **3**(4), 18–27.

Zhong QC (2017a). Cyber synchronous machine (cybersync machine). US Patent 10509373 issued.

Zhong QC (2017b). The ghost operator and its applications to reveal the physical meaning of reactive power for electrical and mechanical systems and others. *IEEE Access* **5**, 13038–13045.

Zhong QC (2017c). Operating doubly-fed induction generators as virtual synchronous generators. US Patent 15727593 pending. UK patent GB2554954 issued.

Zhong QC (2017d). Power electronic converters that take part in the grid regulation without affecting the DC-port operation. UK Patent GB1717573.8 approved.

Zhong QC (2017e). Power electronics-enabled autonomous power systems: Architecture and technical routes. *IEEE Trans. Ind. Electron.* **64**(7), 5907–5918.

Zhong QC (2017f). Synchronized and democratized smart grids to underpin the third industrial revolution. *Proceedings of The 20th IFAC World Congress.*

Zhong QC (2018). Reconfiguration of inertia, damping and fault ride-through for a virtual synchronous machine. UK Patent GB1800572.8 pending.

Zhong QC and Boroyevich D (2013). A droop controller is intrinsically a phase-locked loop. *Proc. of the 39th Annual Conference of the IEEE Industrial Electronics Society, IECON 2013*, pp. 5916–5921, Vienna, Austria.

Zhong QC and Boroyevich D (2016). Structural resemblance between droop controllers and phase-locked loops. *IEEE Access.* **4**, 5733–5741.

Zhong QC and Hornik T (2013). *Control of Power Inverters in Renewable Energy and Smart Grid Integration.* Wiley-IEEE Press.

Zhong QC and Konstantopoulos GC (2017). Current-limiting droop control of grid-connected inverters. *IEEE Trans. Ind. Electron.* **64**(7), 5963–5973.

Zhong QC and Lyu Z (2019a). A 100%-power-electronics-based SYNDEM smart grid testbed. To be published.

Zhong QC and Lyu Z (2019b).Droop-controlled rectifiers that continuously take part in grid regulation. *IEEE Trans. Ind. Electron.* **66**(8), 6516–6526.

Zhong QC and Ming W (2016). A θ-Converter that reduces common mode currents, output voltage ripples, and total capacitance required. *IEEE Trans. Power Electron.* **31**(12), 8435–8447.

Zhong QC and Nguyen PL (2012). Sinusoid-locked loops to detect the frequency, the ampitude and the phase of the fundamental component of a periodic signal. *Proc. of the (24th Chinese Control and Decision Conference (CCDC)*, Taiyuan, China.

Zhong QC and Stefanello M (2017). Port-Hamiltonian control of power electronic converters to achieve passivity. *Proc. of the 56th IEEE Conference on Decision and Control.*

Zhong QC and Stefanello M (2018). Passivity of a power converter enabled by the port-Hamiltonian theory and the Hilbert transform. To be published.

Zhong QC and Weiss G (2009). Static synchronous generators for distributed generation and renewable energy. *Proc. of IEEE PES Power Systems Conference & Exhibition (PSCE)*, pp. 1–6.

Zhong QC and Weiss G (2011). Synchronverters: Inverters that mimic synchronous generators. *IEEE Trans. Ind. Electron.* **58**(4), 1259–1267.

Zhong QC and Zeng Y (2011 Can the output impedance of an inverter be designed capacitive? *Proc. of the 37th Annual IEEE Conference of Industrial Electronics (IECON)*, pp. 1220–1225.

Zhong QC and Zeng Y (2014). Control of inverters via a virtual capacitor to achieve capacitive output impedance. *IEEE Trans. Power Electron.* **29**(10), 5568–5578.

Zhong QC and Zeng Y (2016). Universal droop control of inverters with different types of output impedance. *IEEE Access* **4**, 702–712.

Zhong QC and Zhang X (2019). Impedance-sum stability criterion for power electronic systems with two converters/sources. *IEEE Access.* **7**, 21254–21265.

Zhong QC, Blaabjerg F, Guerrero J and Hornik T (2012a). Improving the voltage quality of an inverter via bypassing the harmonic current components. *Proc. of IEEE Energy Conversion Congress and Exposition (ECCE)*, Raleigh, North Carolina.

Zhong QC, Konstantopoulos GC, Ren B and Krstic M (2018a). Improved synchronverters with bounded frequency and voltage for smart grid integration. *IEEE Trans. Smart Grid.* **9**(2), 786–796.

Zhong QC, Ma Z and Nguyen PL (2012b). PWM-controlled rectifiers without the need of an extra synchronisation unit. *IECON (2012 - 38th Annual Conference on IEEE Industrial Electronics Society*, pp. 691–695.

Zhong QC, Ma Z, Ming WL and Konstantopoulos GC (2015). Grid-friendly wind power systems based on the synchronverter technology. *Energy Conversion and Management.* **89**(1), 719–726.

Zhong QC, Ming WL and Zeng Y (2016). Self-synchronized universal droop controller. *IEEE Access.* **4**, 7145–7153.

Zhong QC, Ming WL, Sheng W and Zhao Y (2017). Beijing converters: Bridge converters with a capacitor added to reduce leakage currents, DC-bus voltage ripples and total capacitance required. *IEEE Trans. Ind. Electron.* **64**(1), 325 –335.

Zhong QC, Nguyen PL, Ma Z and Sheng W (2014). Self-synchronised synchronverters: Inverters without a dedicated synchronisation unit. *IEEE Trans. Power Electron.* **29**(2), 617–630.

Zhong QC, Shao T, Zheng TQ, Li H and Zhang X (2018b). Virtual synchronous machines with reconfigurable inertia and damping. To be published.

Zhong QC, Wang Y and Ren B (2019). Connecting the home grid to the public grid: Field demonstration of virtual synchronous machines. *IEEE Power Electronics Magazine* **6**(4), 41–49.

Zhong QC, Zhao Y and Chai J (2018c). Operation of a doubly-fed induction generator as a virtual synchronous generator (DFIG-VSG). To be published.

Zhou K and Wang D (2003). Digital repetitive controlled three-phase PWM rectifier. *IEEE Trans. Power Electron.* **18**(1), 309–316.

Zhou Z, Unsworth P, Holland P and Igic P (2009). Design and analysis of a feedforward control scheme for a three-phase voltage source pulse width modulation rectifier using sensorless load current signal. *IET Proc. Power Electron.* **2**(4), 421 –430.

Zhu F, Xia M and Antsaklis PJ (2017). On passivity analysis and passivation of event-triggered feedback systems using passivity indices. *IEEE Trans. Autom. Control.* **62**(3), 1397–1402.

Zhu J, Booth CD, Adam GP, Roscoe AJ and Bright CG (2013). Inertia emulation control strategy for VSC-HVDC transmission systems. *IEEE Trans. Power Syst.* **28**(2), 1277–1287.

Zhu X, Wang Y, Xu L, Zhang X and Li H (2011). Virtual inertia control of DFIG-based wind turbines for dynamic grid frequency support *Renewable Power Generation (RPG (2011), IET Conference on*, pp. 1–6 IET.

Ziarani AK and Konrad A (2004). A method of extraction of nonstationary sinusoids. *Signal Processing.* **84**(8), 1323–1346.

Ziarani AK, Konrad A and Sinclair AN (2003). A novel time-domain method of analysis of pulsed sine wave signals. *IEEE Trans. Instrum. Meas.* **52**(3), 809–814.

Index

Power Electronics-Enabled Autonomous Power Systems: Next Generation Smart Grids,
First Edition. Qing-Chang Zhong.
© 2020 John Wiley & Sons Ltd. Published 2020 by John Wiley & Sons Ltd.